ARMY PHYSICAL READINESS TRAINING

with Change 1

Consider Amazon.com for all your Professional Books.
Fast shipping, high quality, and low prices.

3 May 2013

DISTRIBUTION RESTRICTION:
Approved for public release; distribution is unlimited.

United States Government   US Army

Change No. 1

FM 7-22, C1
Headquarters
Department of the Army
Washington, DC, 3 May 2013

Army Physical Readiness Training

1. Change FM 7-22, 28 September 2012, as follows:

Remove old pages:	Insert new pages:
None	Change 1 transmittal
Pages A-19 through A-24	Pages A-19 through A-24

2. A star (*) marks new or changed material.

3. File this transmittal sheet in front of the publication.

DISTRIBUTION RESTRICTION: Approved for public release; distribution is unlimited.

By Order of the Secretary of the Army:

Official:

RAYMOND T. ODIERNO
General, United States Army
Chief of Staff

JOYCE E. MORROW
Administrative Assistant to the
Secretary of the Army
1306702

PIN: 103049-001

*FM 7-22

Field Manual
No. FM 7-22

Headquarters
Department of the Army
Washington, DC, 26 October 2012

Army Physical Readiness Training

Contents

		Page
PREFACE		xv
INTRODUCTION		xvi

PART ONE PHILOSOPHY

Chapter 1	APPROACH	1-1
	Training Program	1-1
	Principles of Training	1-2
Chapter 2	SYSTEM	2-1
	Phases	2-1
	Components	2-3
	Types	2-5
Chapter 3	LEADERSHIP	3-1
	Traits	3-1
	Cooperation	3-2

PART TWO STRATEGY

Chapter 4	TYPES OF PROGRAMS	4-1
	Initial Military Training	4-1
	Advanced Individual Training	4-2
	One Station Unit Training	4-2
	Warrant Officer Candidate School	4-2
	Basic Officer Leader Courses	4-2
	Active and Reserve Components	4-2
	Reserve	4-6
	Unit	4-6
	Individual	4-7
	Scheduling Training	4-7
	Command Responsibilities	4-7
Chapter 5	PLANNING CONSIDERATIONS	5-1
	Goal	5-1
	Session Elements	5-6
	Toughening Phase PRT	5-6
	Toughening Phase PRT Schedule	5-7

DISTRIBUTION RESTRICTION: Approved for public release; distribution is unlimited.

*This publication supersedes TC 3-22.20, dated 20 August 2010.

Contents

	Condensed Time	5-13
	Field Training	5-14
	Sustaining Phase PRT	5-14
	Initial Military Training Sustaining Phase PRT Schedules	5-14
	Condensed Time	5-18
	Field Training	5-18
	PRT in Operational Units	5-18
	Sustaining Phase PRT Schedules	5-20
	Reserve Component	5-32
	Sample Commander's Policy Letter	5-34
Chapter 6	**SPECIAL CONDITIONING PROGRAMS**	**6-1**
	APFT or Unit PRT Goal Failure	6-1
	Army Weight Control Program	6-2
	Reconditioning	6-2
	4 for the Core	6-11
	Exercise 1: Bent-Leg Raise	6-12
	Exercise 2: Side Bridge	6-13
	Exercise 3: Back Bridge	6-14
	Exercise 4: Quadraplex	6-15
	Hip Stability Drill	6-16
	Exercise 1: Lateral Leg Raise	6-16
	Exercise 2: Medial Leg Raise	6-18
	Exercise 3: Bent-Leg Lateral Raise	6-20
	Exercise 4: Single-Leg Tuck	6-22
	Exercise 5: Single-Leg Over	6-24
	Shoulder Stability Drill	6-25
	Exercise 1: "I" Raise	6-26
	Exercise 2: "T" Raise	6-27
	Exercise 3: "Y" Raise	6-28
	Exercise 4: "L" Raise	6-29
	Exercise 5: "W" Raise	6-30
	Strength and Mobility Training	6-31
	Strength Training Machine Drill	6-31
	Exercise 1: Leg Press	6-32
	Modified Exercise 1A: Modified Leg Press	6-33
	Modified Exercise 1B: Single-Leg Press	6-34
	Exercise 2: Leg Curl	6-35
	Modified Eexercise 2A: Modified Leg Curl (Seated)	6-36
	Modified Exercise 2B: Single-Leg Curl (Seated)	6-37
	Modified Exercise 2C: Modified Leg Curl (Prone)	6-38
	Modified Exercise 2D: Single-Leg Curl (Prone)	6-38
	Exercise 3: Heel Raise	6-39
	Modified Exercise 3A: Single-Leg Heel Raise	6-40
	Exercise 4: Chest Press	6-41
	Modified Eexercise 4A: Modified Chest Press	6-42
	Modified Exercise 4B: Single-Arm Chest Press	6-43
	Exercise 5: Seated Row	6-44

Modified Exercise 5A: Straight-Arm Seated Row	6-45
Modified Exercise 5B: Single-Arm Seated Row	6-46
Exercise 6: Overhead Press	6-47
Modified Exercise 6A: Modified Overhead Press	6-48
Modified Exercise 6B: Single-Arm Overhead Press	6-49
Exercise 7: LAT Pull-Down	6-50
Modified Exercise 7A: Straight-Arm LAT Pull-Down	6-51
Modified Exercise 7B: Single-Arm LAT Pull-Down	6-52
Exercise 8: Lateral Raise	6-54
Modified Exercise 8A: Single-Arm Lateral Raise	6-55
Exercise 9: Triceps Extension	6-56
Modified Exercise 9A: Modified Triceps Extension	6-58
Modified Exercise 9B: Single-Arm Triceps Extension	6-59
Exercise 10: Biceps Curl	6-62
Modified Exercise 10A: Modified Biceps Curl	6-63
Modified Exercise 10B: Single-Arm Biceps Curl	6-64
Exercise 11: Trunk Flexion	6-65
Modified Exercise 11: Modified Trunk Flexion	6-66
Exercise 12: Trunk Extension	6-67
Modified Exercise 12: Modified Trunk Extension	6-68
Preparation Drill	6-75
Exercise 1: Bend and Reach	6-75
Modified Exercise 1: Modified Bend and Reach	6-76
Exercise 2: Rear Lunge	6-77
Modified Exercise 2: Modified Rear Lunge	6-78
Exercise 3: High Jumper	6-79
Modified Exercise 3: Modified High Jumper	6-80
Exercise 4: Rower	6-81
Modified Exercise 4: Modified Rower	6-82
Exercise 5: Squat Bender	6-83
Modified Exercise 5: Modified Squat Bender	6-84
Exercise 6: Windmill	6-85
Modified Exercise 6: Modified Windmill	6-87
Exercise 7: Forward Lunge	6-88
Modified Exercise 7: Modified Forward Lunge	6-89
Exercise 8: Prone Row	6-90
Modified Exercise 8: Modified Prone Row	6-91
Exercise 9: Bent-Leg Body Twist	6-92
Modified Exercise 9: Modified Bent-Leg Body Twist	6-93
Exercise 10: Push-Up	6-94
Modified Exercise 10: Modified Push-Up	6-95
Conditioning Drill 1	6-96
Exercise 1: Power Jump	6-96
Modified Exercise 1: Modified Power Jump	6-97
Exercise 2: V-Up	6-98
Modified Exercise 2: Modified V-Up	6-99
Exercise 3: Mountain Climber	6-100
Modified Exercise 3: Modified Mountain Climber	6-101

Contents

Exercise 4: Leg-Tuck and Twist	6-102
Modified Exercise 4: Modified Leg-Tuck and Twist	6-103
Exercise 5: Single-Leg Push-Up	6-104
Modified Exercise 5: Modified Single-Leg Push-Up	6-105
Recovery Drill	6-107
Exercise 1: Overhead Arm Pull	6-107
Modified Exercise 1: Modified Overhead Arm Pull	6-108
Exercise 2: Rear Lunge	6-109
Modified Exercise 2: Modified Rear Lunge	6-110
Exercise 3: Extend and Flex	6-111
Modified Exercise 3: Modified Extend and Flex	6-112
Exercise 4: Thigh Stretch	6-113
Modified Exercise 4: Modified Thigh Stretch	6-114
Exercise 5: Single-Leg Over	6-115
Modified Exercise 5: Modified Single-Leg Over	6-116

PART THREE ACTIVITIES

Chapter 7

EXECUTION OF TRAINING	7-1
Commands	7-1
Platoon Reassembly	7-3
Positions	7-7
Squat Position	7-7
Front Leaning Rest Position	7-8
Six-Point Stance	7-8
Straddle Stance	7-9
Forward Leaning Stance	7-9
Prone Position	7-10
Supine Position	7-10
Cadence	7-11
Commands	7-13
Running Activities	7-14
Recovery Drill	7-14
Mirror Effect	7-15

Chapter 8

PREPARATION AND RECOVERY	8-1
Preparation	8-1
Leadership	8-1
Preparation Drill	8-2
Exercise 1: Bend and Reach	8-4
Exercise 2: Rear Lunge	8-5
Exercise 3: High Jumper	8-6
Exercise 4: Rower	8-7
Exercise 5: Squat Bender	8-8
Exercise 6: Windmill	8-9
Exercise 7: Forward Lunge	8-11
Exercise 8: Prone Row	8-12
Exercise 9: Bent-Leg Body Twist	8-13
Exercise 10: Push-Up	8-14

	Exercise 10A: Push-Up Using the Six-Point Stance	8-15
	Recovery	8-15
	Leadership	8-15
	Commands	8-16
	Recovery Drill	8-16
	Exercise 1: Overhead Arm Pull	8-18
	Exercise 2: Rear Lunge	8-19
	Exercise 3: Extend and Flex	8-20
	Exercise 4: Thigh Stretch	8-21
	Exercise 5: Single-Leg Over	8-22
Chapter 9	**STRENGTH AND MOBILITY ACTIVITIES**	**9-1**
	Exercise Drills	9-1
	Conditioning Drill 1	9-3
	Exercise 1: Power Jump	9-6
	Exercise 2: V-Up	9-7
	Exercise 3: Mountain Climber	9-8
	Exercise 4: Leg Tuck and Twist	9-9
	Exercise 5: Single-Leg Push-Up	9-10
	Conditioning Drill 2	9-11
	Exercise 1: Turn and Lunge	9-13
	Exercise 2: Supine Bicycle	9-14
	Exercise 3: Half Jacks	9-15
	Exercise 4: Swimmer	9-16
	Exercise 5: 8-Count Push-Up	9-17
	Conditioning Drill 3	9-20
	Exercise 1: "Y" Squat	9-21
	Exercise 2: Single-Leg Dead Lift	9-23
	Exercise 3: Side-to-Side Knee Lifts	9-25
	Exercise 4: Front Kick Alternate Toe Touch	9-27
	Exercise 5: Tuck Jump	9-29
	Exercise 6: Straddle-Run Forward and Backward	9-31
	Exercise 7: Half-Squat Laterals	9-32
	Exercise 8: Frog Jumps Forward and Backward	9-33
	Exercise 9: Alternate ¼-Turn Jump	9-34
	Exercise 10: Alternate-Staggered Squat Jump	9-35
	Push-Up and Sit-Up Drill	9-37
	Climbing Drills	9-38
	Climbing Drill 1	9-43
	Exercuse 1: Straight-Arm Pull	9-43
	Exercise 2: Heel Hook	9-44
	Exercise 3: Pull-Up	9-45
	Exercise 4: Leg Tuck	9-46
	Exercise 5: Alternating Grip Pull-Up	9-47
	Climbing Drill 2	9-48
	Exercise 1: Flexed-Arm Hang	9-49
	Exercise 2: Heel Hook	9-50
	Exercise 3: Pull-Up	9-51

Contents

	Exercise 4: Leg Tuck	9-52
	Exercise 5: Alternating Grip Pull-Up	9-53
	Strength Training Circuit	9-54
	Station 1: Sumo Squat	9-56
	Station 2: Straight-Leg Dead Lift	9-58
	Station 3: Forward Lunge	9-59
	Station 4: 8-Count Step-Up	9-60
	Station 5: Pull-Up or Straight-Arm Pull	9-62
	Station 6: Supine Chest Press	9-64
	Station 7: Bent-Over Row	9-65
	Station 8: Overhead Push Press	9-66
	Station 9: Supine Body Twist	9-67
	Station 10: Leg Tuck	9-68
	Guerrilla Drill	9-69
	Exercise 1: Shoulder Roll	9-71
	Exercise 2: Lunge Walk	9-72
	Exercise 3: Soldier Carry	9-73
Chapter 10	**ENDURANCE AND MOBILITY ACTIVITIES**	**10-1**
	Running	10-1
	Leadership	10-4
	Military Movement Drill 1	10-6
	Exercise 1: Verticals	10-8
	Exercise 2: Laterals	10-9
	Exercise 3: Shuttle Sprint	10-10
	Military Movement Drill 2	10-11
	Exercise 1: Power Skip	10-12
	Exercise 2: Crossovers	10-13
	Exercise 3: Crouch Run	10-14
	Speed Running	10-15
	30:60s	10-15
	60:120s	10-15
	300-Yard Shuttle Run	10-16
	Hill Repeats	10-18
	Ability Group Run	10-18
	Unit Formation Run	10-19
	Release Run	10-19
	Terrain Run	10-20
	Foot Marches	10-20
	Conditioning Obstacle Course	10-20
	Endurance Training Machines	10-20
Appendix A	**ARMY PHYSICAL FITNESS TEST (APFT)**	**A-1**
Appendix B	**CLIMBING BARS**	**B-1**
Appendix C	**POSTURE AND BODY MECHANICS**	**C-1**
Appendix D	**ENVIRONMENTAL CONSIDERATIONS**	**D-1**
Appendix E	**OBSTACLE NEGOTIATIONS**	**E-1**
GLOSSARY		**Glossary-1**

Contents

REFERENCES	References-1
INDEX	Index-1

Figures

Figure 1-1. Army PRT System and relationship to ARFORGEN	1-8
Figure 2-1. PRT System	2-2
Figure 2-2. Components of PRT	2-4
Figure 2-3. Types of PRT	2-6
Figure 5-1. Soldier response/adaptation to overreaching, overtraining, and overuse	5-4
Figure 5-2. Sample, commander's policy letter	5-35
Figure 6-1. Army Physical Readiness Training System	6-3
Figure 6-2. Level II reconditioning entry criteria	6-5
Figure 6-3. Level II exit criteria	6-5
Figure 6-4. Rehabilitation and reconditioning responsibilities	6-7
Figure 6-5. Endurance training equipment	6-10
Figure 6-6. Bent-leg raise (4 for the core)	6-12
Figure 6-7. Side bridge	6-13
Figure 6-8. Back bridge	6-14
Figure 6-9. Quadraplex	6-15
Figure 6-10. Lateral leg raise	6-17
Figure 6-11. Medial leg raise	6-19
Figure 6-12. Bent-leg lateral raise (hip stability drill)	6-21
Figure 6-13. Single-leg tuck	6-23
Figure 6-14. Single-leg over	6-24
Figure 6-15. "I" raise	6-26
Figure 6-16. "T" raise	6-27
Figure 6-17. "Y" raise	6-28
Figure 6-18. "L" raise	6-29
Figure 6-19. "W" raise	6-30
Figure 6-20. Leg press	6-32
Figure 6-21. Modified leg press	6-33
Figure 6-22. Single-leg press	6-34
Figure 6-23. Leg curl	6-35
Figure 6-24. Modified leg curl	6-36
Figure 6-25. Single-leg curl	6-37
Figure 6-26. Modified leg curl (prone)	6-38
Figure 6-27. Single-leg curl (prone)	6-38
Figure 6-28. Heel raise	6-39
Figure 6-29. Single-leg heel raise	6-40
Figure 6-30. Chest press	6-41

Contents

Figure 6-31. Modified chest press...6-42
Figure 6-32. Single-arm chest press..6-43
Figure 6-33. Seated row...6-44
Figure 6-34. Straight-arm seated row..6-45
Figure 6-35. Single-arm seated row...6-46
Figure 6-36. Overhead press 6-47
Figure 6-37. Modified overhead press 6-48
Figure 6-38. Single-arm overhead press 6-49
Figure 6-39. Lat pull-down 6-50
Figure 6-40. Straight-arm lat pull-down..6-51
Figure 6-41. Single-arm lat pull-down 6-53
Figure 6-42. Lateral raise...6-54
Figure 6-43. Single-arm lateral raise...6-55
Figure 6-44. Triceps extension 6-57
Figure 6-45. Modified triceps extension using a high pulley 6-58
Figure 6-46. Modified triceps extension using a triceps extension machine 6-58
Figure 6-47. Single-arm triceps extension using a high pulley 6-60
Figure 6-48. Single-arm triceps extension using a triceps extension machine........6-61
Figure 6-49. Biceps curl 6-62
Figure 6-50. Modified biceps curl 6-63
Figure 6-51. Single-arm biceps curl 6-64
Figure 6-52. Trunk flexion 6-65
Figure 6-53. Modified trunk flexion...6-66
Figure 6-54. Trunk extension 6-67
Figure 6-55. Modified trunk extension..6-68
Figure 6-56. Bend and reach 6-75
Figure 6-57. Modified bend and reach 6-76
Figure 6-58. Rear lunge..6-77
Figure 6-59. Modified rear lunge 6-78
Figure 6-60. High jumper 6-79
Figure 6-61. Modified high jumper (remaining on the ground)...................................6-80
Figure 6-62. Rower 6-81
Figure 6-63. Modified rower (limited range of movement)...6-82
Figure 6-64. Modified rower (without use of arms) 6-82
Figure 6-65. Squat bender 6-83
Figure 6-66. Modified squat bender..6-84
Figure 6-67. Windmill...6-85
Figure 6-68. Modified windmill (body twist)..6-87
Figure 6-69. Modified windmill (hands on hips) 6-87
Figure 6-70. Modified windmill (single arm) 6-87
Figure 6-71. Forward lunge...6-88
Figure 6-72. Modified forward lunge 6-89

Figure 6-73. Prone row ... 6-90
Figure 6-74. Modified prone row (assuming starting position) ... 6-91
Figure 6-75. Modified prone row (using the arms) ... 6-91
Figure 6-76. Bent-leg body twist ... 6-92
Figure 6-77. Modified bent-leg body twist (head on the ground and arms at 45 degrees) .. 6-93
Figure 6-78. Modified bent-leg body twist (head elevated and arms at 90 degrees) ... 6-93
Figure 6-79. Push-up ... 6-94
Figure 6-80. Push-up in the 6-point stance ... 6-94
Figure 6-81. Modified push-up variation for assuming the 6-point stance ... 6-95
Figure 6-82. Modified push-up ... 6-95
Figure 6-83. Power jump ... 6-96
Figure 6-84. Modified power jump ... 6-97
Figure 6-85. V-up ... 6-98
Figure 6-86. Modified V-up ... 6-99
Figure 6-87. Mountain climber ... 6-100
Figure 6-88. Modified mountain climber ... 6-101
Figure 6-89. Leg-tuck and twist ... 6-102
Figure 6-90. Modified leg-tuck and twist ... 6-103
Figure 6-91. Single-leg push-up ... 6-104
Figure 6-92. Variation for assuming the 6-point stance ... 6-105
Figure 6-93. Modified single-leg push-up ... 6-105
Figure 6-94. Overhead arm pull ... 6-107
Figure 6-95. Modified overhead arm pull and front arm pull ... 6-108
Figure 6-96. Rear lunge ... 6-109
Figure 6-97. Modified rear lunge ... 6-110
Figure 6-98. Extend and flex ... 6-111
Figure 6-99. Modified extend and flex (standing) ... 6-112
Figure 6-100. Stepping into the modified extend and flex (prone) ... 6-112
Figure 6-101. Modified extend and flex (prone) starting position ... 6-112
Figure 6-102. Thigh stretch ... 6-113
Figure 6-103. Modified thigh stretch (assuming the seated position) ... 6-114
Figure 6-104. Modified thigh stretch starting positions ... 6-114
Figure 6-105. Single-leg over ... 6-115
Figure 6-106. Modified single-leg over ... 6-116
Figure 7-1. Platoon rectangular formation ... 7-2
Figure 7-2. Platoon rectangular formation extended and uncovered ... 7-3
Figure 7-3. Forming a company, company in line with platoons in column ... 7-3
Figure 7-4. Company extended and uncovered, company in line with platoons in column ... 7-4
Figure 7-5. Formation of company en masse ... 7-4
Figure 7-6. Company en masse extended and uncovered ... 7-5
Figure 7-7. Platoon formation en masse ... 7-6
Figure 7-8. Platoon formation extended and covered ... 7-6

Contents

Figure 7-9. Squat position 7-7
Figure 7-10. Front leaning rest position 7-8
Figure 7-11. Six-point stance 7-8
Figure 7-12. Straddle stance 7-9
Figure 7-13. Forward leaning stance 7-9
Figure 7-14. Prone position 7-10
Figure 7-15. Supine position 7-10
Figure 7-16. Hands down assist to supine position 7-11
Figure 8-1. Bend and reach 8-4
Figure 8-2. Rear lunge 8-5
Figure 8-3. High jumper 8-6
Figure 8-4. Rower 8-7
Figure 8-5. Squat bender 8-8
Figure 8-6. Windmill 8-9
Figure 8-7. Forward lunge 8-11
Figure 8-8. Prone row 8-12
Figure 8-9. Bent-leg body twist 8-13
Figure 8-10. Push-up 8-14
Figure 8-11. Push-up using the six-point stance 8-15
Figure 8-12. Overhead arm pull 8-18
Figure 8-13. Rear lunge 8-19
Figure 8-14. Extend and flex 8-20
Figure 8-15. Thigh stretch 8-21
Figure 8-16. Single-leg over 8-22
Figure 9-1. Strength and mobility-related WTBDs 9-1
Figure 9-2. Power jump 9-6
Figure 9-3. V-up 9-7
Figure 9-4. Mountain climber 9-8
Figure 9-5. Leg tuck and twist 9-9
Figure 9-6. Single-leg push-up 9-10
Figure 9-7. Turn and lunge 9-13
Figure 9-8. Supine bicycle 9-14
Figure 9-9. Half jacks 9-15
Figure 9-10. Swimmer 9-16
Figure 9-11. 8-count push-up 9-18
Figure 9-12. "Y" squat 9-21
Figure 9-13. Single-leg dead lift 9-24
Figure 9-14. Side-to-side knee lifts 9-26
Figure 9-15. Front kick alternate toe touch 9-28
Figure 9-16. Tuck jump 9-29
Figure 9-17. Straddle-run forward and backward 9-31
Figure 9-18. Half-squat laterals 9-32

Figure 9-19. Frog jumps forward and backward ... 9-33
Figure 9-20. Alternate ¼-turn jump ... 9-34
Figure 9-21. Alternate-staggered squat jump ... 9-35
Figure 9-22. Climbing pod ... 9-40
Figure 9-23. Hand positions ... 9-41
Figure 9-24. Straight-arm pull ... 9-43
Figure 9-25. Heel hook ... 9-44
Figure 9-26. Pull-up ... 9-45
Figure 9-27. Leg tuck ... 9-46
Figure 9-28. Alternating grip pull-up ... 9-47
Figure 9-29. Flexed-arm hang ... 9-49
Figure 9-30. Heel hook ... 9-50
Figure 9-31. Pull-up ... 9-51
Figure 9-32. Leg tuck ... 9-52
Figure 9-33. Alternating grip pull-up ... 9-53
Figure 9-34. Strength training circuit ... 9-55
Figure 9-35. Sumo squat ... 9-57
Figure 9-36. Straight-leg dead lift ... 9-58
Figure 9-37. Forward lunge ... 9-59
Figure 9-38. 8-count step-up ... 9-61
Figure 9-39. Pull-up ... 9-62
Figure 9-40. Straight-arm pull ... 9-63
Figure 9-41. Supine chest press ... 9-64
Figure 9-42. Bent-over row ... 9-65
Figure 9-43. Overhead push press ... 9-66
Figure 9-44. Supine body twist ... 9-67
Figure 9-45. Leg tuck ... 9-68
Figure 9-46. Shoulder roll ... 9-71
Figure 9-47. Lunge walk ... 9-72
Figure 9-48. Soldier carry ... 9-73
Figure 10-1. Moving under direct and indirect fire ... 10-1
Figure 10-2. Sustained running form ... 10-5
Figure 10-3. Military movement drill 1 ... 10-7
Figure 10-4. Verticals ... 10-8
Figure 10-5. Laterals ... 10-9
Figure 10-6. Shuttle sprint ... 10-10
Figure 10-7. Power skip ... 10-12
Figure 10-8. Crossovers ... 10-13
Figure 10-9. Crouch run ... 10-14
Figure 10-10. Speed running on a straight course ... 10-16
Figure 10-11. 300-yard shuttle run ... 10-17
Figure 10-12. Endurance training machines ... 10-21

Contents

Figure A-1. Push-up event narrative A-6
Figure A-2. Push-up additional checkpoints A-7
Figure A-3. Sit-up event narrative A-8
Figure A-4. Sit-up additional checkpoints A-9
Figure A-5. Sit-up hand and feet position A-9
Figure A-6. 2-mile run event narrative A-10
Figure A-7. 800-yard swim test narrative A-13
Figure A-8. 6.2-mile stationary cycle ergometer test narrative A-14
Figure A-9. 6.2-mile bicycle test narrative A-16
Figure A-10. 2.5-mile walk narrative A-17
Figure A-11A. DA Form 705 sample (page 1) A-19
Figure A-11B. DA Form 705 sample (page 2) A-20
Figure A-11C. DA Form 705 sample (page 3) A-21
Figure A-11D. DA Form 705 sample (page 4) A-22
Figure A-11E. DA Form 705 sample (page 5) A-23
Figure A-11F. DA Form 705 sample (page 6) A-24
Figure B-1. Climbing bars B-1
Figure B-2. Climbing bars, dimensions, top view B-2
Figure B-3. Climbing bar dimensions, side view B-3
Figure B-4. Multiple climbing bar pods B-4
Figure C-1. Poor posture limits range of motion C-2
Figure C-2. Good posture allows better range of motion C-2
Figure C-3. Rear lunge C-3
Figure C-4. Good (left) and poor (center and right) sitting posture C-4
Figure C-5. Good (left) and poor (right) standing posture C-5
Figure C-6. Soldiers in the flexed (right) and extended (left) postures C-6
Figure C-7. Performing extension to compensate for flexion C-6
Figure C-8. Performing decompression to compensate for compression C-7
Figure C-9. Soldiers moving under load C-8
Figure C-10. Set the hips and tighten the abdominal muscles C-9
Figure C-11. Power position C-10
Figure C-12. Lifting from the ground C-11
Figure C-13. Lifting overhead C-11
Figure C-14. Pushing C-12
Figure C-15. Pulling/climbing C-12
Figure C-16. Rotation C-13
Figure C-17. Jumping and landing C-13
Figure C-18. Lunging C-14
Figure C-19. Marching and foot marching C-14
Figure C-20. Changing direction C-15
Figure D-1. Wind chill chart D-3
Figure D-2. Clothing recommendations for PRT D-4

Figure E-1. Obstacles in combat ... E-1
Figure E-2. Jumping obstacles ... E-6
Figure E-3. Dodging obstacles ... E-6
Figure E-4. Climbing obstacles ... E-7
Figure E-5. Horizontal traversing obstacles ... E-7
Figure E-6. Crawling obstacles ... E-8
Figure E-7. Vaulting obstacles ... E-8
Figure E-8. Balancing obstacles ... E-9
Figure E-9. Black quadrant CFOC ... E-13
Figure E-10. Blue quadrant CFOC ... E-15
Figure E-11. White quadrant CFOC ... E-17
Figure E-12. Red quadrant CFOC ... E-19
Figure E-13. Tough one (course sketch) ... E-21
Figure E-14. Slide for life (course sketch) ... E-22
Figure E-15. Confidence climb (course sketch) ... E-23
Figure E-16. Skyscraper (course sketch) ... E-24
Figure E-17. Belly robber (course sketch) ... E-25
Figure E-18. Tarzan (course sketch) ... E-25
Figure E-19. Low belly over (course sketch) ... E-26
Figure E-20. Dirty name (course sketch) ... E-27
Figure E-21. Tough nut (course sketch) ... E-28
Figure E-22. Belly crawl (course sketch) ... E-28
Figure E-23. Inclining wall (course sketch) ... E-29
Figure E-24. High step over (course sketch) ... E-29
Figure E-25. Swing, stop, and jump (course sketch) ... E-30
Figure E-26. Six vaults (course sketch) ... E-31
Figure E-27. Easy balancer (course sketch) ... E-31
Figure E-28. Belly buster (course sketch) ... E-32
Figure E-29. Low wire (course sketch) ... E-32
Figure E-30. Hip-hip (course sketch) ... E-33
Figure E-31. Reverse climb (course sketch) ... E-34
Figure E-32. Weaver (course sketch) ... E-35
Figure E-33. Balancing logs (course sketch) ... E-35
Figure E-34. Island hopper (course sketch) ... E-36

Tables

Table 1-1. Principles of training ... 1-2
Table 1-2. Warrior tasks and battle drills, physical requirements for performance ... 1-4
Table 1-3. Warrior tasks and battle drills to components matrix ... 1-5
Table 1-4. Warrior tasks and battle drills to activities matrix ... 1-6
Table 5-1. Symptoms of overtraining ... 5-2

Table 5-2. Toughening phase PRT daily session overview (BCT and OSUT-R/W/B phases) 5-8
Table 5-3. Toughening phase PRT schedule (BCT and OSUT-R/W/B phases) 5-9
Table 5-4. Condensed sessions (toughening phase) 5-13
Table 5-5. Field training sessions (toughening phase) 5-14
Table 5-6. Sustaining phase PRT daily session overview (AIT and OSUT-B/G phases) 5-17
Table 5-7. Condensed sessions (sustaining phase) 5-18
Table 5-8. Field training sessions (sustaining phase) 5-18
Table 5-9. Sustaining phase PRT daily session overview (ARFORGEN) 5-21
Table 5-10. Unit PRT reset schedule, Month 1 5-22
Table 5-11. Unit PRT train/ready schedule, Month 1 5-28
Table 5-12. Unit PRT, available schedule 5-30
Table 5-13. Deployed PRT, collective schedule 5-31
Table 5-14. Deployed PRT, individual schedule 5-31
Table 5-15. RC quarterly and annual PRT schedule 5-32
Table 5-16. RC annual collective PRT schedule 5-33
Table 5-17. RC individual PRT schedule 5-34
Table 6-1. Reconditioning Level I training schedule 6-9
Table 6-2. Shoulder stability drill (SSD) 6-25
Table 6-3. Reconditioning Level II training schedule 6-69
Table 6-4. Reconditioning walk-to-run progression 6-69
Table 6-5. Reconditioning phase level II exit criteria 6-116
Table 8-1. Preparation drill 8-2
Table 8-2. Body segments trained in the conduct of the preparation drill 8-3
Table 8-3. Recovery drill 8-16
Table 8-4. Body segments trained in the conduct of the recovery drill 8-17
Table 9-1. Strength and mobility drills and activities 9-2
Table 9-2. Strength and mobility activity prescription 9-3
Table 9-3. Body segments trained in the conduct of CD 1 9-5
Table 9-4. Body segments trained in the conduct of CD 2 9-12
Table 9-5. Body segments trained in the conduct of CD 3 9-20
Table 9-6. Body segments trained in the conduct of PSD 9-38
Table 9-7. Body segments trained in the conduct of CL1 9-42
Table 9-8. Body segments trained in CL 2 9-48
Table 9-9. Body segments trained in the conduct of the STC 9-54
Table 9-10. Equipment required to conduct platoon-size STC 9-56
Table 9-11. Body segments trained in the guerrilla drill 9-70
Table 10-1. Endurance and mobility activities 10-2
Table 10-2. Endurance and mobility activity prescription 10-3
Table 10-3. Ability group assignment 10-18
Table 10-4. Quarter-mile split times based on AGR pace 10-19
Table A-1. Alternate aerobic event standards A-12

Preface

Soldier physical readiness is acquired through the challenge of a precise, progressive, and integrated physical training program. A well-conceived plan of military physical readiness training must be an integral part of every unit training program. This field manual prescribes doctrine for the execution of the Army Physical Readiness Training System.

This publication comprises the print portion of Army physical readiness training. Companion portions are available on the internet.

Terms that have joint or Army definitions are identified in both the Glossary and the text.

This publication prescribes DA Form 705 (Army Physical Fitness Test Scorecard).

The proponent for this publication is the United States Army Training and Doctrine Command (TRADOC). The preparing agency is the United States Army Physical Fitness School. Submit comments and recommendations for improvement of this field manual on DA Form 2028 (Recommended Changes to Publications and Blank Forms). To contact the United States Army Physical Fitness School, write—

DCG-IMT-TSSD

U.S. Army Physical Fitness School

ATTN: Director

4325 Jackson Blvd

Fort Jackson, SC 29207-5015

This regulation applies to the Active Army, the Army National Guard(ARNG)/Army National Guard of the United States(ARNGUS), and the United States Army Reserve(USAR), unless otherwise stated.

Unless this publication states otherwise, masculine nouns and pronouns refer to both men and women.

Introduction

The Army assesses, plans, prepares, and executes training and leader development through training based on tasks, conditions, and standards. Knowing the task, assessing the level of proficiency against the standard and developing a sustained or improved training plan is the essence of all Army training.

Army training overall prepares Soldiers, leaders, and units to fight in the full spectrum of operations. Combat readiness is the Army's primary focus as it transitions to a more agile, versatile, lethal, and survivable force.

Physical readiness training prepares Soldiers and units for the physical challenges of fulfilling the mission in the face of a wide range of threats, in complex operational environments, and with emerging technologies.

- Part I, Philosophy, covers approach, system, and leadership.
- Part II, Strategy, covers types of programs, planning considerations, and special conditioning programs.
- Part III, Activities, covers execution of training, preparation and recovery, strength and mobility, and endurance and mobility.
- Appendix A is the Army Physical Fitness Test.
- Appendix B discusses climbing bars.
- Appendix C discusses posture and body mechanics.
- Appendix D discusses environmental considerations.
- Appendix E discusses obstacle negotiation.

This field manual—

- Provides Soldiers and leaders with the doctrine of Army physical readiness training.
- Reflects lessons learned in battles past and present, time-tested theories, and principles and emerging trends in physical culture.
- Helps ensure the continuity of our nation's strength and security.
- Prepares Soldiers physically for full spectrum operations.
- Explains training requirements and objectives.
- Provides instructions, required resources, and reasons why physical fitness is a directed mandatory training requirement as specified in AR 350-1, *Army Training and Leader Development*.
- Allows leaders to adapt physical readiness training to unit missions and individual capabilities.
- Guides leaders in the progressive conditioning of Soldier strength, endurance, and mobility.
- Provides a variety of physical readiness training activities that enhance military skills needed for effective combat and duty performance.

PART ONE
Philosophy

This part discusses the philosophy of Army physical readiness training.

Chapter 1
Approach

Military leaders have always recognized that the effectiveness of Soldiers depends largely on their physical condition. Full spectrum operations place a premium on the Soldier's strength, stamina, agility, resiliency, and coordination. Victory—and even the Soldier's life—so often depend upon these factors. To march long distances in fighting load through rugged country and to fight effectively upon arriving at the area of combat; to drive fast-moving tanks and motor vehicles over rough terrain; to assault; to run and crawl for long distances; to jump in and out of craters and trenches; and to jump over obstacles; to lift and carry heavy objects; to keep going for many hours without sleep or rest—all these activities of warfare and many others require superb physical conditioning. Accordingly, this chapter links Army physical readiness training (PRT) to Army Force Generation (ARFORGEN).

TRAINING PROGRAM

1-1. This chapter introduces the elements and resources used in the Army Physical Readiness Training Program.

AR 350-1, ARMY TRAINING AND LEADER DEVELOPMENT

1-2. This regulation prescribes policy and procedure for the conduct of the Army Physical Fitness Training Program.

Who does AR 350-1 Apply to?

1-3. AR 350-1 applies to all Soldiers, functional branches, units, and operating agencies.
- **Physical readiness is the ability to meet the physical demands of any combat or duty position, accomplish the mission, and continue to fight and win.**
- Physical readiness training provides the physical component that contributes to tactical and technical competence, and **forms the physical foundation** for all training. Commanders and supervisors must establish PRT programs consistent with the requirements in AR 350-1, with their unit missions, and with this field manual (FM). Soldiers must meet the physical fitness standards set forth in AR 350-1 and in the Army Physical Fitness Test (APFT) provided in Appendix A.
- AR 350-1 specifies that physical fitness training is one of the Army's mandatory training requirements.

Why is PRT a mandatory training requirement?

1-4. Physical readiness training is a mandatory training requirement because it is—
- Considered by senior leaders to be essential to individual, unit, and force readiness.
- Required by law for all individuals and units.

ADP 7-0, TRAINING UNITS AND DEVELOPING LEADERS

1-5. This FM provides the training and leader development methodology that forms the foundation for developing competent and confident Soldiers in the conduct of full spectrum operations. The tasks, conditions, and standards of PRT activities derive from the mission analysis of the physical demands of unit mission, core mission essential task list (C-METL) or directed mission essential task list (D-METL), and warrior tasks and battle drills (WTBDs). The Army PRT System is performance-based, incorporating physically demanding activities that prepare Soldiers and units to accomplish the physical requirements of WTBDs. As Soldiers' physical performance levels increase, standards remain constant, but conditions become more demanding. To ensure the generation of superior combat power, the end state requires leaders to integrate the relative physical performance capabilities of every Soldier. Soldiers and leaders must execute the planned training, assess performance, and retrain until they meet Army Physical Fitness Training Program standards as specified in AR 350-1, *Training and Leader Development*. Conditions should replicate wartime conditions as nearly as possible.

PRINCIPLES OF TRAINING

1-6. The Army's approach to PRT links directly to its seven principles of training (ADP 7-0). Leaders must understand how these Army training principles (see Table 1-1) and PRT relate to improving war-fighting capabilities.

Table 1-1. Principles of training

1	Commanders and Other Leaders are Responsible for Training
2	Noncommissioned Officers Train Individuals, Crews, and Small Teams
3	Train as You Will Fight
4	Train to Standard
5	Train to Sustain
6	Conduct Multiechelon and Concurrent Training
7	Train to Develop Agile Leaders and Organizations

COMMANDERS AND OTHER LEADERS ARE RESPONSIBLE FOR TRAINING

1-7. Physical readiness training is the commander's program. Chapter 3, Leadership, discusses this principle of training in detail. Commanders are the primary training managers and trainers for their organization. Senior noncommissioned officers (NCOs) at every level of command are vital to helping commanders meet their training responsibilities. Senior NCOs are often the most experienced trainers in the unit; they are, therefore, essential to a successful PRT program. Leaders should emphasize the value of PRT by clearly explaining the objectives and benefits of the program. They must also use the time allotted for PRT effectively.

1-8. Each PRT session has specific tasks, conditions, and standards that support the physical requirements needed to accomplish the unit's C- and D-METLs. As the unit's primary training manager, commanders must do the following to optimize the effect of PRT:
- Incorporate mission command in PRT.
- Supervise the planning, preparation, execution, and assessment of PRT.
- Align PRT with mission/METL (mission-essential task list) requirements in support of full spectrum operations.
- Train to standard according to this FM.

- Assess individual and unit physical readiness according to this FM.
- Provide resources required to execute PRT.
- Incorporate safety and composite risk management (CRM).
- Ensure training is realistic and performance-oriented.
- Ensure training replicates the operational environment as closely as possible.

NONCOMMISSIONED OFFICERS TRAIN INDIVIDUALS, CREWS, AND SMALL TEAMS

1-9. Noncommissioned officers serve as the primary trainers for enlisted Soldiers, crews, and small teams. Noncommissioned officers must conduct standards-based, performance-oriented, mission- and METL-focused PRT. To accomplish the PRT mission, NCOs—
- Identify specific tasks that PRT enhances in support of the unit's C- or D-METL.
  - Individual.
  - Crew.
  - Small team.
- Prepare, rehearse, and execute PRT.
- Evaluate PRT and conduct AARs to provide feedback to the commander.

1-10. Senior NCOs train junior NCOs and aid in developing junior officers, ensuring mastery of PRT drills, exercise activities, and assessments.

1-11. This FM discusses these tenets of training in—
- Chapter 3, Leadership.
- Chapter 5, Planning Considerations.
- Chapter 7, Execution of Training.
- Appendix A, Army Physical Fitness Test.

TRAIN AS YOU WILL FIGHT

1-12. All Army training is based on the principle "Train as you will fight;" therefore, the primary focus of PRT goes far beyond preparation for the APFT. Soldiers improve their physical readiness capabilities through PRT. For Soldiers to achieve the desired standard of physical readiness, every unit training program must include a well-conceived plan of PRT. Training must be both realistic and performance-oriented to ensure physical readiness to meet mission/METL requirements.

Train the Fundamentals First

1-13. Toughening phase training provides foundational fitness and fundamental motor skills, which lay the foundation for all other activities in the sustaining phase. Once Soldiers are able to perform all of the exercises, drills, and activities to standard in this FM, they should be prepared to perform most physical challenges and advanced PRT.

Tenets

1-14. The eight tenets of train as you will fight, as they relate to PRT, are—
- PRT must support full spectrum operations and promote quick transitions between missions.
- PRT must support proficiency in combined arms operations and unified actions.
- PRT focus is on training the fundamentals first.
- PRT must be performance-oriented, conducted under realistic conditions, and mission focused.
- PRT should incorporate challenging, complex, ambiguous, and uncomfortable situations.
- PRT must incorporate safety and CRM.
- PRT must be conducted under conditions that replicate the operational environment.
- PRT must be conducted during deployments.

Chapter 1

Realism

1-15. Army PRT should be tough, realistic, and physically challenging, yet safe in its execution. The objective is to develop Soldiers' physical capabilities to perform their duty assignments and combat roles. Army PRT incorporates those types of training activities that directly support war-fighting tasks within full spectrum operations. Physical readiness training activities include such fundamental skills as climbing, crawling, jumping, landing, and sprinting, because all contribute to success in the more complex skills of obstacle negotiation, combatives, and military movement.

Performance-Oriented Training

1-16. Performance-oriented training involves performing tasks physically. The focus is on results, not process. Soldiers and units need to be proficient in the WTBDs required to perform their missions during duty and wartime conditions; therefore, Army PRT must be performance-based, incorporating physically demanding exercises, drills, and activities that prepare Soldiers and units to accomplish the physical requirements associated with the successful accomplishment of WTBDs. The tasks, conditions, and standards of PRT activities derive from the mission analysis of the physical demands of WTBDs. Table 1-2 shows examples of physical requirements for the performance of WTBDs.

Table 1-2. Warrior tasks and battle drills, physical requirements for performance

Shoot	Physical Requirements
Employ hand grenades	Run under load, jump, bound, high/low crawl, climb, push, pull, squat, lunge, roll, stop, start, change direction, get up/down, and throw.
Move	**Physical Requirements**
Perform individual movement techniques	March/run under load, jump, bound, high/low crawl, climb, push, pull, squat, lunge, roll, stop, start, change direction, and get up/down.
Navigate from one point to another	March/run under load, jump, bound, high/low crawl, climb, push, pull, squat, lunge, roll, stop, start, change direction, and get up/down.
Move under fire	Run fast under load, jump, bound, crawl, push, pull, squat, roll, stop, start, change direction, and get up/down.
Survive	**Physical Requirements**
Perform Combatives	React to man-to-man contact: push, pull, run, roll, throw, land, manipulate body weight, squat, lunge, rotate, bend, block, strike, kick, stop, start, change direction, and get up/down.
Adapt	**Physical Requirements**
Assess and Respond to Threats (Escalation of Force)	React to man-to-man contact: push, pull, run, roll, throw, land, manipulate body weight, squat, lunge, rotate, bend, block, strike, kick, stop, start, change direction, and get up/down. Run under load, jump, bound, high/low crawl, climb, push, pull, squat, lunge, roll, stop, start, change direction, get up/down, and throw.
Battle Drills	**Physical Requirements**
React to contact	Run fast under load, jump, bound, crawl, push, pull, squat, roll, stop, start, change direction, and get up/down.
Evacuate a casualty	Squat, lunge, flex/extend/rotate trunk, walk/run, lift, and carry.

Integrated Approach

1-17. The Army PRT System employs an integrated approach to physical conditioning by training the critical components of strength, endurance, and mobility. Table 1-3 and Table 1-4 show the correlation between WTBDs and PRT components and activities. Standards remain constant as Soldier physical performance levels increase, but conditions become more demanding. Soldiers and leaders execute the planned training, assess performance, and retrain until they meet Army PRT System standards under conditions that try to replicate wartime conditions. The end state requires leaders to integrate the relative physical performance capabilities of

every Soldier to generate superior combat power. Critical to successful individual and unit performance is the ability to develop the physical potential of all Soldiers for maximum performance in the accomplishment of the WTBDs. The tenets of this principle of training are discussed in detail in—
- Chapter 2, System.
- Chapter 4, Types of Programs.
- Chapter 5, Planning Considerations.

Table 1-3. Warrior tasks and battle drills to components matrix

PRT Components	Warrior Tasks					Battle Drills	
	Employ hand grenades	Perform individual movement techniques	Navigate from one point to another	Move under fire	Perform Combatives	Assess and Respond to Threats (Escalation of Force)	React to contact	Evacuate a casualty
Strength								
Muscular Strength	X	X	X	X	X	X	X	X
Muscular Endurance	X	X	X	X	X	X		X
Endurance								
Anaerobic Endurance	X	X	X	X	X	X	X	X
Aerobic Endurance		X	X			X		X
Mobility								
Agility	X	X	X	X	X	X	X	X
Balance	X	X	X	X	X	X	X	X
Coordination	X	X	X	X	X	X	X	X
Flexibility	X	X	X	X	X	X	X	X
Posture	X	X	X	X	X	X	X	X
Stability	X	X	X	X	X	X	X	X
Speed	X	X	X	X	X	X	X	X
Power	X	X	X	X	X	X	X	X

Table 1-4. Warrior tasks and battle drills to activities matrix

PRT Activities	Employ hand grenades	Perform individual movement techniques	Navigate from one point to another	Move under fire	Perform Combatives	Assess and Respond to Threats (Escalation of Force)	React to contact	Evacuate a casualty
Conditioning Drill 1	X	X	X	X	X	X	X	X
Conditioning Drill 2	X	X	X	X	X	X	X	X
Conditioning Drill 3	X	X	X	X	X	X	X	X
Guerrilla Drill	X	X	X	X	X	X	X	X
Climbing Drill 1	X	X	X	X	X	X		X
Climbing Drill 2	X	X	X	X	X	X		X
Strength Training Circuit	X	X	X	X	X	X	X	X
Military Movement Drill 1	X	X	X	X	X	X	X	X
Military Movement Drill 2	X	X	X	X	X	X	X	X
30:60s		X	X	X	X	X	X	X
60:120s		X	X	X	X	X	X	X
300-yd Shuttle Run	X	X	X	X	X	X	X	X
Ability Group Run		X	X					
Unit Formation Run		X	X					
Release Run		X	X	X	X	X	X	X
Terrain Run	X	X	X	X	X	X	X	X
Hill Repeats	X	X	X	X	X	X	X	X
Foot Marching		X	X	X				X
Obstacle Course Negotiation	X	X	X	X	X	X	X	X
Combatives	X	X	X	X	X	X	X	X

TRAIN TO STANDARD

1-18. Training to standard using appropriate doctrine prepares Soldiers to fight and sustain in the fight during full spectrum operations; therefore to be most effective, standards and doctrine must be uniformly known, understood, replicable, and accepted. Doctrine represents a professional Army's collective thinking about how it intends to fight, train, equip, and modernize. It is the condensed expression of the Army's approach to warfighting. The tactics, techniques, procedures, organizations, support structures, equipment, and training must all derive from it. In accordance with ADP 7-0, *Training Units and Developing Leaders,* mastery, not just proficiency, should be the goal of all training. Leaders should continually challenge Soldiers and units by

varying the conditions to make successful achievement of the standard more challenging. The tenets of standards-based training are—
- Leaders know and enforce standards.
- Leaders define success in the absence of standards.
- Leaders train to standard, not time.

1-19. Physical readiness training doctrine applies Army-wide. It includes all Soldiers, functional branches, units, and operating agencies. Physical readiness training provides a foundation for combat readiness and must be an integral part of every Soldier's life. Unit readiness begins with the physical fitness of Soldiers and the NCOs and officers who lead them. Physical readiness training must be conducted according to the Army Physical Fitness Training Program, as prescribed in AR 350-1, and conform to the Army doctrine prescribed in this FM. Army doctrine continues to evolve to reflect lessons learned in major periods of armed conflict.

1-20. Commanders train and develop Soldiers and leaders to adapt, preparing their subordinates to operate in positions of increased responsibility. Commanders intensify training experiences by varying training conditions. Activities must impose both physical and metabolic demands on the Soldier. For example, requiring the Soldier to surmount a ledge, climb stairs, sprint between covered and concealed positions, and evacuate casualties all challenge the Soldier to overcome an ever changing set of physical demands. To prepare Soldiers to meet the physical demands of their profession, a system of training must focus on the development of strength, endurance and mobility, plus the enhancement of the body's metabolic pathways. Developing the ability of Soldiers to meet the changing physical demands that are placed upon them without undue fatigue or risk of injury is woven into the fabric of the PRT System. Standards are achieved through precise control of the following:
- Prescribe appropriate intensity and duration to which Soldiers perform PRT.
- Properly distribute external loads across the major joints of the body.
- Integrate and balance the components of strength, endurance, and mobility.
- Provide adequate rest, recovery, and nutrition.

1-21. Every PRT session emphasizes the performance-related factors for the successful accomplishment of WTBDs. The systematic stress of each Soldier's metabolic system substantially influences their ability to perform physically at an optimum level. Competence in individual Soldier performance of all PRT activities instills confidence in the ability to perform. It also gives personnel the confidence that all Soldiers in the unit have similar physical capabilities and the mental and physical discipline needed to adapt to changing situations and physical conditions. Commanders at every echelon integrate training events in their training plans to develop and train imaginative, adaptive leaders, and units. Commanders should understand the fundamental doctrinal training principles described in this FM and apply them accurately. This ensures Soldiers are physically prepared to accomplish the unit mission/C- and/D-METLs.

TRAIN TO SUSTAIN

1-22. Units must be able to operate continuously while deployed. Physical readiness training provides a foundation for combat readiness and must be an integral part of every Soldier's life. Soldiers and leaders are responsible for maintaining a high state of physical readiness to support training and operational missions. Units need to be capable of fighting for sustained periods. Soldiers should therefore become experts in the conduct and performance of PRT. This link between training and sustainment is vital to mission success. Once Soldiers and units train to the required level of proficiency, leaders structure individual and collective training plans to retrain critical tasks at the minimum frequency needed to sustain proficiency. Sustainment training is the key to maintaining unit proficiency despite personnel turbulence and operational deployments. Army units train to accomplish their missions by frequent sustainment training on critical tasks.

CONDUCT MULTIECHELON AND CONCURRENT TRAINING

1-23. Multi-echelon training is the simultaneous training of more than one echelon on different tasks. It is the most effective and efficient way of sustaining proficiency on mission-essential tasks with limited time and resources. All multi-echelon training techniques have these distinct characteristics:
- They require detailed planning and coordination by commanders and leaders at each echelon.

Chapter 1

- They maintain battle focus by linking individual and collective battle tasks with unit METL tasks and within large-scale training event METL tasks.
- They habitually train at least two echelons simultaneously on selected METL tasks and require maximum use of allocated resources and available time.

1-24. Concurrent training occurs when a leader conducts training within another type of training. It complements the execution of primary training objectives by allowing leaders to make the most efficient use of available time. Similarly, while Soldiers are waiting their turn on the firing line at a range, their leaders can train them on other tasks. Leaders look for ways to use all available training time. Concurrent training can occur during multi echelon training. In PRT, for example, concurrent training occurs when part of the unit is conducting climbing drills (CLs) while the others are performing conditioning drills (CDs). Upon completion, the groups will change in order to optimize the use of limited equipment.

Army Force Generation Model

1-25. Prior to the conduct of multi-echelon training, commanders assess their units' proficiency levels to determine the appropriate tasks to be trained. The same is true for commanders in the execution of PRT. The commander plans PRT based on the assessed level of physical readiness of his Soldiers. An example is the ARFORGEN model that utilizes the reset, train/ready, and available phases. (Figure 1-1, Army PRT System and relationship to ARFORGEN.) The PRT System consists of three training phases: initial conditioning, toughening, and sustaining. These three phases align with Soldiers' current career paths (future Soldier, initial military training [IMT], and unit PRT) within the operational, institutional, and self-development domains of the Army training system.

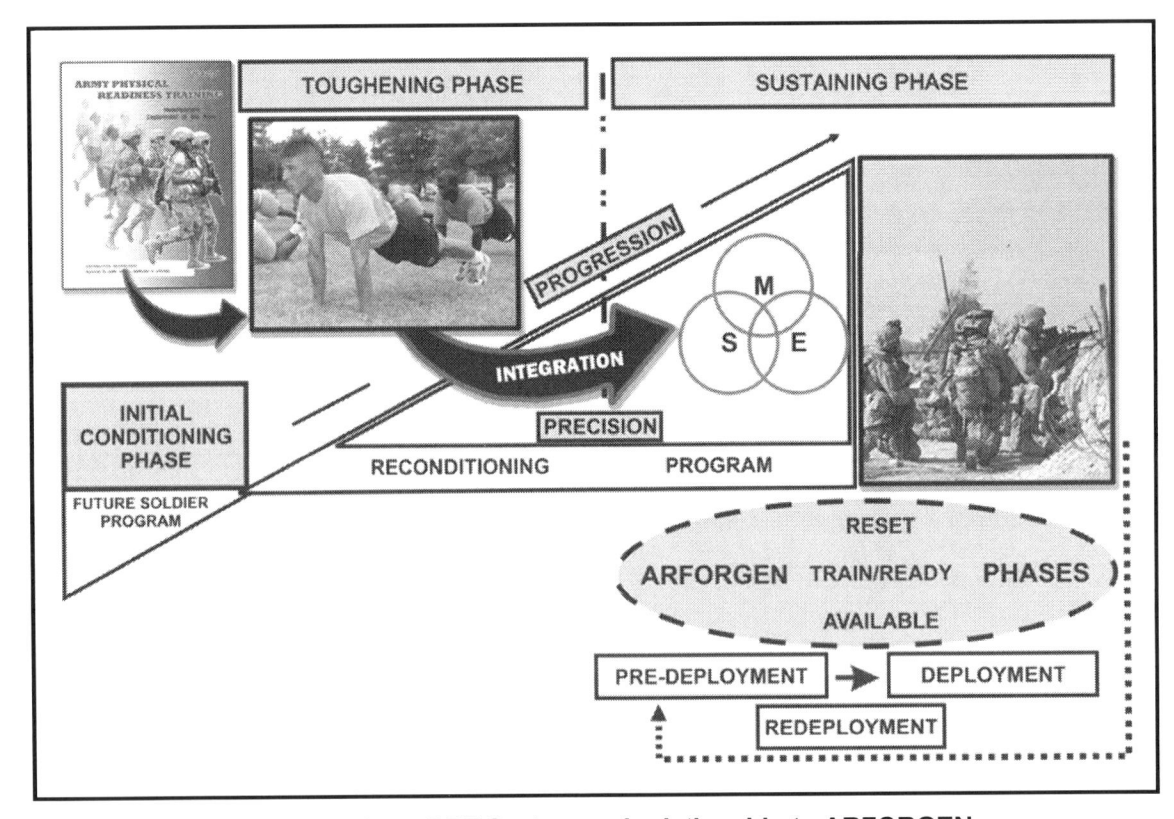

Figure 1-1. Army PRT System and relationship to ARFORGEN

TRAIN TO DEVELOP AGILE LEADERS AND ORGANIZATIONS

1-26. In accordance with FM 7-0, the Army trains and educates its Soldiers to develop agile leaders and units to be successful in any operational environment. Training and developing leaders is an embedded component of

Approach

every training event, especially in PRT. Noncommissioned officers are responsible for conducting standards-based, performance-oriented, and realistic training. Senior NCOs train junior NCOs and assist in the development of junior officers in their mastery of PRT drills, exercises, activities, and assessments. Noncommissioned officers have an opportunity to lead everyday during PRT. Nothing is more important to the Army than building confident, competent, adaptive leaders for tomorrow. See ADP 7-0, for the tenets that underlie the development of agile and competent leaders and organizations.

1-27. Physical readiness is a mandatory training requirement that requires synchronization of the Army Physical Fitness Training Program strategy across the training domains of the Army Training System: the operational domain, the institutional domain, and the self-development domain. The objective of PRT is to prepare Soldiers to meet the physical demands related to mission and C- or D-METL. This occurs through an organized schedule of prescribed PRT drills and activities. These exercises, drills, and activities are methodically sequenced to adequately challenge all Soldiers through progressive conditioning of the entire body while controlling injuries. Commanders execute a vital role in PRT leader training and development in the operational and self-development domains. They plan training in detail, prepare for training thoroughly, execute training to standard and evaluate short-term training proficiency in terms of desired long-term results.

> *"Military physical training should build Soldiers up physically, wake Soldiers up mentally, fill Soldiers with enthusiasm, and discipline them."*
> Koehler's West Point Manual of Disciplinary Physical Training (1919)

Summary

This FM provides Soldiers and leaders with the doctrine of Army PRT. It is a product of our history, forged out of the great battles from the past to the present. Its doctrinal concepts also reflect emerging trends in current physical culture. This FM will impact the Army in a manner of importance toward the continuation of our national strength and security. The purpose of Army PRT is not merely to make our Soldiers look fit, but to actually make them physically ready for the conduct of full spectrum operations.

Chapter 2
System

Army physical readiness is defined as the ability to meet the physical demands of any combat or duty position, accomplish the mission, and continue to fight and win.

The goal of the Army Physical Fitness Training Program is to develop Soldiers who are physically capable and ready to perform their duty assignments or combat roles. To reach this goal, leaders use the PRT System to aim first at developing strength, endurance, and mobility. Soldiers must be able to perform required duties and sustain activity during full spectrum operations. Soldiers trained through PRT demonstrate the mobility to apply strength and endurance to the performance of basic military skills such as marching, speed running, jumping, vaulting, climbing, crawling, combatives, and water survival. These skills are essential to personal safety and effective Soldier performance—not only in training, but also, and more importantly, during combat operations.

Physical fitness and health form the basis of physical readiness. Physical readiness is in turn essential to combat readiness. Physical readiness training prepares Soldiers and units physically to be successful in the conduct of full spectrum operations. Secondary goals of PRT are to instill confidence and the will to win; develop teamwork and unit cohesion; and integrate aggressiveness, resourcefulness, and resilience. The PRT System brings Soldiers to a state of physical readiness through a systematic program of drills and activities specifically designed to enhance performance of WTBDs. Army PRT seeks to attain the development of all Soldiers' physical attributes to the fullest extent of their given potential. This will instill confidence in their ability to perform their duties under all circumstances.

"Soldiers should train to become stronger, faster, mobile, lethal, resilient, and smarter."
Frank A. Palkoska, Director USAPFS

PHASES

2-1. Commanders face the continual challenge of training Soldiers with different physical capabilities. Training to the level of the least fit removes rigor from the program, while excessive rigor places less fit Soldiers at risk of injury. Most commanders recognize this dilemma and try to occupy a reasonable middle ground. This chapter guides commanders in the implementation of safe and challenging PRT. It should be applied according to Chapters 5 and 6.

2-2. The initial conditioning phase prepares future Soldiers to learn and adapt to Army PRT. Toughening phase activities develop foundational fitness and fundamental movement skills that prepare Soldiers to transition to the sustaining phase. Sustaining phase activities develop a higher level of physical readiness required by duty position and C- or D-METL. Reconditioning restores Soldiers' physical fitness levels that enable them to safely re-enter the toughening or sustaining phase and progress to their previous level of conditioning. See Chapter 6 for more information on reconditioning. Types of PRT training include on-ground, off-ground, and combatives. Within these types of training are three fundamental components: strength, endurance, and mobility. Phased training follows the principles of precision, progression, and integration. Finally, Army PRT optimizes physical performance within an environment of injury control. Figure 2-1 shows the PRT System's phases, types of training, components, principles, and reconditioning as they apply to ARFORGEN.

Chapter 2

INITIAL CONDITIONING PHASE

2-3. The purpose of the initial conditioning phase is to establish a safe starting point for people considering entering the Army. This includes those individuals enrolled in the Army's Future Soldier Program and in the Reserve Officer Training Corps. This phase of training is conducted before enlistment or pre-commissioning.

TOUGHENING PHASE

2-4. The purpose of the toughening phase is to develop foundational fitness and fundamental movement skills. A variety of training activities with precise standards of execution ensures that bones, muscles, and connective tissues gradually toughen, rather than break. In the toughening phase, Soldiers gradually become proficient at managing their own body weight. Toughening phase activities develop essential skills associated with critical Soldier tasks such as jumping, landing, climbing, lunging, bending, reaching, and lifting. Physical readiness improves through progression in these activities. The toughening phase occurs during IMT, basic combat training (BCT), one station unit training (OSUT) (red/white/blue phases), and Basic Officer Leader Course A (BOLC A). The toughening phase prepares Soldiers to move to the sustaining phase.

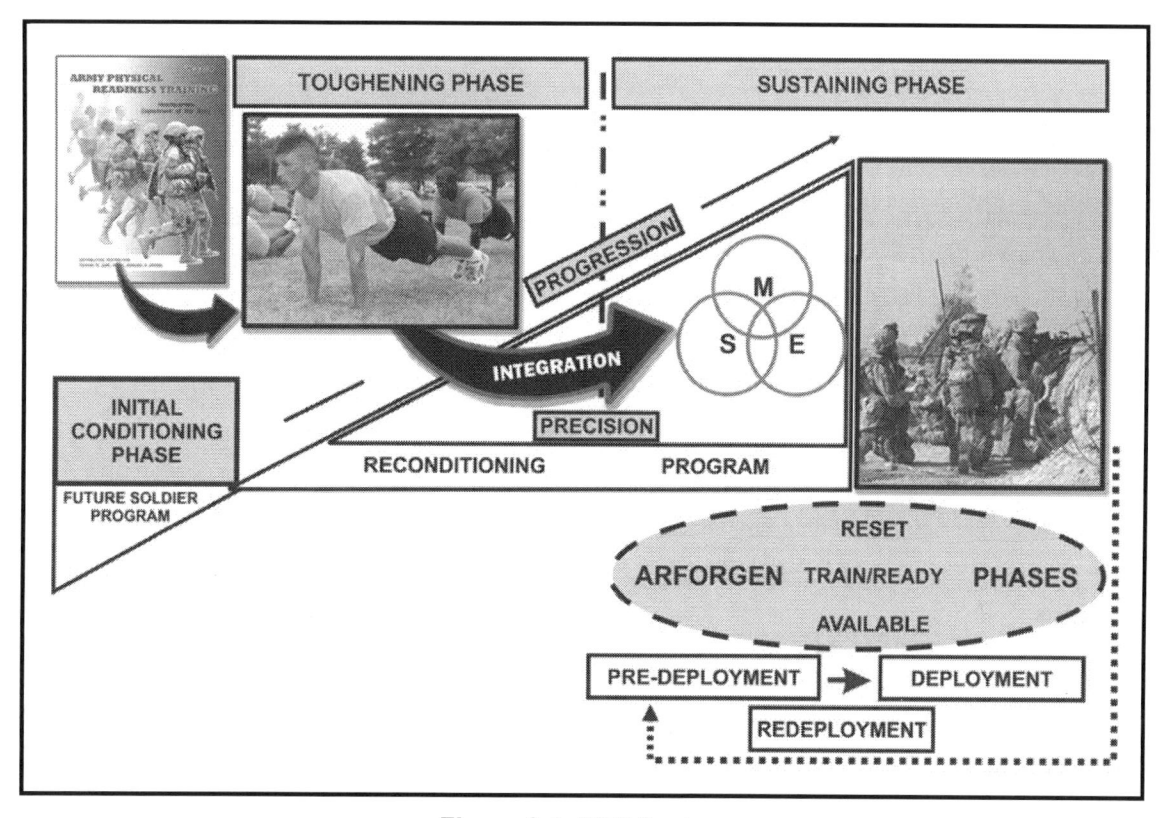

Figure 2-1. PRT System

SUSTAINING PHASE

2-5. The purpose of the sustaining phase is to continue physical development and maintain a high level of physical readiness appropriate to duty position and the requirements of the unit's C- or D-METL as it applies to ARFORGEN. See AR 350-1 to reference ARFORGEN. Sustaining phase activities are conducted in unit PRT throughout the Army. In this phase, activities become more demanding. Exercises, drills, and activities such as advanced calisthenics, military movement, kettlebell, and CLs are performed with increasing resistance. Endurance and mobility activities such as foot marching, speed running, and sustained running increase in intensity and duration. Activities that directly support unit mission and C- or D-METL, such as individual movement techniques, casualty carries, obstacle courses, and combatives are integrated into PRT sessions.

RECONDITIONING

2-6. The objective of reconditioning is to restore physical fitness levels that enable Soldiers to reenter the toughening or sustaining phase safely, and then progress to their previous levels of conditioning. See Chapter 6, Special Conditioning Programs, for more information on rehabilitation and reconditioning PRT. Soldiers may participate in reconditioning after rehabilitation and recovery from injury or illness, and then re-enter training in the toughening or sustaining phases.

2-7. Factors such as extended deployment, field training, block leave, and recovery from illness or injury can cause Soldiers to move from the toughening or sustaining phases to reconditioning. Once Soldiers meet the transition criteria for re-entry into unit training, they may do so. Units usually conduct either reconditioning and toughening or reconditioning and sustaining phases at the same time.

PRINCIPLES

2-8. The conduct of Army PRT follows the principles of precision, progression, and integration. These principles ensure that Soldiers perform all PRT sessions, activities, drills, and exercises correctly, within the appropriate intensity and duration for optimal conditioning and injury control.

PRECISION

2-9. Precision is the strict adherence to optimal execution standards for PRT activities. Precision is based on the premise that the quality of the movement or form is just as important as the weight lifted, repetitions performed or speed of running. It is important not only for improving physical skills and abilities, but to decrease the likelihood of injury due to the development of faulty movement patterns. Adhering to precise execution standards in the conduct of all PRT activities ensures the development of body management and fundamental movement skills.

PROGRESSION

2-10. Progression is the systematic increase in the intensity, duration, volume, and difficulty of PRT activities. The proper progression of PRT activities allows the body to positively adapt to the stresses of training. When progression is violated by too rapid an increase in intensity, duration, volume or difficulty the Soldier is unable to adapt to the demands of training. The Soldier is then unable to recover, which leads to overtraining or the possibility of injury. Phased training ensures appropriate progression.

INTEGRATION

2-11. Integration uses multiple training activities to achieve balance and appropriate recovery between activities in the PRT program. Because most WTBDs require a blend of strength, endurance, and mobility, PRT activities are designed to challenge all three components in an integrated manner. The principle of integration is evident when WTBDs and their component movements are incorporated in PRT. For example, CDs and CLs develop the strength, mobility, and physical skills needed to negotiate obstacles. Military movement drills (MMDs) improve running form and movement under direct or indirect fire. The guerrilla drill (GD) develops the strength and skill associated with casualty evacuation and combatives. The drills, exercises, and activities in this FM integrate essential Soldier tasks, making PRT a critical link in the chain of overall Soldier physical readiness.

COMPONENTS

2-12. The PRT System incorporates the three components of training shown in Figure 2-2.

Chapter 2

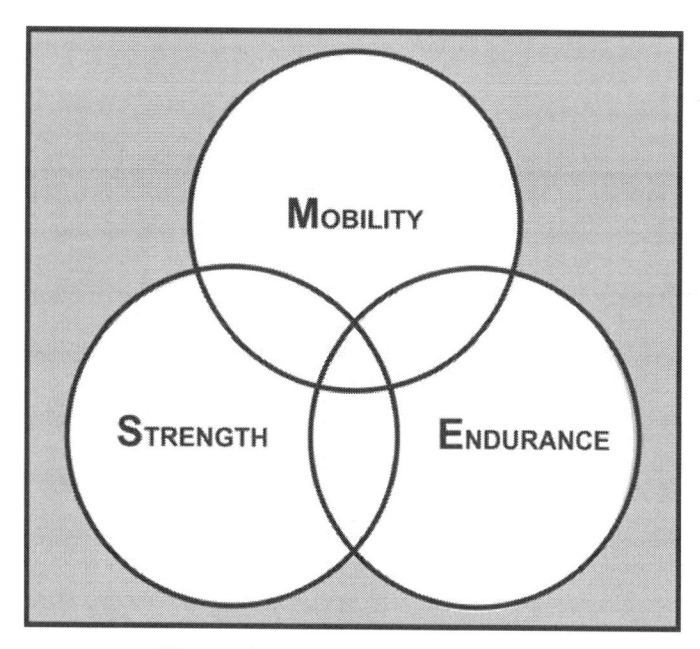

Figure 2-2. Components of PRT

STRENGTH

2-13. Strength is the ability to overcome resistance. Strength runs a continuum between two subcomponents: absolute muscular strength (the capacity of a muscle/muscle group to exert a force against a maximal resistance) and muscular endurance (the capacity of a muscle/muscle group to exert a force repeatedly or to hold a fixed or static contraction over a period time). Soldiers need strength to foot march under load; enter and clear a building or trench line; repeatedly load heavy rounds; lift equipment; transport a wounded Soldier to the casualty collection point; and most of all, to be able to withstand the rigors of continuous operations while under load. A well-designed, strength-training program improves performance and appearance and controls injuries. The Army's approach to strength training is performance-oriented. The goal is to attain the muscular strength required to perform functional movements against resistance. Calisthenics are the foundation of Army strength training and body management. They develop the fundamental movement skills needed for Soldiers to manipulate their own body weight and exert force against external resistance. Strength is further developed through the performance of advanced calisthenics, resistance training, CL, and the GD.

ENDURANCE

2-14. This is the ability to sustain activity. The component of endurance, like strength, also runs a continuum between the ability to sustain high-intensity activity of short duration (anaerobic), and low-intensity activity of long duration (aerobic).

2-15. A properly planned and executed endurance training program balances anaerobic and aerobic training. Analysis of the mission and C- or D-METL for nearly all units shows a significant need for anaerobic endurance. Anaerobic training has a crossover value in improvement of aerobic capability. However, aerobic training alone does little to improve anaerobic capacity. To enhance effectiveness and survivability, Soldiers must train to perform activities of high intensity and short duration efficiently. Endurance programs based solely on sustained running, while likely to improve aerobic endurance, fail to prepare units for the type of anaerobic endurance they will need for the conduct of full spectrum operations.

- Examples of anaerobic training are speed running, individual movement techniques, and negotiation of obstacles.
- Examples of aerobic training are foot marching, sustained running, cycling, and swimming.

MOBILITY

2-16. This is the functional application of strength and endurance. It is movement proficiency. Strength with mobility allows a Soldier to squat and lift an injured Soldier. Without sufficient mobility, a strong Soldier may have difficulty executing the same casualty transport technique. Endurance without mobility may be acceptable to a distance runner, but for Soldiers performing individual movement techniques, both components are essential for optimal performance.

"Movement, as such, may replace by its effect any remedy, but all the remedies in the world cannot take the place of movement."

Tissot, XVIII Century

QUALITATIVE PERFORMANCE FACTORS

2-17. Performing movements with correct posture and precision improves physical readiness while controlling injuries. Qualitative performance factors for improved mobility include:

Agility is the ability to stop, start, change direction, and efficiently change body position. Performing the GD, the shuttle run (SR), and negotiating obstacles all improve agility.

Balance is the ability to maintain equilibrium. The drills in this FM are designed to challenge and improve balance. Balance is an essential component of movement. External forces such as gravity and momentum act upon the body at any given time. Sensing these forces and responding appropriately leads to quality movements.

Coordination is the ability to perform multiple tasks. Driving military vehicles and operating various machinery and weaponry requires coordination. Coordination of arm, leg, and trunk movement is essential in climbing and individual movement techniques.

Flexibility is the range of movement at a joint and its surrounding muscles. Flexibility is essential to performing quality movements safely. Regular, progressive, and precise performance of calisthenics and resistance exercises promote flexibility. Spending time on slow, sustained stretching exercises during the recovery drill (RD) may also help to improve flexibility.

Posture is any position in which the body resides. Posture constantly changes as the body shifts to adapt to forces of gravity and momentum. Good posture is important to military bearing and optimal body function. Proper carriage of the body while standing, sitting, lifting, marching, and running is essential to movement quality and performance.

Stability is the ability to maintain or restore equilibrium when acted on by forces trying to displace it. Stability depends on structural strength and body management. It is developed through regular precise performance of PRT drills. Quality movements through a full range of motion, such as lifting a heavy load from the ground to an overhead position, require stability to ensure optimal performance without injury.

Speed is rate of movement. Many Soldier tasks require speed. Speed improves through better technique and conditioning. Lengthening stride (technique) and increasing pace (conditioning) improve running speed.

Power is the product of strength and speed. Throwing, jumping, striking, and moving explosively from a starting position require both speed and strength. Power is generated in the trunk (hips and torso). Developing trunk strength, stability, and mobility is important to increasing power. Soldiers, as tactical athletes, are power performers.

TYPES

2-18. The PRT System incorporates the three types of training shown in Figure 2-3.

Chapter 2

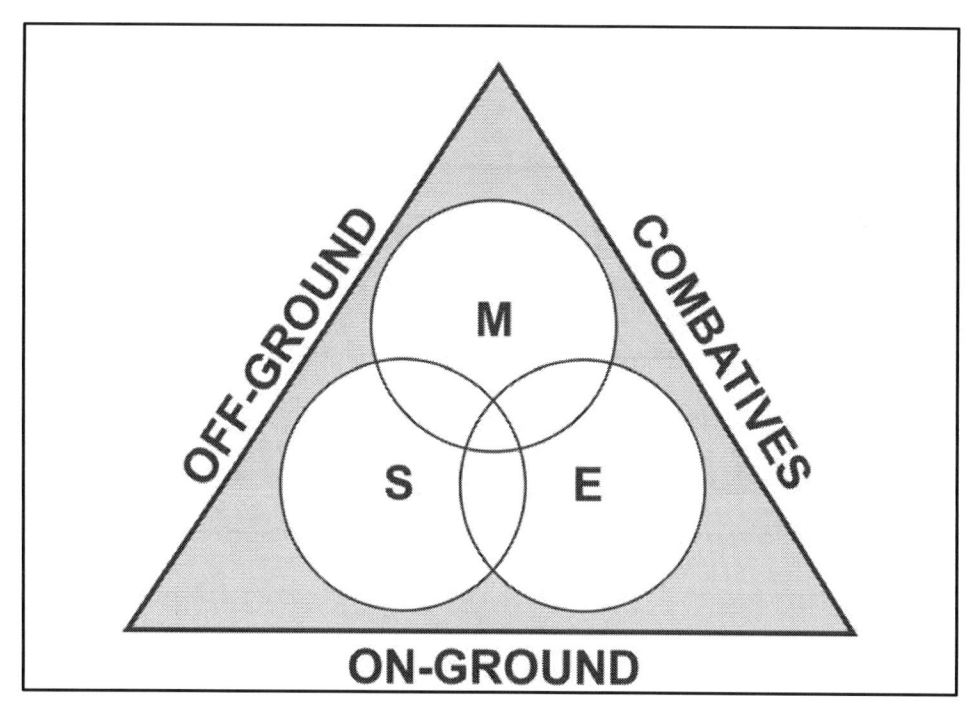

Figure 2-3. Types of PRT

ON-GROUND TRAINING

2-19. On-ground training includes activities in which Soldiers maintain contact with the ground. Activities such as marching, speed running, sustained running, calisthenics, and resistance training create a foundation for physical fitness and movement skills.

OFF-GROUND TRAINING

2-20. Off-ground training includes activities that take place off the ground briefly (jumping and landing) or while suspended above ground for longer periods (climbing bar and negotiation of high obstacles). Examples of jumping and landing exercises are high jumper, power jump, and verticals. Negotiation of high obstacles (reverse climb and cargo net) and exercises using the climbing pod (pull-up and leg tuck) require manipulation of the body and specific movement skills while suspended above ground.

COMBATIVES TRAINING

2-21. This includes techniques that deter or defeat opponents using projectile (weapons), striking and/or close range (grappling). (See FM 3-25.150.)

Summary

The Army's PRT System consists of three phases: the initial conditioning phase, the toughening phase, and the sustaining phase. The initial conditioning phase prepares future Soldiers to learn and adapt to Army PRT. Toughening phase activities develop foundational fitness and fundamental movement skills that prepare Soldiers to transition to the sustaining phase. Activities in the sustaining phase develop a higher level of physical readiness required by duty position and/or C- or D-METL. Reconditioning restores Soldiers to physical readiness levels that allow them to safely re-enter the toughening or sustaining phase. Types of PRT include on-ground, off-ground, and combatives. Within these types of training are three fundamental components: strength, endurance, and mobility. Phased training of these components is guided by the overarching principles of precision, progression, and integration. Finally, Army PRT optimizes physical performance within an environment of injury control. Figure 2-1 illustrates the PRT System's phases, types of training, components, principles, and reconditioning as they apply to ARFORGEN.

Chapter 3

Leadership

"The American Soldier...demands professional competence in his leaders in battle; he wants to know that the job is going to be done right, with no unnecessary casualties. The noncommissioned officer wearing the chevron is supposed to be the best Soldier in the platoon, and he is supposed to know how to perform all duties expected of him. The American Soldier expects his sergeant to be able to teach him how to do his job, and expects even more from his officers."

General of the Army Omar N. Bradley

Throughout history, the Army has had confident leaders of character and competence. Leaders develop through a dynamic process consisting of three equally important training domains: operational, institutional, and self-development according to AR 350-1. The process incorporating these domains provides the following key leadership elements: fundamental military specialty experience; education that instills key competencies; personal and professional development goals that enable leaders to develop the skills, the knowledge, and the attitudes needed for success. Leaders at all levels should understand that PRT improves Soldier resiliency, which is a vital component of a combat-ready force. This chapter addresses the importance of leadership as it applies to PRT.

TRAITS

3-1. The success or failure of the PRT program depends upon the quality of its leadership. Leadership is the process of influencing Soldiers by providing purpose, direction, and motivation. The best outcome results only when Soldiers extend themselves completely in strenuous physical activities and perform all exercises in the prescribed form. Officers and NCOs lead, train, motivate, and inspire their Soldiers. Only the best leadership can inspire Soldiers to cooperate to this extent. For these reasons, only the best qualified NCOs in the unit should lead PRT. The leader must exemplify the Army adage: *Be, Know, Do.*

COMPETENCE

3-2. All officers, NCOs, and PRT leaders must set and enforce standards through complete mastery of this TFM. They must not only be able to explain and demonstrate all activities, but also must know the best methods of presenting and conducting them. Leaders set the example. The PRT leader demonstrates tactical and technical competence through a mastery of PRT subject matter. Mastery is the first step in developing confidence, assurance, and poise. Thorough knowledge of this FM allows the PRT leader to apply the training principles of precision, progression, and integration needed to attain Soldier physical readiness. Skill in demonstrating and leading all PRT exercises, drills, and activities is essential to teaching technique and is invaluable to the PRT leader. The unprepared, hesitant leader loses the confidence and respect of Soldiers almost immediately. The well-prepared, confident leader gains the respect and cooperation of all Soldiers at the outset.

PHYSICAL QUALIFICATIONS AND APPEARANCE

3-3. The personal appearance and physical qualifications of the PRT leader affect his effectiveness. He should exemplify the things he is seeking to teach. It is a great advantage if the leader himself can do all and more than he asks of his men. He must be physically fit because PRT leadership is so strenuous that considerable strength, endurance, and mobility are essential prerequisites for success.

Chapter 3

KNOWLEDGE OF HUMAN BEHAVIOR

3-4. Successful leadership in PRT requires the leader to know and appreciate the individual physical and mental differences of his Soldiers. He must get to know his Soldiers as individuals and be quick to recognize signs indicating their reactions to his instruction. The successful PRT leader ensures that his subordinates understand the critical importance of PRT to the successful accomplishment of WTBDs in support of the unit's C- or D-METL. This is accomplished by understanding Soldiers, knowing how to lead and motivate them, understanding how they learn, and using this knowledge in PRT sessions. To succeed, PRT leaders must have the confidence of the Soldiers. He gains their confidence by winning their respect. He wins their respect by his sincerity, integrity, determination, sense of justice, energy, self-confidence, and force of character. A leader who has the admiration and respect of his Soldiers easily secures their cooperation. The leader treats the Soldiers with consideration and avoids imposing unreasonable physical demands on them. If Soldiers are exercised too violently, they become so stiff and sore that they look upon the next PRT session with apprehension. When this happens, Soldiers can develop an antagonistic attitude toward the leader and the program. Instead of cooperating, they will malinger at every opportunity.

ENTHUSIASM

3-5. Another essential quality of the PRT leader is enthusiasm. Successful Army PRT activities must be carried on in a continuous and vigorous manner. Soldiers reflect the attitude of the PRT leader. If the leader is enthusiastic, his instructed Soldiers will be enthusiastic. If the leader is apathetic, his instructed Soldiers will be apathetic. The enthusiasm of a leader springs from the realization of the importance of the mission. Leaders must be inspired by the thought that what they do every minute of every day may mean the difference between life and death. There is no more effective method of obtaining the energetic, wholehearted participation of Soldiers in the PRT program than by providing skilled, enthusiastic leadership.

"The instructor must lose himself in his work, must demand precision, encourage here, correct there, reprove one man, and boost another. In fact, he must so strive himself, that his men will be proud of their leader in every way, proud of his appearance, proud of his ability, proud of his fairness, and proud because their instructor is helping to make their organization the best in the Army."

LTC Herman J. Koehler, First Master of the Sword, United States Military Academy

COOPERATION

3-6. A successful PRT program requires the full cooperation of all Soldiers. Orderly movement of Soldiers and units requires a precise and unified effort. A Soldier belongs to a team that works smoothly when every Soldier plays his part. Each Soldier knows what to do in response to a command as well as what his fellow Soldiers must do. The Soldier's confidence in the team grows until he feels as sure of them as he does of himself. The final result is teamwork, and teamwork is attained though the medium of drills.

3-7. A drill consists of certain movements that allow the unit to conduct an activity with order and precision. Drills train Soldiers to do their parts exactly so that, on command, the unit moves instantly and smoothly. Drill training starts the day a Soldier enters the Army. In the beginning, he is taught the movements of his feet and arms used in PRT, marching, and handling the weapon. He is trained in all these activities until he reaches a point where he does them automatically in response to a command. He is then placed in a unit and trained to do all these activities with other Soldiers. Squads, platoons, and companies drill with the smoothness of machinery. The result is cooperative, unified action—teamwork. Soldiers are at their best when inspired to have pride in themselves and their organization. This pride finds expression in perfect response to command.

MOTIVATION

3-8. Commanders and leaders at all levels may provide one of the best incentives for their Soldiers when they are visible and actively participate in PRT. When Soldiers feel their chain of command believes in PRT to the extent that they themselves regularly engage in the activities, they are motivated to greater effort. Troops also develop a greater esprit de corps and respect for their officers and NCOs when all actively participate. Finally, the frequent use of Soldiers as assistant instructors (AIs) also serves as an incentive. Soldiers will work hard for this honor and positively respond to AI responsibilities.

RESPONSIBILITIES

3-9. Leaders must provide facilities and funds to support a PRT program that will develop physical readiness in all Soldiers.

FACILITIES

3-10. Exercise drill activities require flat, grassy areas. The GD, speed, and sustained running require well lighted running routes, tracks, and marked fields. Strength development requires kettlebells, step-up benches, and climbing bars.

SPECIAL PROGRAMS

3-11. Leaders must follow training guidelines for individual, reconditioning, pregnancy, and post-partum weight control, APFT failure, and new Soldier programs.

EXERCISE NAMES

3-12. Soldiers learn all exercises by name, sequence, and movement. This ensures efficient use of time and precision of execution.

CORRECTIVE TRAINING

3-13. Assistant instructors must remove Soldiers who need corrective training from the formation. This applies to Soldiers not performing exercises, drills or activities to standard. The AI corrects all mistakes and ensures proper execution.

SCHEDULING AND SUPERVISING

3-14. Leaders responsible for scheduling and supervising PRT should take the following actions:
- Make PRT as important as any other training activity.
- Dedicate sufficient time for PRT (60 to 90 minutes).
- Avoid substituting other training or routine duties during scheduled PRT.
- Schedule and conduct PRT when it makes the most sense. Physical readiness training should not be reserved only for the early morning hours and may run during or at the end of the duty day.
- Prevent the misuse of allotted PRT time by using qualified personnel to supervise and lead.
- Provide for mass participation regardless of rank, age or gender during every PRT session.
- Adhere to PRT schedules for the toughening and the sustaining phases.
- Use appropriate PRT formations.
- Use preparatory commands and commands of execution.
- Use cadence appropriate for planned activities.
- Require PRT leaders to lead and conduct activities with the Soldiers to determine appropriate intensity levels.
- Require one AI for every 15 Soldiers.
- Require AIs to supervise the execution of all PRT activities and make appropriate corrections.

3-15. Leaders have the latitude to adjust the PRT schedule to balance it with other training to avoid conflicts with physically demanding events that can lead to overtraining. For example, if the confidence obstacle course (CFOC) is the day's main physical training event, leaders should not schedule strength training for PRT (unless it is conducted later in the training day). If conflicts cannot be resolved, PRT should be performed after a physically demanding event (later in the duty day), rather than before the event (in the morning) for safety reasons. It is also acceptable to not conduct the scheduled PRT session in order to provide adequate rest and recovery.

> **Summary**
> Leaders are challenged with scheduling and executing PRT programs that ensure Soldiers and units are prepared to successfully perform their wartime mission. Effective leadership is essential to the success of any program. Successful leaders possess qualities that gain the confidence and respect of their Soldiers.

PART TWO

Strategy

This part discusses the strategy of Army physical readiness training.

Chapter 4

Types of Programs

> "The quality of the unit is determined by the overall picture of physical condition and total military fitness of all its members. It is more important that all men in a unit receive the benefits of a balanced and well-directed program of physical training than that a few members achieve record performances. The physical training program, therefore, is directed toward the total conditioning of all men."
>
> FM 21-20, Physical Training (1946)

Army PRT achieves other valuable outcomes in addition to developing and maintaining a high level of individual and unit readiness. These outcomes include: basic military skills and survivability along with their intangible benefits. The basic military skills associated with PRT include foot marching, running, swimming, jumping, vaulting, climbing, crawling, lifting, and load carrying. Survivability is often dependent upon maneuverability and mental alertness. Intangible benefits include teamwork, aggressiveness, confidence, resourcefulness, a will to win, discipline, and adaptability. Physical resilience is also a gained attribute.

INITIAL MILITARY TRAINING

4-1. Initial military training has the following elements: BCT, advanced individual training (AIT), OSUT, and Basic Officer Leader Courses A and B (BOLC A and B).

BASIC COMBAT TRAINING

4-2. The training program in BCT provides foundational fitness and fundamental motor skill development. New Soldiers report to BCT at various levels of physical readiness and ability. During the first weeks of training, the focus is on progressive training of the whole body. To minimize the risk of injury, Soldiers must perform exercises precisely. Also, their intensity must progress gradually. The toughening phase BCT training schedules in Chapter 5, Planning Considerations, when executed to standard, provide the proper training intensity and exercise volume and gradual progression appropriate to improving physical fitness and controlling injuries. Commanders should evaluate each new Soldier who falls below the BCT standard and give special assistance to improve deficiencies. Supplemental training should not punish a new Soldier for the inability to perform well. Commanders and PRT leaders need to realize that it takes at least six to eight weeks to begin positive changes in physical fitness levels; therefore, some Soldiers may require additional time to make the improvements required to meet Army standards.

Chapter 4

"More PRT does not equal better PRT. Training quality is more important than the number of repetitions performed."

William R. Rieger, National Strength and Conditioning Association
Certified Strength and Conditioning Specialist

ADVANCED INDIVIDUAL TRAINING

4-3. Advanced individual training focuses on technical and MOS-oriented (military occupational specialty) subjects; therefore, PRT should continue to prepare these Soldiers to meet the physical requirements of their first unit of assignment. It is recommended that commanders should continue conducting toughening phase activities until Soldiers meet Army standards before transitioning to sustaining phase activities. See Chapter 5 for AIT planning considerations.

ONE STATION UNIT TRAINING

4-4. Physical readiness training in OSUT brings Soldiers through the toughening phase and prepares them for the rigors of their first unit of assignment. New Soldiers follow the same progression as BCT during the red/white/blue phases of OSUT.

4-5. Commanders should continue conducting toughening phase activities until Soldiers meet Army standards. Soldiers can then transition to sustaining phase activities during the black/gold phases of OSUT. These activities are more difficult and complex and prepare Soldiers to perform the physical requirements of their duty assignments as their units prepare for full spectrum operations. See Chapter 5 for OSUT planning considerations.

WARRANT OFFICER CANDIDATE SCHOOL

4-6. Physical readiness training in warrant officer candidate school employs sustaining phase exercises, drills, and activities to prepare Soldiers for the rigors of warrant officer candidate school and their first unit of assignment.

BASIC OFFICER LEADER COURSES

4-7. The training program in BOLC A brings Soldiers up to a level of physical readiness that prepares them for the rigors of BOLC B. Cadets and officer candidates report to BOLC A at various levels of physical readiness and ability. During the first weeks of training, the focus is on progressive training of the whole body. It is recommended that Soldiers in BOLC A perform toughening phase activities during PRT sessions. Soldiers in BOLC B transition to performing sustaining phase activities during PRT sessions. To minimize the risk of injury, Soldiers perform exercises precisely and the intensity progresses gradually. Commanders should evaluate each new Soldier who falls below the BOLC A standard and give special assistance to improve deficiencies. Again, more PRT is not necessarily better. Instructors emphasize quality of the training, not quantity of exercises performed. Commanders and PRT leaders need to realize that it takes at least six to eight weeks to begin positive changes in physical fitness levels; therefore, some Soldiers may require additional time to make the improvements required to meet Army standards.

ACTIVE AND RESERVE COMPONENTS

4-8. This section covers PRT programs for the active and reserve component (RC) forces. It also provides an overview of the Army training management process and its relationship to the development of individual and unit PRT programs.

ACTIVE

4-9. Active component PRT includes unit, individual, reconditioning, and special conditioning programs.

Unit

4-10. The goal of Army PRT is to improve each Soldier's physical ability to survive and win in any operational environment. Physical readiness includes all aspects of physical performance and requires training well above that of simple preparation for the APFT. Commanders are responsible for the training, performance, and readiness of their Soldiers. Physical readiness training is a commander's program; therefore, commanders should employ the Army training management process specified in FM 7-0. The Army training management process provides a systematic way to manage time and resources to meet training objectives through purposeful training activities. Commanders use this process to identify training requirements and to subsequently plan, prepare, execute, and assess all training.

Mission-Essential Task List

4-11. The Army's training management model provides the framework for commanders to achieve proficiency in their unit's mission-essential task list (METL). The unit METL drives training. Key to the success of this process is the inclusion of bottom-up feedback. This approach applies mission command to the training process. With this approach, senior leaders provide training focus, direction, and resources. Subordinate leaders develop objectives and training requirements specific to the unit and provide feedback on training proficiency. They also identify unit needs and train to standard according to the unit training schedule or the event training plan. Senior leaders provide guidance based on mission and priorities, requiring subordinate leaders to conduct analysis to identify both collective and individual tasks that support the higher headquarters mission-essential tasks. This is the top-down approach to training. Input provided by subordinate leaders identifies critical training needs in order to achieve task proficiency. This is the bottom-up approach to training. This process, when combined, creates the top-down/bottom-up approach to training. This ensures effective communication of the requirements and of the planning, preparing, executing, and assessing of training. This process is essential to ensure proper conduct and execution of the unit PRT program.

Operating Tempo

4-12. Well-planned PRT maximizes physical performance in the completion of critical Soldier and leader tasks that support the unit's mission and C- and/or D-METL. It must reflect the commander's training objectives and goals and must reflect the principles of precision, progression, and integration. With ever changing operating tempo (OPTEMPO), units and Soldiers must continue to train to improve/sustain METL performance. Training priorities dictate how often and how rigorously PRT occurs.

Army Force Generation

4-13. Army Force Generation is the driving force behind training management. The Army provides campaign capable expeditionary forces through ARFORGEN. Army Force Generation applies to both regular Army and RC (Army National Guard and U.S. Army Reserve) units. Unit commanders and PRT leaders can plan PRT based on the specific requirements addressed in each of the ARFORGEN phases. Chapter 5 provides commanders and PRT leaders with example training schedules based on the three phases of ARFORGEN. Each phase has a specific focus.

Reset

4-14. The reset phase focuses on individual and collective training tasks that support their C- and/or D-METL.

Train/Ready

4-15. The train/ready phase focuses on higher level collective tasks associated specifically with deployment.

Available

4-16. The available phase continues focus on higher-level collective tasks. The unit achieves trained status and becomes available for immediate alert and deployment.

Chapter 4

Top Down/Bottom Up

4-17. Units not involved in ARFORGEN should still follow the Army training management process. The unit's mission and METL still drive training. The top-down/bottom-up approach to training mentioned previously ensures effective communication of the requirements and of the planning, preparing, executing, and assessing of training. Senior leaders continue to provide training focus, direction, and resources. Subordinate leaders continue to develop objectives and training requirements specific to the unit and provide feedback on training proficiency. They also identify unit needs and train to standard according to the unit training schedule or event training plan. Senior leaders provide guidance based on mission and priorities, requiring subordinate leaders to conduct analysis to identify both collective and individual tasks that support the higher headquarters mission-essential tasks. Well-planned PRT maximizes physical performance in the completion of critical Soldier and leader tasks that support the unit's mission and METL. It must reflect the commander's training objective and goals and be based on the principles of precision, progression, and integration. With ever changing OPTEMPO, units and Soldiers must continue to train to improve or sustain METL performance. Training priorities dictate how often and how rigorously PRT occurs. Professional development schools, hospitals, military police, communication centers, and Department of the Army staff have various challenges in planning and conducting PRT. Leaders should make every effort to conduct phased unit or individual PRT five times a week. See Chapter 5 for unit PRT schedules.

INDIVIDUAL

4-18. Commanders that authorize the use of individual training programs for their Soldiers should follow the same training management principles outlined in the previous paragraphs. Army Force Generation is the driving force behind training management. The Army provides campaign-capable expeditionary forces through ARFORGEN. Army Force Generation also applies to RC (Army National Guard and U.S. Army Reserve) units; therefore, leaders and individual Soldiers need to use the PRT system outlined in this FM to help them achieve and sustain high levels of physical readiness required in the conduct of duty position or full spectrum operations. Individual PRT programs must be designed to improve the individual's contribution to the unit's physical readiness. Conditioning, CLs, and GDs, foot marching, and running activities employed in unit PRT can be performed individually or with a partner. Individual and small group PRT should develop and maintain a level of physical readiness equivalent to that required for success in performance of the unit mission and C-METL or D-METL. Chapter 5, Planning Considerations, provides commanders and PRT leaders with examples of collective and individual training schedules based on the three phases of ARFORGEN.

4-19. All Soldiers must understand that it is their personal responsibility to achieve and sustain a high level of physical readiness and resilience. Individual physical readiness includes all aspects of physical performance and requires training well above that of simple preparation for the APFT. Many Soldiers are assigned to duty positions that restrict participation in collective unit PRT programs. Examples include Army staff, hospitals, service-school staff and faculty, recruiting, Reserve Officer Training Corps, Reserve and National Guard units. In such units, commanders must develop leadership environments that encourage and motivate Soldiers to accept individual responsibility for their own physical readiness. Physical readiness and resilience requirements are the same for these personnel as for others.

RECONDITIONING PROGRAM

4-20. As mentioned in Chapter 2, System, the objective of the reconditioning program is to restore physical fitness levels of Soldiers on medical profile that enable them to re-enter the toughening or sustaining phase. Commanders and health care personnel provide special aid to Soldiers assigned to reconditioning PRT for one or more of the following medical conditions: injury, illness, or surgery. Chapter 6, Special Conditioning Programs, provides more information on reconditioning.

PREGNANCY AND POSTPARTUM TRAINING

4-21. The U.S. Army Medical Command has responsibility for the Army Pregnancy Postpartum Physical Training (PPPT) Program. The Army PPPT Program is designed to maintain health and fitness levels of pregnant Soldiers and to assist them in returning to pre-pregnancy fitness levels after the end of their pregnancy. The goal is to integrate the Soldier into her unit PRT program with an emphasis on meeting the standards for the

Army Weight Control Program (AWCP) and APFT. Pregnancy postpartum physical training program standards, policies, procedures, and responsibilities are set forth in the United States Army Public Health Command (USAPHC), Technical Guide Series 255A-E, U.S. Army Pregnancy Post Partum Physical Training Program. The USAPHC is responsible to ensure that the Technical Guide Series 255A-E manuals are updated periodically and made available in a web-based format. USAPHC is responsible for training PPPT instructor trainers and health care experts who provide training for the PPPT program as specified in the Technical Guide Series 255A-E.

Senior Commanders

4-22. Senior commanders have responsibility for PPPT program execution and will ensure the following:
- All eligible Soldiers will participate in the installation level PPPT program.
- Soldiers maintain health and fitness levels throughout their pregnancy and return to pre-pregnancy fitness levels.
- Soldiers will safely reintegrate into their unit's PRT program.
- Soldiers meet AWCP and APFT standards.
- Medical consultation and support are provided.
- Healthcare instruction is available for the local PPPT program.
- Facilities and equipment are available for conducting the PPPT.
- Personnel are designated to conduct the PT portion of the PPPT program.

Publications

4-23. Adhere to the content, standards, policies, procedures, and responsibilities in the guide series and regulation.
- AR 350-1, *Army Training and Leader Development.*
- USAPHC Technical Guide Series 255A-E, U.S. Army Pregnancy/Postpartum Physical Training Program. The USAPHC provides and updates this series of guides, which provides the standards, policies, procedures, and responsibilities that Medical Command must follow in administering the PPPT program.

Reserve Component and Remotely Located Soldiers

4-24. Reserve component Soldiers, geographically remote Soldiers, and those assigned to installations with a small population of pregnant Soldiers may use the materials designed for an individualized PPPT program. These are available from USAPHC.

Eligibility

4-25. Soldiers diagnosed as pregnant or who are recovering from childbirth are exempt from regular unit physical training and APFT for the duration of the pregnancy and 180 days past the end of their pregnancy. These Soldiers are required to enroll in the Army PPPT Program. Before they may participate in the physical training portions of the PPPT program, they must receive clearance to do so from their health care provider. Before they start convalescent leave, postpartum Soldiers receive a postpartum profile. This 45-day temporary profile starts the day of the birth or end of the pregnancy. It specifies that the Soldier may engage in physical training at her own pace. Soldiers are encouraged to use the at-home component of the Army PPPT Program while on convalescent leave. Postpartum Soldiers may return to regular unit physical training before 180 days after the end of their pregnancy, but must receive health care provider clearance to do so.

FAILURE TO PERFORM TO STANDARD

4-26. Most units are diverse in physical readiness levels due to injuries, illnesses, deployments, and new Soldiers. This diversity may affect the number of APFT and unit physical readiness standard failures. Over time, a solid PRT program allows Soldiers to achieve the Army and unit standards. Performing high-quality training once per day is a better approach than conducting additional high-volume training that could lead to overuse injuries. Additional reinforcement training, if determined appropriate by the commander, should focus on identified weaknesses and sustain strengths. **Do not use supplemental training to punish a Soldier for the inability to perform well.**

NEW SOLDIERS ENTERING UNITS

4-27. The new Soldier's threshold level of physical performance may fall below the minimum for his gaining unit. He may be considered a borderline APFT performer or be borderline overweight. He may be fresh out of BCT, AIT, or OSUT, or may have just completed a permanent change of station move or returned from an extended deployment. These Soldiers are facing new conditions relating to physical performance (acclimatization to altitude, temperature, and humidity), which may take them up to four weeks to adapt. Although Soldiers leave IMT prepared to transition to the sustaining phase, they may de-train due to leave, transit, and in-processing at their new duty assignments just like Soldiers in operational units. New Soldiers need to train in the unit for 90 days before PRT leaders or AIs assess the Soldiers' fitness levels. This timeframe allows them to acclimatize, assimilate into a unit PRT program, and adapt physiologically and psychologically.

WEIGHT CONTROL

4-28. Overweight Soldiers need not perform PRT with a special group. Instead, they should participate in unit PRT and continue to train with their units; however, they may require supplemental PRT, plus education on diet and exercise (Chapter 6 and AR 600-9). The supplemental PRT session focus for overweight Soldiers who perform unit PRT is on low-impact activities and resistance training to achieve caloric expenditure, build lean muscle mass, and promote optimal fat loss. Aim for 20 to 60 minutes of exercise by either walking or splitting the session between machines (15 minutes each on the bike, stepper, and rower). Leaders synchronize additional resistance training activities with strength and mobility sessions conducted during unit PRT. These additional training sessions should focus on total body strength development.

4-29. Overweight Soldiers not performing unit PRT should follow the activities on the unit schedule and supplement with further aerobic exercise. Resistance training for overweight Soldiers should be initially limited to normal PRT activities such as CDs and CLs. Resistance exercise can stimulate muscle growth and aid fat loss. The more lean mass is present, the more calories are needed to sustain it. Weight loss may not occur if lean mass is added through resistance training. In this case, Soldiers will have a lower body fat percentage, but not a lower weight. Because AR 600-9 specifies that satisfactory progress for this program is measured in pounds, not body fat, reassessment of the Soldier's progress should include both weigh-ins and circumference measurements.

RESERVE

4-30. Today's Soldier understands the critical importance of individual physical readiness. This is especially true for RC Soldiers whose collective training periods are spread throughout the training year. Reserve component units must meet the challenge of physical readiness for mission performance often with less collective training time than regular Army units; therefore, it is critical for RC commanders to apply the Army training management process. Using this process, the commander can systematically manage time and resources to meet training objectives through purposeful training activities. He also uses the process to identify training requirements and subsequently plan, prepare, execute, and assess all training. The Army's training management model provides the framework for commanders to achieve proficiency in their unit's mission-essential task list (METL). The unit METL drives training.

UNIT

4-31. Army Force Generation is the driving force behind training management. The Army provides campaign capable, expeditionary forces through ARFORGEN. Army Force Generation also applies to Army National Guard and U.S. Army Reserve units. Army Force Generation is based on a three-phase readiness cycle. The three phases of ARFORGEN are:
- Reset.
- Train/Ready.
- Available.

4-32. Each phase has a specific focus. The reset phase focuses on individual and collective training tasks that support their C- and/or D-METL. The train/ready phase focuses on higher level collective tasks associated specifically with deployment. The available phase continues focus on higher level collective tasks as the unit is considered trained and available for immediate alert and deployment to a specified contingency. Unit commanders and PRT leaders can plan PRT based on the specific requirements addressed in each of the ARFORGEN phases. Chapter 5 provides commanders and PRT leaders with training schedules based on the three phases of ARFORGEN.

4-33. Unit PRT activities should be incorporated into individual duty for training (IDT) periods. Commanders must determine how much emphasis to place on PRT activities and allocate time and resources accordingly. At a minimum, one hour of PRT activities should be incorporated into every sixteen hours of unit training during IDT periods. During annual training (AT) periods, units should try to conduct PRT five times per week.

4-34. Valuable RC collective PRT time should not be focused on preparing Soldiers to take the APFT; nor should the focus of PRT during IDT periods be on achieving a training effect. The focus should be on precisely teaching and leading the activities in this FM. On some occasions, Soldiers might have to perform at near-maximal effort during training, such as in the conduct of a unit foot march or other training activities. This should be the exception, not the norm. A training program in which Soldiers are expected to perform at near-maximal effort once a month will not achieve the desired physiological changes, no matter how intense. This type of program probably causes more harm than good and typically violates the commander's CRM.

INDIVIDUAL

4-35. An ideal unit PRT program strives to give Soldiers the knowledge they need to conduct their own quality PRT sessions between unit assemblies. The program should increase Soldier motivation so they habitually train on their own. Incorporating the PRT activities in this FM into IDT periods is one way to effect motivation with the added benefit of providing commanders a physical readiness snapshot. Most of the exercises, drills, and activities in this FM support the type of RC unit PRT program described in this section. For example, Soldiers would collectively learn CD 1 during the unit assembly—then train on their own between unit assemblies—raising their proficiency and readiness level at the same time. Soldiers are then prepared for PRT sessions conducted during subsequent IDT and AT periods. Few of the exercises, drills, and activities in this FM require expensive or hard-to-obtain equipment so they can easily be performed individually.

SCHEDULING TRAINING

4-36. Use USAR troop program unit and Army National Guard mobilization day Soldiers who have civilian health and fitness experience to assist in conducting the program, especially the reconditioning program (Chapter 6). All NCOs should learn and be able to teach the exercises, drills, and activities in this FM.

4-37. Chapter 5, Planning Considerations, covers how PRT activities can be integrated into an example RC yearly training cycle. The focus of collective PRT during unit AT should be on increasing the unit physical readiness level. For this to be effective, PRT activities on the example unit AT schedule must be introduced during IDT periods and trained individually before AT. Chapter 5 also provides 5-day PRT schedules that can be used during AT periods or by RC Soldiers for individual training sessions.

COMMAND RESPONSIBILITIES

4-38. Effective leadership is critical to the success of a PRT program. History has taught us that often Soldiers and units may not be afforded the time to develop an appropriate level of physical readiness and resilience during mobilization. Commanders can reduce this risk by applying the following strategies to meet individual and unit goals and objectives.

GUIDANCE

4-39. Clearly explaining the objectives and benefits of the program ensures that the time allotted for PRT is used effectively; therefore, leaders must constantly emphasize the value of PRT and commanders must provide resources to support a program that will improve each Soldier's level of physical readiness. Mandatory participation is essential. All individuals, regardless of rank, age or gender benefit from regular exercise. In

Chapter 4

some instances, leaders will need to make special efforts to overcome recurring problems that interfere with regular training. To foster a positive attitude, unit and PRT leaders must be knowledgeable, understanding, and fair, but demanding. A high level of physical readiness and resilience cannot be attained by simply going through the motions. Smart, realistic, and challenging training to standard is essential. Leaders should not punish Soldiers who fail to perform to standard, because this often does more harm than good. They must recognize individual differences and motivate Soldiers to put forth their best efforts. The application of reconditioning PRT will progressively return Soldiers with medical profiles to the unit. It also allows them to train with the unit whenever possible, within the limits of their profiles.

LEADERSHIP BY EXAMPLE

4-40. Leaders must understand and practice Army physical readiness doctrine. Their example will emphasize the importance of PRT and highlight it as a key element of the unit's training mission. Command presence and participation at PRT formations and assessments will set a positive example for subordinates.

Leadership Training

4-41. Commanders must ensure that leaders are trained to supervise and conduct PRT. The doctrinal concepts and unit program models presented in this FM are starting points for commanders and PRT leaders to optimize unit PRT and assessment.

Evaluation And Standards

4-42. Commanders must use the unit's mission and C- or D-METL as criteria for evaluating PRT program effectiveness.

DISCIPLINE

4-43. Highly disciplined and physically fit Soldiers make for a corps spirit that inspires organizations to dare because of their ability to do. PRT programs must therefore develop every Soldier's physical potential to the fullest. When PRT is executed precisely, Soldiers develop discipline; disciplined Soldiers perform all duties with greater confidence and success. Well-run programs also enhance physical resilience.

"Such discipline may therefore be defined as the voluntary, intelligent, and cheerful subordination of every individual in an equal degree with every other individual of the mass to which he belongs, and of which he is an interdependent and not independent unit, through which the object of the mass can alone be attained."

LTC Herman J. Koehler

SAFETY

4-44. Safety is a major consideration when planning and evaluating PRT programs. Commanders should use the CRM process for all PRT activities to ensure they do not place their Soldiers at undue risk for injury or accident. The commander should address:
- Environmental conditions.
- Emergency procedures.
- Facilities.
- Differences in age.
- Gender.
- Level of conditioning of each Soldier in the unit.

"The best form of welfare for the troops is first-class training."

B. H. Liddell Hart, British Military Tactician

Summary

PRT is the commander's program. It must reflect his training goals and be based on the principles of precision, progression, and integration. The purpose of the PRT program is to develop and maintain a high level of unit physical readiness appropriate to duty position and for the conduct of full spectrum operations. The goal is to improve each Soldier's physical ability to survive, be resilient and win on the battlefield. Well-planned PRT optimizes physical performance in the completion of the critical Soldier and leader tasks that support the unit's mission, C-METL and/or D-METL. The unit METL drives training. Army Force Generation is the driving force behind training management. The Army provides campaign capable, expeditionary forces through ARFORGEN. Army Force Generation applies to both regular Army and RS (Army National Guard and U.S. Army Reserve) units.

Chapter 5
Planning Considerations

"Physical fitness is the basis for all other forms of excellence."

John Fitzgerald Kennedy

This chapter provides commanders and PRT leaders a template for efficiently implementing Army PRT doctrine into the unit training plan. Specifically, it provides training guidance for the toughening and sustaining phases.

GOAL

5-1. The overall goal of the Army Physical Fitness Training Program is to develop Soldiers who are physically capable, ready to perform their duty assignments or combat roles and to promote good health, resiliency and physical readiness through exercise. To best plan PRT to achieve this goal, leaders must know the PRT system. Chapter 2 explains the Army PRT goal. Adherence to the exercise principles of precision, progression, and integration are key to program effectiveness and injury control. These principles of exercise should be used in developing all PRT schedules.

PRECISION

5-2. This is strict adherence to the best execution standards for PRT activities. Precision assumes that the quality of movement is just as important as the amount of weight lifted, number of repetitions performed or distance run. For example, when a Soldier can no longer maintain the PRT leader's push-up cadence speed or the correct form while performing push-ups, he will get into the six-point stance and continue his push-ups. This allows precision and completion of the specified number of repetitions.

5-3. Precision is essential in resistance training to develop strength and mobility whether the Soldier is performing CDs, CLs, the strength training circuit, or using strength training machines (STMs). Precision also depends on the use of stable body positions, appropriate range of motion, proper speed, and proper breathing. Too little stability, too much weight, exceeding the appropriate range of motion, improper speed, and improper breathing technique reinforce faulty motor patterns. Over time, these practices could lead to injury. Chapter 9, Strength and Mobility, provides safe and effective resistance training techniques. When a Soldier fails to maintain proper running form or speed during speed running, he should slow down to regain proper running form. Typically, Soldiers perform the first two repetitions of speed running intervals (30:60s and 60:120s) or the 300-yard SR too quickly. When this happens, it causes form to break down and affects the ability to maintain speed for the specified number of repetitions. Soldiers should be instructed to pay attention to their speed in order to maintain precision.

PROGRESSION

5-4. This is the systematic increase in the intensity or duration of PRT activities. Proper progression allows the body to positively adapt to the stresses of training. When intensity or duration is increased too rapidly, the Soldier cannot adapt to the demands of training, and is unable to recover, leading to overtraining and possible injury. The following are gradually increased to produce the desired physiological effect:
- Intensity (resistance and pace).
- Exercise volume (number of sets and repetitions).
- Duration (time).

5-5. In the toughening phase, the duration of the ability group runs (AGRs) progress from 10 minutes to 30 minutes gradually over the training cycle. The pace of individual Soldiers or the group also gradually increases

Chapter 5

over time. For example, in the sustaining phase during the conduct of CLs and speed running, Soldiers progress from wearing the individual physical fitness uniform (IPFU) to Army combat uniforms (ACUs), boots, advanced combat helmet (ACH), and improved outer tactical vest (IOTV). Progression in strength is achieved by increasing one or more of the following for each exercise when using strength training equipment.

- Resistance (weight).
- Number of sets.
- Number of repetitions.

5-6. Adhering to the scheduled intensity and duration prevents the Soldier from progressing too fast. How fast the Soldier should progress also depends on how regularly he performs challenging activities and how much rest and recovery time he gets. PRT time is a valuable resource, especially during the toughening phase. Every PRT session develops strength, endurance, and mobility. To ensure improvement, PRT sessions in IMT occur 5 or 6 times a week and last 45 to 60 minutes. PRT sessions in the sustaining phase last 60 minutes or more and occur 4 to 5 times a week. If PRT cannot be conducted first thing in the morning, it should be conducted at some other time during the duty day. Training sessions should be sequenced to ensure adequate recovery.

5-7. Commanders and PRT leaders must avoid overtraining syndrome during the planning and conduct of the PRT program. Overtraining significantly impacts Soldier resiliency through the degradation of physical performance, as well as, behavioral and emotional well-being. Through a proper ramp of progression (intensity, duration, and type of exercise), PRT exercises, drills, and activities provide a demanding physical challenge that leads to improvements in affective, cognitive, and psychomotor Soldier performance.

Overtraining

5-8. Overtraining occurs when training involves excessive frequency, intensity and/or duration of training that may result in extreme fatigue, illness or injury. This may occur within a short period of time (days) or cumulatively (weeks/months) over the length of the training cycle and beyond. Overtraining often results from a lack of adequate recovery, rest or in some cases, a lack of nutrient intake. Thus, too much training, too little recovery, and/or poor nutrient intake (fueling) may elicit both the physical and psychological symptoms associated with overtraining syndrome. Refer to Table 5-1 for the symptoms associated with overtraining syndrome.

Table 5-1. Symptoms of overtraining

SYMPTOMS OF OVERTRAINING SYNDROME		
Performance Issues	*Physiological Symptoms*	*Psychological Symptoms*
<ul><li>Early Fatigue</li><li>Increased Heart Rate w/less Effort</li><li>Decreased Strength, Endurance, Speed, and Coordination</li><li>Decreased Aerobic Capacity</li><li>Delayed Recovery</li></ul>	<ul><li>Persistent Fatigue</li><li>On-going Muscle Soreness</li><li>Loss of Appetite</li><li>Excessive Weight Loss</li><li>Excessive Loss of Body Fat</li><li>Irregular Menses</li><li>Increased Resting Heart Rate</li><li>Chronic Muscle Soreness</li><li>Increase in Overuse Injuries</li><li>Difficulty Sleeping</li><li>Frequent Colds or Infections</li></ul>	<ul><li>Irritation or Anger</li><li>Depression</li><li>Difficulty in Concentration</li><li>Increased Sensitivity to Emotional Stress</li><li>Loss of Competitive Drive</li><li>Loss of Enthusiasm</li></ul>

Overreaching

5-9. The term "overreaching" refers to the earliest phase of overtraining. Overreaching consists of extreme muscle soreness that occurs as a result of excessive training with inadequate rest/recovery between hard training sessions. This process of overreaching occurs quickly after several consecutive days of hard training. Overreaching has both positive and negative results. When planned as part of the periodized training program, overreaching allows for the suppression of performance while developing tolerance. For highly conditioned Soldiers, overreaching is a planned component of their training for peak performance. Their higher fitness levels allows for a tolerance to this more intense training with proper rest/recovery and nutrient intake. Short-term overreaching followed by an appropriate tapering period can elicit significant strength and power gains. Muscle soreness and general fatigue are normal outcomes following a series of intense workouts; however, if these outcomes are never completely resolved and performance continues to decline, these may be the first indicators of overtraining syndrome. Commanders and PRT leaders need to recognize these symptoms, especially in IMT and need to adjust training and recovery for these Soldiers. Performance indicators and physiological symptoms of overtraining are listed in Table 5-1.

5-10. Continued overreaching will lead to overtraining and elicit negative results. In many instances, Soldiers that experience a degradation of performance (a loss of strength or speed) feel the need to train even harder. Contrary to their belief, pushing harder not only decreases the chance of improved performance, but increases the risk of injury. Recovery, rest, and proper nutrient intake will elicit more improvement than training harder. When the volume and intensity of exercise exceeds Soldiers' capacity to recover, they cease making progress and may even lose strength and endurance. Overtraining is a common problem in both resistance training and running activities. Improvements in strength and endurance occur only during the rest period following hard training. This process, referred to as supercompensation, takes 12 to 24 hours for the body to completely rebuild. If sufficient rest is not available, then complete recovery cannot occur. Overreaching as a training practice is not appropriate, nor is it recommended for Soldiers in IMT, especially for those who have low fitness levels, high foot time, and high training OPTEMPO. Overreaching may lead to overtraining syndrome and overuse injuries when hard training continues beyond a reasonable period of time.

Overuse

5-11. Continued overreaching without adequate rest/recovery and nutrient intake leads to overtraining and eventually overuse injuries. The effects of overtraining syndrome may last weeks or months, inhibiting Soldier performance and possibly causing acute or chronic injuries that may limit or end a Soldier's term of service. Specific examples include rhabdomyolysis, pubic ramus stress fractures, compartment syndrome, and femoral neck stress fractures. Commanders and PRT leaders must be cognizant of overtraining symptoms listed in Table 5-1. Figure 5-1 is a graphic description of Soldier response/adaptation to overreaching, overtraining, and overuse.

Chapter 5

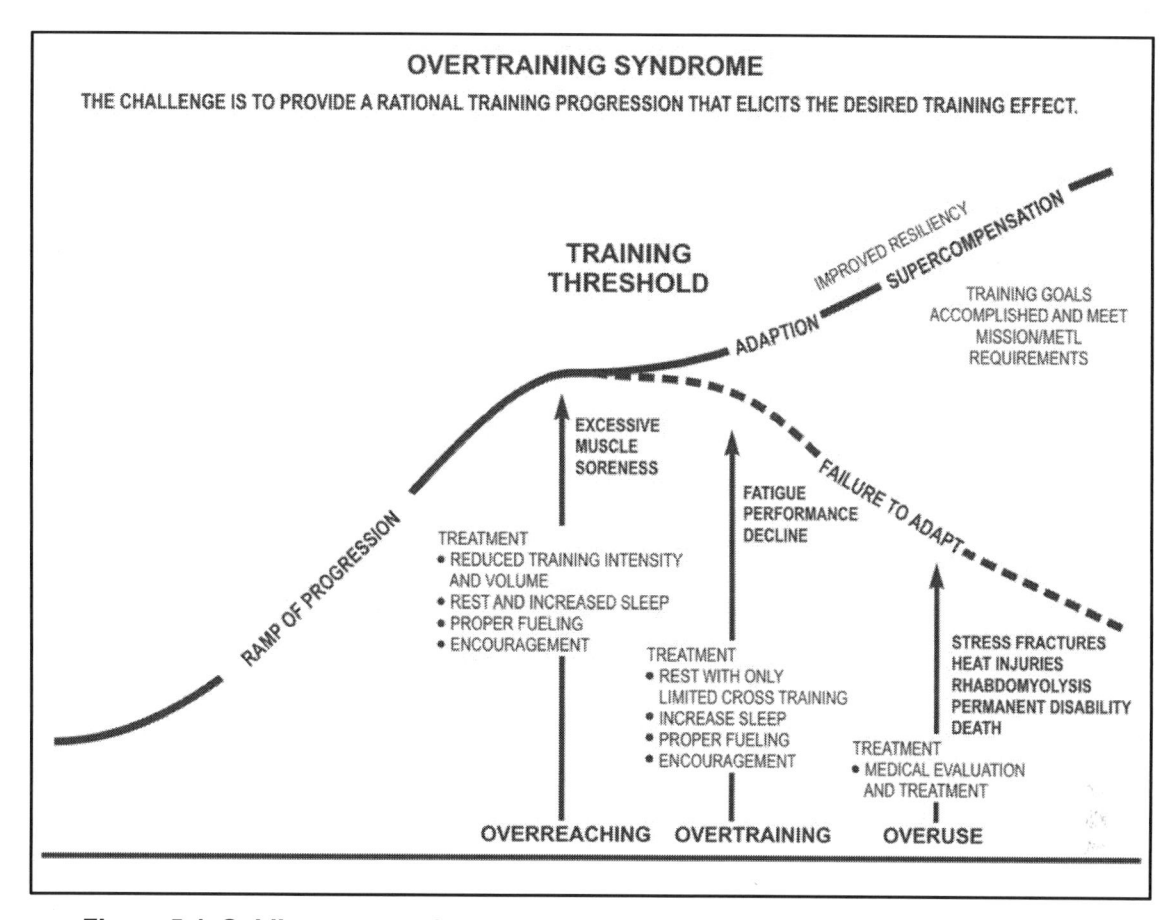

Figure 5-1. Soldier response/adaptation to overreaching, overtraining, and overuse

Causes Of Overtraining Syndrome And Overuse Injuries

5-12. Safe progression for performance improvement is complex, involving many variables that impact success (entry fitness level, ramp of progression, total volume of activity, rest/recovery, and nutrient intake). Many of these variables can be controlled following the principles of precision, progression, and integration, as well as, monitoring Soldiers in training and making training adjustments as required. Common mistakes to compensate for low performance and rate of improvement are the conduct of multiple training sessions, high intensity "smoke sessions," and/or excessive corrective action using exercise. All of these are detrimental to performance improvement and lead to overuse injury.

Multiple Training Sessions

5-13. Multiple training sessions per day have both positive and negative effects as they relate to performance improvement and injury control. Highly conditioned Soldiers may respond well to an additional daily training session that challenges them differently than the one conducted earlier that same day. For example, on strength and mobility days, the morning PRT session may consist of CD and CL exercises, while the second PRT session may target specific muscle groups using resistance training equipment. On endurance and mobility days, speed training may be conducted during the morning PRT session and during activities such as aquatics. Use of endurance training machines (ETM) and agility exercises may be conducted in the afternoon. Soldiers with lower fitness levels, such as those entering IMT, recovering from injury, and those returning from extended deployment (RESET), are better served with a second training session of lower intensity that targets specific needs for improvement, but does not lead to overtraining. For example, Soldiers in BCT and OSUT (red, white, and blue phases [R/W/B]) typically perform a challenging PRT session in the early morning. If a second session is conducted in the afternoon, it should consist of activities that address specific needs and/or technique and mobility improvement, but not be so intense that they cause undue fatigue that may lead to overtraining. For example, a second PRT session should consist of activities that promote stability, mobility, and proper body

Planning Considerations

mechanics, such as, 4 for the core (4C), hip stability drill (HSD), and RD. Commanders and PRT leaders should understand that "***more is not better***" and additional recovery time (rest) may elicit higher performance than the conduct of additional PRT sessions. In IMT, PRT two-a-days are highly discouraged and should be treated as the exception rather than the rule.

High Intensity/Volume Training Sessions

5-14. Soldiers commonly refer to these training sessions as "smoke sessions." Many times in these types of sessions, the difficulty, intensity, and volume of exercise is too high and the purpose may be to punish Soldiers by bringing them to the point of exhaustion. This type of training is a dangerous practice that inhibits building resiliency because performance is degraded, motivation is lowered, and risk of injury is high. Thus, training sessions for the sole purpose of "smoking" Soldiers have no place in the PRT system. Many times, these sessions produce life-threatening conditions for Soldiers, such as, heat fatalities, debilitating overuse injuries, and rhabdomyolysis and may lead to permanent disability or death.

Corrective Action

5-15. When exercise is used for corrective action, it is often performed incorrectly, promoting overtraining syndrome, and overuse injuries. Often corrective action mimics "smoke sessions," punishing Soldiers with little or no corrective value. Consideration must be given to the number of times per day exercises are used for corrective action for individual Soldiers and groups of Soldiers to avoid the cumulative effect and limit the potential for overtraining syndrome. The following guidelines should be followed when employing exercise as corrective action.

- Only the following exercises should be selected for performance of corrective action.
  - Rower.
  - Squat bender.
  - Windmill.
  - Prone row.
  - Push-up.
  - V-up.
  - Leg tuck and twist.
  - Supine bicycle.
  - Swimmer.
  - 8-count push-up.
- Only one of the above exercises may be selected for each corrective action.
- The number of repetitions should not exceed FIVE for any one of the exercises listed above.

INTEGRATION

5-16. Integration is the use of multiple training activities to achieve balance in the PRT program and enhance appropriate recovery between PRT activities. Because most WTBDs require a blend of strength, endurance, and mobility, PRT activity schedules are designed to progress Soldiers in their physical activity in an integrated manner. Several different exercises and activities are employed to develop all three components. Leaders should balance the PRT schedule with other training to avoid conflicts with physically demanding events that can lead to overtraining. For example, if the CFOC course is the day's main physical training event, leaders should not schedule strength training for PRT, unless it is conducted later in the training day. If conflicts cannot be resolved, PRT should be performed after a physically demanding event (later in the duty day), rather than before the event (in the morning). The PRT schedule provides a well-rounded program that develops all of the components of physical readiness equally. PRT drills and activities include exercises that condition all major muscle groups for a total body workout. Failure to adhere to the training schedule as written will result in an emphasis on one component at the expense of another. The activities in PRT schedules should allow Soldiers to improve their overall physical fitness, combat readiness, and achieve the standard of the APFT.

5-17. Commanders and PRT leaders schedule PRT sessions based on the number of days available for each week of training. For example, if only three PRT days are available in the toughening phase, then the toughening phase schedule is followed, and those days where PRT is not conducted are omitted. Omitted training days are missed and should not be made up. The same principle applies to training schedules in the sustaining phase.

SESSION ELEMENTS

5-18. PRT sessions consist of the elements of preparation, activities, and recovery. Each element includes the exercises needed to conduct performance-oriented PRT sessions that effectively address physical readiness components.

PREPARATION

5-19. The preparation drill (PD) is a dynamic warm-up consisting of ten exercises that appropriately prepare Soldiers for more intense PRT activities. Conduct the PD before all PRT activities.

ACTIVITIES

5-20. Activities address specific PRT goals in the areas of strength, endurance, and mobility. They take most of the training time (30 to 60 minutes). Conduct at least two strength and mobility days and two endurance and mobility days each week, with one endurance and mobility training session consisting of speed running. Follow the guidelines listed below:
- Conduct strength and mobility training every other day.
- Conduct endurance and mobility training (running) every other day. This also applies to foot marches more than 5 km in the toughening phase.
- Avoid conducting foot marches and endurance and mobility training on the same or consecutive days.
- Perform speed running once per week, preferably in the middle of the week. In the sustaining phase, speed running may be conducted twice per week for well-conditioned Soldiers.
- A typical five-day training week will include two or three strength and mobility days that alternate with two or three endurance and mobility days.
- Conduct the PD before the APFT. If required, Soldiers may perform push-ups in CD 1 on their knees. After the conclusion of the AFPT, the RD is conducted.
- Schedule APFTs so Soldiers have advance notice. Preferably, the APFT should be scheduled on Monday to allow for recovery provided by the weekend. If the APFT is not conducted on a Monday, no strenuous PRT should be conducted on the day before the APFT. The conduct of the PD, 4C, HSD, and RD provide an active recovery day before the APFT (refer to Table 5-3, Session 2-5).

RECOVERY

5-21. This includes walking (after running activities) and the performance of the RD at the end of all PRT sessions. Recovery gradually and safely tapers off activities to bring the body back to its pre-exercise state. The element of recovery carries over until the next exercise session is performed. Restoring adequate hydration and energy balance through proper nutrition and ensuring adequate sleep allows the body to refuel and rest. This results in a positive adaptation to the stress of training, improves Soldier resiliency, and optimizes gains in strength, endurance, and mobility while controlling injuries.

TOUGHENING PHASE PRT

5-22. As described in Chapter 2, the purpose of the toughening phase is to develop foundational fitness and fundamental skills. Soldiers in BCT, one station unit training (R/W/B phases), and BOLC A are in the toughening phase. The following PRT drills and activities are scheduled during the toughening phase:
- Preparation drill (PD).
- 4 for the core (4C).
- Hip stability drill (HSD).
- Conditioning drills 1 and 2 (CD 1 and CD 2).

- Climbing drill 1 (CL 1).
- Strength training circuit (STC).
- Push-up and sit-up drill (PSD).
- Military movement drill 1 (MMD1).
- 30:60s.
- 60:120s.
- 300-yard shuttle run (SR).
- Ability group run (AGR).
- Release run (RR).
- Foot march with fighting load (FM-fl).
- Conditioning obstacle course (CDOC).
- Confidence obstacle course (CFOC).
- Combatives (CB).
- Recovery drill (RD).

TOUGHENING PHASE PRT SCHEDULE

5-23. The toughening phase PRT schedule is used in BCT and OSUT (R/W/B phases). The BOLC A leaders can use this schedule as a guide for developing PRT in their course program of instruction (POI). Physical readiness training should be conducted five to six days per week depending on the POI and course training schedule. When following this schedule, all PRT sessions occur in order, regardless of the off day(s). Each day's PRT activities also occur in the order listed. Not every toughening phase activity is listed in this schedule. To achieve optimal progression while controlling injuries, toughening phase PRT activities are specifically ordered and sessions sequenced according to the system described in Chapter 2. The activities and sessions should therefore be performed in the order listed on the schedule. Refer to Table 5-2, Toughening Phase PRT Daily Session Overview, for an example of activity sequencing and session purpose. Higher level activities such as the CDOC, confidence obstacle course, and combatives are most appropriate when performed in the sustaining phase; however, Soldiers are introduced to these activities while still in the toughening phase. Thus, PRT leaders must focus their instruction of these activities on proper technique and lead-up skills to ensure safety and successful execution. Table 5-3, Toughening Phase PRT Schedules (BCT and OSUT-R/W/B phases) describe the ordered sequence of training to be used for PRT in BCT and OSUT-R/W/B phases.

Table 5-2. Toughening phase PRT daily session overview (BCT and OSUT-R/W/B phases)

Monday	Tuesday	Wednesday	Thursday	Friday	Saturday
Preparation: PD Activities: HSD, MMD 1, AGR Recovery: RD	Preparation: PD Activities: 4C, CD1, CL1, PSD Recovery: RD	Preparation: PD Activities: HSD, MMD 1, 30:60s Recovery: RD	Preparation: PD Activities: 4C, CD1, CL1, PSD Recovery: RD	Preparation: PD Activities: HSD, MMD 1, AGR Recovery: RD	Preparation: PD Activities: 4C, CD1, CL1, PSD Recovery: RD

Monday

Preparation: PD Activities: HSD, MMD 1, AGR Recovery: RD	The purpose of this session is to improve the endurance and mobility needed for the successful performance of WTBDs. Preparation and The Hip Stability Drill ready the body for a variety of activities that develop body management competencies. The Military Movement Drill 1 helps improve running form while preparing the Soldier for sustained running. The AGR improves aerobic endurance through sustained running at an appropriate pace. Recovery safely returns Soldiers to a pre-exercise state while improving mobility.

Tuesday

Preparation: PD Activities: 4C, CD1, CL1, PSD Recovery: RD	The purpose of this session is to improve the strength and mobility needed for the successful performance of WTBDs. Preparation and Four for the Core ready the body for a variety of activities that develop body management competencies. Conditioning Drill 1 improves total body muscular strength, endurance, and mobility. Climbing Drill 1 increases upper body strength, trunk strength, and creates muscle balance. The Push-up and Sit-up Drill improves APFT performance. Recovery safely returns Soldiers to a pre-exercise state while improving mobility.

Wednesday

Preparation: PD Activities: HSD, MMD 1, 30:60s, 300-yd SR Recovery: RD	The purpose of this session is to improve the conditioning required to successfully perform critical WTBDs such as Individual Movement Techniques and move under direct and indirect fire. Preparation and The Hip Stability Drill ready and condition the body for a variety of body management competencies. Military Movement Drill 1 helps improve running form while preparing the Soldier for speed running. 30:60s enhance anaerobic power through sustained repeats of high intensity running with intermittent periods of recovery. The 300-yard Shuttle Run develops speed, agility, and anaerobic power. Recovery safely returns Soldiers to a pre-exercise state while improving mobility.

Thursday

Preparation: PD Activities: 4C, CD1, CL1, PSD Recovery: RD	The purpose of this session is to improve the strength and mobility needed for the successful performance of WTBDs. Preparation and Four for the Core ready the body for a variety of activities that develop body management competencies. Conditioning Drill 1 improves total body muscular strength, endurance, and mobility. Climbing Drill 1 increases upper body strength, trunk strength, and creates muscle balance. The Push-up and Sit-up Drill improves APFT performance. Recovery safely returns Soldiers to a pre-exercise state while improving mobility.

Friday

Preparation: PD Activities: HSD, MMD 1, AGR Recovery: RD	The purpose of this session is to improve the endurance and mobility needed for the successful performance of WTBDs. Preparation and the Hip Stability Drill ready the body for a variety of activities that develop body management competencies. Military Movement Drill 1 helps improve running form while preparing the Soldier for sustained running. The AGR improves aerobic endurance through sustained running at an appropriate pace. Recovery safely returns Soldiers to a pre-exercise state while improving mobility.

Saturday

Preparation: PD Activities: 4C, CD1, CL1, PSD Recovery: RD	The purpose of this session is to improve the strength and mobility needed for the successful performance of WTBDs. Preparation and Four for the Core ready the body for a variety of activities that develop body management competencies. Conditioning Drill 1 improves total body muscular strength, endurance, and mobility. Climbing Drill 1 increases upper body strength, trunk strength, and creates muscle balance. The Push-up and Sit-up Drill improves APFT performance. Recovery safely returns Soldiers to a pre-exercise state while improving mobility.

Abbreviations	PD – Preparation Drill RD – Recovery Drill PSD – PU/SU Drill FM – Foot March STC – Strength Training Circuit	4C – Four for the Core CD – Conditioning Drill SR – Shuttle Run RR – Release Run	HSD – Hip Stability Drill CL – Climbing Drill AGR – Ability Group Run TR – Terrain Run

Table 5-3. Toughening phase PRT schedule (BCT and OSUT-R/W/B phases)

Session	Week 1 Schedule
1-1	Preparation: PD (INSTRUCTION)
1-2	Preparation: PD (5 reps) Activities: HSD and MMD 1 (INSTRUCTION) and 1-mile run Assessment (ability group placement) Recovery: RD (INSTRUCTION)
1-3	Preparation: PD (5 reps) Activities: 4C and CD1 (INSTRUCTION) Recovery: RD (20 seconds)
1-4	Preparation: PD (5 reps) Activities: HSD (5 reps), MMD 1 (1 rep) and 30:60s x 6 reps (INSTRUCTION) Recovery: RD (20 seconds)
1-5	Preparation: PD (5 reps) Activities: 4C (60 seconds), CD 1 (5 reps) and CD 2 (INSTRUCTION) Recovery: RD (20 seconds)
1-6	Preparation: PD (5 reps) Activities: HSD (5 reps), MMD 1 (1 rep), AGR (A 15 min @ 7:30; B 15 min @ 9:00; C 10 min @ 10:30; D 10 min @ 12:00) Recovery: RD (20 seconds)
Session	**Week 2 Schedule**
2-1	Preparation: PD (5 reps) Activities: HSD (5 reps), MMD 1 (1 rep), AGR (A 15 min @ 7:15; B 15 min @ 8:30; C 12 min @ 10:00; D 12 min @ 11:00) Recovery: RD (20 seconds)
2-2	Preparation: PD (5 reps) Activities: 4C (60 seconds), CD 1 (5 reps) and CL 1 (INSTRUCTION) Recovery: RD (20 seconds)
2-3	Preparation: PD (5 reps) Activities: HSD (5 reps), MMD 1 (1 rep) and 30:60s (6 reps) Recovery: RD (20 seconds)
2-4	Preparation: PD (5 reps) Activities: 4C (60 seconds), CD 1 (5 reps), CD 2 (5 reps) and CL 1 (5 reps) Recovery: RD (20 seconds)
2-5	Preparation: PD (5 reps) Activities: HSD (5 reps) and 4C (60 seconds) Recovery: RD (20 seconds)
2-6	Preparation: PD (5 reps) Activity: Practice APFT Recovery: RD (20 seconds)
Abbreviations	PD – Preparation Drill 4C – Four for the Core HSD – Hip Stability Drill RD – Recovery Drill CD – Conditioning Drill CL – Climbing Drill PSD – PU/SU Drill SR – Shuttle Run AGR – Ability Group Run FM – Foot March RR – Release Run STC – Strength Training Circuit

Chapter 5

Table 5-3. Toughening phase PRT schedule (BCT and OSUT-R/W/B phases) continued

Session	Week 3 Schedule
3-1	Preparation: PD (5 reps) Activities: HSD (5 reps) MMD 1 (1 rep), AGR (A 20 min @ 7:15; B 20 min @ 8:30; C 14 min @ 9:30; D 14 min @ 10:30) Recovery: RD (20 seconds)
3-2	Preparation: PD (5 reps) Activities: 4C (60 seconds), CD 1 (5 reps), CD 2 (5 reps), CL 1 (5 reps) and PSD (INSTRUCTION) Recovery: RD (20 seconds)
3-3	Preparation: PD (5 reps) Activities: HSD (5 reps), MMD 1 (1 rep), 30:60s (8 reps) and 300-yd SR (INSTRUCTION) Recovery: RD (20 seconds)
3-4	Preparation: PD (5 reps) Activities: 4C (60 seconds), CD 1 (5 reps), CD 2 (5 reps), CL 1 (5 reps) and PSD (2 x 30 seconds) Recovery: RD (20 seconds)
3-5	Preparation: PD (5 reps) Activities: HSD (5 reps), MMD 1 (1 rep), AGR (A 20 min @ 7:15; B 20 min @ 8:30; C 14 min @ 9:30; D 14 min @ 10:30) Recovery: RD (20 seconds)
3-6	Preparation: PD (5 reps) Activities: 4C (60 seconds), CD 1 (5 reps), CD 2 (5 reps), CL 1 (5 reps) and PSD (2 x 30 seconds) Recovery: RD (20 seconds)
Session	Week 4 Schedule
4-1	Preparation: PD (5 reps) Activities: HSD (5 reps), MMD 1 (1 rep), AGR (A 25 min @ 7:15; B 25 min @ 8:15; C 16 min @ 9:30; D 16 min @ 10:00) Recovery: RD (20 seconds)
4-2	Preparation: PD (5 reps) Activities: 4C (60 seconds), CD 1 (5 reps), CD 2 (5 reps), CL 1 (5 reps) and PSD (2 x 45 seconds) or STC (INSTRUCTION) Recovery: RD (20 seconds)
4-3	Preparation: PD (5 reps) Activities: HSD (5 reps), MMD 1 (1 rep), 60:120s x 6 reps (INSTRUCTION) and 300-YD SR (1 rep) Recovery: RD (20 seconds)
4-4	Preparation: PD (5 reps) Activities: 4C (60 seconds), CD 1 (5 reps), CD 2 (5 reps), CL 1 (5 reps) and PSD (2 x 45 seconds) or STC (2 rotations x 60 seconds @ each exercise-includes movement) Recovery: RD (20 seconds)
4-5	Preparation: PD (5 reps) Activities: HSD (5 reps), MMD 1 (1 rep), AGR (A 25 min @ 7:15; B 25 min @ 8:15; C 16 min @ 9:30; D 16 min @ 10:00) Recovery: RD (20 seconds)
4-6	Preparation: PD (5 reps) Activities: 4C (60 seconds), CD 1 (5 reps), CD 2 (5 reps), CL 1 (5 reps) and PSD (2 x 45 seconds) or STC (2 rotations x 60 seconds @ each exercise-includes movement) Recovery: RD (20 seconds)
Abbreviations	PD – Preparation Drill 4C – Four for the Core HSD – Hip Stability Drill RD – Recovery Drill CD – Conditioning Drill CL – Climbing Drill PSD – PU/SU Drill SR – Shuttle Run AGR – Ability Group Run FM – Foot March RR – Release Run STC – Strength Training Circuit

Table 5-3. Toughening phase PRT schedule (BCT and OSUT-R/W/B phases) continued

Session	Week 5 Schedule
5-1	Preparation: PD (5 reps) Activities: HSD (5 reps), MMD 1 (1 rep), AGR (A 30 min @ 7:30; B 25 min @ 8:00; C 18 min @ 9:00; D 18 min @ 10:00) Recovery: RD (20 seconds)
5-2	Preparation: PD (5 reps) Activities: 4C (60 seconds), CD 1 (5 reps), CD 2 (5 reps), CL 1 (5 reps) and PSD (2 x 60 seconds) or STC (2 rotations x 60 seconds @ each exercise-includes movement) Recovery: RD (20 seconds)
5-3	Preparation: PD (5 reps) Activities: HSD (5 reps), MMD 1 (1 rep), 60:120s (8 reps) and 300-yd SR (1 rep) Recovery: RD (20 seconds)
5-4	Preparation: PD (5 reps) Activities: 4C (60 seconds), CD 1 (5 reps), CD 2 (5 reps), CL 1 (5 reps) and PSD (2 x 60 seconds) or STC (2 rotations x 60 seconds @ each exercise-includes movement) Recovery: RD (20 seconds)
5-5	Preparation: PD (5 reps) Activities: HSD (5 reps) and 4C (60 seconds) Recovery: RD (20 seconds)
5-6	Preparation: PD (5 reps) Activity: Practice APFT Recovery: RD (20 seconds)
Session	**Week 6 Schedule**
6-1	Preparation: PD (5 reps) Activities: HSD (5 reps), MMD 1 (1 rep), AGR (A 30 min @ 7:30; B 30 min @ 8:00; C 20 min @ 8:30; D 20 min @ 9:30) or RR (20 minutes) Recovery: RD (20 seconds)
6-2	Preparation: PD (5 reps) Activities: 4C (60 seconds), CD 1 (5 reps), CD 2 (5 reps), CL 1 (5 reps) and PSD (2 x 60 seconds) or STC (3 rotations x 60 seconds @ each exercise-includes movement) Recovery: RD (20 seconds)
6-3	Preparation: PD (5 reps) Activities: HSD (5 reps), MMD 1 (1 rep), 60: 120s (8 reps) and 300-yd SR (1 rep) Recovery: RD (20 seconds)
6-4	Preparation: PD (5 reps) Activities: 4C (60 seconds), CD 1 (5 reps), CD 2 (5 reps), CL 1 (5 reps) and PSD (2 x 60 seconds) or STC (2 rotations x 60 seconds @ each exercise-includes movement) Recovery: RD (20 seconds)
6-5	Preparation: PD (5 reps) Activities: HSD (5 reps), MMD 1 (1 rep), AGR (A 30 min @ 7:30; B 30 min @ 8:00; C 20 min @ 8:30; D 20 min @ 9:30) Recovery: RD (20 seconds)
6-6	Preparation: PD (5 reps) Activities: 4C (60 seconds), CD 1 (5 reps), CD 2 (5 reps), CL 1 (5 reps) and PSD (2 x 60 seconds) or STC (2 rotations x 60 seconds @ each exercise-includes movement) Recovery: RD (20 seconds)
Abbreviations	PD – Preparation Drill 4C – Four for the Core HSD – Hip Stability Drill RD – Recovery Drill CD – Conditioning Drill CL – Climbing Drill PSD – PU/SU Drill SR – Shuttle Run AGR – Ability Group Run FM – Foot March RR – Release Run STC – Strength Training Circuit

Table 5-3. Toughening phase PRT schedule (BCT and OSUT-R/W/B phases) continued

Session	Week 7 Schedule
7-1	Preparation: PD (5 reps) Activities: HSD (5 reps), MMD 1 (1 rep), AGR (A 30 min @ 7:15; B 30 min @ 7:45; C 20 min @ 8:15; D 20 min @ 9:30) or RR (20 minutes) Recovery: RD (20 seconds)
7-2	Preparation: PD (5 reps) Activities: 4C (60 seconds), CD 1 (5 reps), CD 2 (5 reps), CL 1 (5 reps) and PSD (2 x 60 seconds) or STC (3 rotations x 60 seconds @ each exercise-includes movement) Recovery: RD (20 seconds)
7-3	Preparation: PD (5 reps) Activities: HSD (5 reps), MMD 1 (1 rep) and 60:120s (10 reps) Recovery: RD (20 seconds)
7-4	Preparation: PD (5 reps) Activities: 4C (60 seconds), CD 1 (5 reps), CD 2 (5 reps), CL 1 (5 reps) and PSD (2 x 60 seconds) or STC (3 rotations x 60 seconds @ each exercise-includes movement) Recovery: RD (20 seconds)
7-5	Preparation: PD (5 reps) Activities: 4C (60 seconds) and HSD (5 reps) Recovery: RD (20 seconds)
7-6	Preparation: PD (5 reps) Activity: Record APFT Recovery: RD (20 seconds)
Session	Week 8 Schedule
8-1	Preparation: PD (5 reps) Activities: HSD (5 reps), MMD 1 (1 rep), AGR (A 30 min @ 7:15; B 30 min @ 7:45; C 20 min @ 8:15; D 20 min @ 9:30) or RR (20 minutes) Recovery: RD (20 seconds)
8-2	Preparation: PD (5 reps) Activities: 4C (60 seconds), CD 1 (5 reps), CD 2 (5 reps), CL 1 (5 reps) and PSD (2 x 60 seconds) or STC (3 rotations x 60 seconds @ each exercise-includes movement) Recovery: RD (20 seconds)
8-3	Preparation: PD (5 reps) Activities: HSD (5 reps), MMD 1 (1 rep) and 60:120s (10 reps) Recovery: RD (20 seconds)
8-4	Preparation: PD (5 reps) Activities: 4C (60 seconds), CD 1 (5 reps), CD 2 (5 reps), CL 1 (5 reps) and PSD (2 x 60 seconds) or STC (3 rotations x 60 seconds @ each exercise-includes movement) Recovery: RD (20 seconds)
8-5	Preparation: PD (5 reps) Activities: HSD (5 reps), MMD 1 (1 rep) AGR (A 30 min @ 7:15; B 30 min @ 7:45; C 20 min @ 8:15; D 20 min @ 9:30) or RR (20 minutes) Recovery: RD (20 seconds)
8-6	Preparation: PD (5 reps) Activities: 4C (60 seconds), CD 1 (5 reps), CD 2 (5 reps), CL 1 (5 reps) and PSD (2 x 60 seconds) or STC (3 rotations x 60 seconds @ each exercise-includes movement) Recovery: RD (20 seconds)
Abbreviations	PD – Preparation Drill 4C – Four for the Core HSD – Hip Stability Drill RD – Recovery Drill CD – Conditioning Drill CL – Climbing Drill PSD – PU/SU Drill SR – Shuttle Run AGR – Ability Group Run FM – Foot March RR – Release Run STC – Strength Training Circuit

Table 5-3. Toughening phase PRT schedule (BCT and OSUT-R/W/B phases) continued

Session	Week 9 Schedule
9-1	Preparation: PD (5 reps) Activities: HSD (5 reps), MMD 1 (1 rep), AGR (A 30 min @ 7:15; B 30 min @ 7:45; C 20 min @ 8:15; D 20 min @ 9:30) or RR (20 minutes) Recovery: RD (20 seconds)
9-2	Preparation: PD (5 reps) Activities: 4C (60 seconds), CD 1 (5 reps), CD 2 (5 reps), CL 1 (5 reps) and PSD (2 x 60 seconds) or STC (3 rotations x 60 seconds @ each exercise-includes movement) Recovery: RD (20 seconds)
9-3	Preparation: PD (5 reps) Activities: HSD (5 reps), MMD 1 (1 rep), 60:120s (10 reps) and 300-yd SR (1 rep) Recovery: RD (20 seconds)
9-4	Preparation: PD (5 reps) Activities: 4C (60 seconds), CD 1 (5 reps), CD 2 (5 reps), CL 1 (5 reps) and PSD (2 x 60 seconds) or STC (3 rotations x 60 seconds @ each exercise-includes movement) Recovery: RD (20 seconds)
9-5	Preparation: PD (5 reps) Activities: HSD (5 reps), MMD 1 (1 rep), AGR (A 30 min @ 7:15; B 30 min @ 7:45; C 20 min @ 8:15; D 20 min @ 9:30) or RR (20 minutes) Recovery: RD (20 seconds)
9-6	Preparation: PD (5 reps) Activity: 4C (60 seconds), CD 1 (5 reps), CD 2 (5 reps), CL 1 (5 reps) and PSD (2 x 60 seconds) or STC (3 rotations x 60 seconds @ each exercise-includes movement) Recovery: RD (20 seconds)
Session	Week 10 Schedule
10-1	Preparation: PD (5 reps) Activities: HSD (5 reps), MMD 1 (1 rep), AGR (A 30 min @ 7:15; B 30 min @ 7:45; C 20 min @ 8:15; D 20 min @ 9:30) or RR (20 minutes) Recovery: RD (20 seconds)
10-2	Preparation: PD (5 reps) Activity: 4C (60 seconds), CD 1 (5 reps), CD 2 (5 reps), CL 1 (5 reps) and PSD (2 x 60 seconds) or STC (3 rotations x 60 seconds @ each exercise-includes movement) Recovery: RD (20 seconds)
10-3	Preparation: PD (5 reps) Activities: HSD (5 reps), MMD 1 (1 rep), 60:120s (10 reps) and 300-yd SR (1 rep) Recovery: RD (20 seconds)
10-4	GRADUATION
Abbreviations	PD – Preparation Drill 4C – Four for the Core HSD – Hip Stability Drill RD – Recovery Drill CD – Conditioning Drill CL – Climbing Drill PSD – PU/SU Drill SR – Shuttle Run AGR – Ability Group Run FM – Foot March RR – Release Run STC – Strength Training Circuit

CONDENSED TIME

5-24. When scheduled training requirements in the training POI conflict with the designated time available for PRT, commanders, and PRT leaders may choose to perform one of the two sessions shown in Table 5-4.

Table 5-4. Condensed sessions (toughening phase)

Session	Toughening Phase
Strength and Mobility	Preparation: PD (5 reps) Activities: CD 1&2 (5 reps ea) and PSD 2x30 sec Recovery: RD (20 seconds)
Endurance and Mobility	Preparation: PD (5 reps) Activities: MMD1 (1 rep) and 30:60s (8-10 reps) Recovery: RD (20 seconds)

Chapter 5

FIELD TRAINING

5-25. Toughening phase PRT should be conducted whenever possible within the constraints of the environment, whether on a range or during a field training exercise (FTX). The example schedule shown in Table 5-5 may be conducted anywhere and is not resource intensive.

Table 5-5. Field training sessions (toughening phase)

Session	Schedule
Endurance and Mobility	Preparation: PD (5 reps) Activities: MMD1 (2 reps) and 300-yd SR (2 reps) Recovery: RD (20 seconds)
Strength and Mobility	Preparation: PD (5 reps) Activities: CD 1&2 (5 reps ea) and PSD (4x30 sec) Recovery: RD (20 seconds)

SUSTAINING PHASE PRT

5-26. As described in Chapter 2, the purpose of the sustaining phase is to develop a high level of physical readiness in Soldiers. Training results enable Soldiers to perform WTBDs and those physical tasks associated with the performance of their duty positions and operational missions. Soldiers in AIT, one station unit training (black and gold phases [B/G]), BOLC B, and Soldiers assigned to operational units are in the sustaining phase. The following PRT drills and activities are scheduled during the sustaining phase.

- Preparation drill (PD).
- 4 for the core (4C).
- Hip stability drill (HSD).
- Conditioning drills 1, 2, and 3 (CD 1, CD 2, and CD 3).
- Climbing drills 1 and 2 (CL 1 and CL 2).
- Guerrilla drill (GD).
- Push-up and sit-up drill (PSD).
- Strength training circuit (STC).
- Military movement drills 1 and 2 (MMD 1 and 2).
- 30:60s.
- 60:120s.
- Ability group run (AGR).
- Release runs (RR).
- Foot march with fighting load (FM-fl).
- Foot march with approach march load (FM-aml).
- Foot march with emergency approach march load (FM-eaml).
- Terrain running (TR).
- Hill repeats (HR).
- 300-yard shuttle run (SR).
- Conditioning obstacle course (CDOC).
- Confidence obstacle courses (CFOC).
- Combatives (CB).
- Recovery drill (RD).

INITIAL MILITARY TRAINING SUSTAINING PHASE PRT SCHEDULES

5-27. Sustaining phase PRT activities should be used in AIT, OSUT (B/G phases), and BOLC B.

Planning Considerations

ADVANCED INDIVIDUAL TRAINING

5-28. Training schedule development in AIT is a complex process. Several variables impact the ability to apply one training schedule across all of AIT. These variables include: how units fill, length of the training cycle, student to leader ratio, training conducted by shift, availability of PRT training areas, MOS specific training requirements, equipment, and facilities; therefore, commanders and PRT leaders should apply the following doctrinal guidelines when developing their unit PRT schedules:

- PRT sessions should be scheduled for four-to-five days per week, depending on the POI and course training schedule.
- Alternate strength and mobility emphasis weeks with endurance and mobility emphasis weeks on five-day per week training schedules throughout the length of the training cycle. A strength and mobility emphasis week contains three strength and mobility training days and two endurance and mobility training days. An endurance and mobility emphasis week contains three endurance and mobility training days and two strength and mobility training days.
- During four-day per week training schedules, alternate strength and mobility days with endurance and mobility days.
- Utilize exercises, drills, and activities listed in paragraph 5-26 when developing AIT PRT schedules. See Chapters 9, Strength and Mobility Activities, and Chapter 10, Endurance and Mobility Activities, for a detailed description of strength and mobility and endurance and mobility drills and activities.
- Supplemental PRT exercises, drills, and activities found on the USAPFS website may be integrated into sustaining phase PRT schedules. Commanders and PRT leaders are responsible for ensuring Soldiers have achieved the appropriate level of fitness and motor skill development before conducting these more complex activities.
- Strength and mobility exercise progression is accomplished by progressing from 5 repetitions of each exercise to a maximum of 10 repetitions per exercise. If more than 10 repetitions per exercise are desired, repeat the drill in its entirety. Examples of a rational progression include the following ranges of repetitions performed: PD (5-10 reps), CDs 1, 2, and 3 (5-10 reps), CLs 1 and 2 (5-10 reps), and the GD (1-3 reps).
- For those strength and mobility exercises and drills that use time, a rational progression involves increasing the amount of time allocated for each exercise. Examples of a rational progression include: PSD (2-4 sets @ 30-60 seconds) and the strength training circuit (2-3 rotations @ 60 seconds).
- The initial assessment for new fills is the 1-mile run assessment. This assessment is used to assign Soldiers to the appropriate running ability groups.
- Sustaining phase ability group run (AGR) times are different from toughening phase AGR times. Refer to Chapter 10, Endurance and Mobility Activities, Tables 10-3 and 10-4 for placing Soldiers into AGR groups, pacing, and split times.
- Speed running is the most important endurance and mobility activity; therefore, speed running is scheduled at least one time per week. If there is only one endurance and mobility activity session scheduled per week, it will be speed running. Speed running includes the following activities: 30:60s, 60:120s, and the 300-yd SR. Speed running progression for 30:60s and 60:120s ranges from 6 to 10 repetitions. The 300-yd SR progresses from 1 to 3 repetitions.
- Release runs and HR are a combination of sustained and speed running; however, these activities will not replace 30:60s, 60:120s, and/or the 300-yd SR. Release run progression should not exceed 30 minutes total running time. The progression for HR is running up or down gentle slopes progressing to steeper hills, distances of 40 to 60 yards, and increasing from 5 to 10 repetitions.
- Foot marching is a movement component of maneuver and is a critical Soldier physical requirement. Regular foot marching helps to avoid the cumulative effects of lower-body injury trauma and prepares Soldiers to successfully move under load. Refer to FM 21-18, *Foot Marching*, for specific guidance on foot marching variables such as: terrain, frequency, load, rate of march, distance, visibility, halts, and rest. Length of the training cycle, MOS requirements, and POI will determine how these variables are applied to the PRT schedule.

- If combatives training is conducted as a PRT session, it should be conducted only one time per week. Preferably, combatives should only replace a sustained running or foot march session during an endurance emphasis week and one of the three strength training sessions during a strength emphasis week.
- Army Physical Fitness Tests will be conducted according to Appendix A of this FM and the course POI. Preferably, the APFT should be scheduled on Monday. If the APFT is not conducted on a Monday, then no PRT is scheduled on the day before the APFT.

5-29. AR 350-1 acknowledges that specified units and schools have separate physical fitness standards. Within AIT, examples include diving, parachute, and parachute rigger military occupational specialties (MOSs). Commanders that train MOSs that have separate PRT and testing requirements will request approval from the DCG-IMT to implement PRT exercises, drills, and activities to meet these higher physical fitness standards. The USAPFS can assist commanders with the development of PRT programs identified as having separate PRT and testing requirements.

5-30. Refer to Table 5-6, Sustaining Phase PRT Daily Session Overview, for an example of activity sequencing and session purpose. These activities increase in difficulty, complexity, intensity, and/or duration.

Planning Considerations

Table 5-6. Sustaining phase PRT daily session overview (AIT and OSUT-B/G phases)

Monday	Tuesday	Wednesday	Thursday	Friday
Preparation: PD Activities: MMD 1&2, AGR or Release Run Recovery: RD	Preparation: PD Activities: GD, CD 1&2, CL 1, PSD or STC and PSD Recovery: RD	Preparation: PD Activities: MMD 1&2, 60:120s, 300 yd SR Recovery: RD	Preparation: PD Activities: GD, CD 3, CL 2, PSD or STC and PSD Recovery: RD	Preparation: PD Activities: MMD 1&2, Hill Repeats or Terrain Run or 10K FM w/fl Recovery: RD

Monday

Preparation: PD Activities: MMD 1&2, AGR or Release Run Recovery: RD	The purpose of this session is to improve the endurance and mobility needed for the successful performance of WTBDs. Preparation readies and conditions the body for a variety of body management competencies. Military Movement Drills 1 and 2 help improve running form while preparing the Soldier for sustained running. The Ability Group Run or Release Run improves aerobic endurance and speed through sustained running. Recovery safely returns Soldiers to a pre-exercise state while improving mobility.

Tuesday

Preparation: PD Activities: GD, CD 1&2, CL 1, PSD or STC and PSD Recovery: RD	The purpose of this session is to improve the strength and mobility needed for the successful performance of WTBDs. Preparation readies and conditions the body for a variety of body management competencies. The Guerrilla Drill develops functional mobility for the performance of combatives and the ability to carry another Soldier. Conditioning Drills 1 and 2 consist of calisthenics that are designed to functionally train upper body and trunk muscular strength and endurance needed to successfully perform WTBDs. Climbing Drill 1 improves the upper body and trunk strength needed for manipulating body weight. The Push-up and Sit-up Drill provide upper-body strength and APFT improvement. The Strength Training Circuit develops total body strength and movement proficiency. Recovery safely returns Soldiers to a pre-exercise state while improving mobility.

Wednesday

Preparation: PD Activities: MMD 1&2, 60:120s and 300 YD SR Recovery: RD	The purpose of this session is to improve the conditioning required to successfully perform WTBDs such as IMT and move under direct and indirect fire. Preparation readies and conditions the body for a variety of body management competencies. The Military Movement Drills 1 and 2 help improve running form while preparing the Soldier for speed running. 60:120s enhance speed and anaerobic power through sustained repeats of high intensity running with intermittent periods of recovery. The 300-yd Shuttle Run develops anaerobic endurance and functional mobility. Recovery safely returns Soldiers to a pre-exercise state while improving mobility.

Thursday

Preparation: PD Activities: GD, CD 3, CL 2, PSD or STC and PSD Recovery: RD	The purpose of this session is to improve the functional strength and mobility needed for the successful performance of WTBDs. Soldiers perform these drills wearing ACUs, boots, and ACH. Preparation readies and conditions the body for a variety of body management competencies. The Guerrilla Drill develops functional mobility for the performance of combatives and the ability to carry another Soldier. Conditioning Drill 3 consists of advanced calisthenics that improve power, coordination and agility. Climbing Drill 2 improves the upper body and trunk strength needed for manipulating body weight under load. The Push-up and Sit-up Drill provide upper-body strength and APFT improvement. The Strength Training Circuit develops total body strength and movement proficiency. Recovery safely returns Soldiers to a pre-exercise state while improving mobility.

Friday

Preparation: PD Activities: MMD 1&2, Hill Repeats or Terrain Run or 10K FM (aml) Recovery: RD	The purpose of this session is to improve the strength, endurance, and mobility needed for the successful performance of foot marching and running over various terrains. Preparation readies and conditions the body for a variety of body management competencies. The foot march improves the muscular and aerobic endurance needed for foot marching. Hill Repeats and Terrain Running improve the Soldier's ability to move quickly with agility over various terrains with or without a load. Recovery safely returns Soldiers to a pre-exercise state while improving mobility.
Abbreviations	PD – Preparation Drill 4C – Four for the Core HSD – Hip Stability Drill RD – Recovery Drill CD – Conditioning Drill CL – Climbing Drill PSD – PU/SU Drill SR – Shuttle Run AGR–Ability Group Run FM – Foot March RR – Release Run TR – Terrain Run STC – Strength Training Circuit FM-aml (foot march-approach march load)

Chapter 5

OSUT

5-31. Physical readiness training in OSUT consists of a combination of toughening and sustaining phase exercises, drills, and activities. Commanders and PRT leaders should follow the toughening phase PRT schedule during the R/W/B phases of OSUT. Refer to Table 5-2 for the BCT and OSUT-R/W/B phase daily overview and for the toughening phase BCT and OSUT-R/W/B phase PRT schedule. During the B/G phases of OSUT, PRT schedules will be developed using sustaining phase exercises, drills, and activities found in paragraph 5-26. Refer to paragraph 5-28 and Table 5-6 for doctrinal guidelines when developing OSUT (B/G phase) PRT schedules.

BOLC B

5-32. The PRT schedule development for BOLC B is based upon adapting the conduct of sustaining phase exercises, drills, and activities to the course POIs and training schedules. Refer to paragraph 5-26 for PRT drills and activities selection, paragraph 5-28 for PRT scheduling guidelines, and Table 5-6 for a daily session overview. An example of a weekly PRT schedule and the purpose behind each session can be found in Table 5-6 also.

CONDENSED TIME

5-33. When scheduled training requirements in the training POI conflict with the designated time available for PRT, commanders and PRT leaders may choose to perform one of the two sessions shown in Table 5-7.

Table 5-7. Condensed sessions (sustaining phase)

Session	Sustaining Phase
Strength and Mobility	Preparation: PD (5-10 reps) Activities: GD (1 rep), CD 1&2 (5-10 reps ea) Recovery: RD (20 seconds)
Endurance and Mobility	Preparation: PD (5-10 reps) Activities: MMD2 (1 rep) and 60:120s (8-10 reps) Recovery: RD (20 seconds)

FIELD TRAINING

5-34. Sustaining phase PRT should be conducted whenever possible within the constraints of the environment, whether on a range or during an FTX. The example schedule shown in Table 5-8 may be conducted anywhere and is not resource intensive.

Table 5-8. Field training sessions (sustaining phase)

Session	Schedule
Endurance and Mobility	Preparation: PD (5 reps) Activities: MMD1 (2 reps) and 300-yd SR (2 reps) Recovery: RD (20 seconds)
Strength and Mobility	Preparation: PD (5 reps) Activities: CD 1&2 (5 reps ea) and PSD (4x30 sec) Recovery: RD (20 seconds)

PRT IN OPERATIONAL UNITS

5-35. Sustaining phase PRT supports ARFORGEN. See AR 350-1 for a detailed description of the ARFORGEN model.

ARMY FORCE GENERATION

5-36. Army Force Generation uses a structured progression of increased unit readiness over time. This results in recurring periods of availability of trained, ready, and cohesive units prepared for operational deployment as specified in the Army Campaign Plan. The recurring structured progression of increasing unit readiness focuses on reset, train/ready, and available phases according to operational readiness cycles. With the potential to have shortened ARFORGEN cycles, commanders need to stay vigilant in planning and programming PRT that supports full spectrum operations.

Reset Phase

5-37. Units returning from deployment are placed in the reset phase. Units usually remain in the reset phase for up to 180 days (6 months) for active component units and up to 365 days (12 months) for RC units. The goal is to achieve readiness status level of C1 as soon as possible. Typically, Soldiers in these units experience detraining or injury and may return less fit than before deployment. Special consideration must be given to this when planning PRT. Once the unit has stabilized and recovered, commanders and PRT leaders should conduct PRT assessments using foot marches, APFT, or unit readiness standards. This suggests an appropriate start point for regular PRT. For example, exercise sessions should first be conducted at a lower intensity, duration, and exercise volume. Sessions should increase progressively as Soldiers improve and regain their previous fitness levels. Initial PRT sessions should be no longer than 60 minutes in duration and progress to 90 minutes. See the sample reset schedule in this chapter for appropriate progression; sets and repetitions for strength and mobility activities; and sets, repetitions, pace, recovery, and total time in endurance and mobility activities.

Train/Ready Phase

5-38. Once units complete the reset phase, PRT leaders should continue to conduct strength and mobility activities and endurance and mobility activities two to three times per week each. The leaders select activities and drills under sustaining phase activities or supplemental PRT exercises, drills, and activities from the USAPFS website. Commanders and PRT leaders should continue to schedule PRT sessions that specifically enhance mission and C- or D-METL task performance. For example, emphasis should be placed on activities that involve wearing ACUs, boots, IOTV, ACH, and individual weapons. These activities include military movement drills 1 and 2, speed running, GD, CL 2, foot marching, combatives, and obstacle course negotiation. See the sample train/ready phase schedule in this chapter. Units identified within the train/ready phase have no set duration.

Available Phase

5-39. Units in the available phase should focus on activities and drills that support operational missions according to their C-METL or D-METL. See the sample available phase schedule in this chapter. This schedule can be repeated throughout the available phase until units are deployed. Once deployed to the theater of operations, units should continue to conduct PRT activities that are safe and appropriate to the operating environment. Commanders and PRT leaders may select activities and drills to ensure a balanced, progressive, integrated program that can be conducted safely within the constraints of the operating environment. Refer to paragraph 5-26 for a list of sustaining phase PRT exercises, drills, and activities.

Deployment

5-40. Deployment to the theatre of operations may present limitations and constraints on the conduct of PRT. For this reason, special considerations must be taken when planning and conducting individual and collective PRT sessions. Typically, endurance and mobility activities such as sustained running are more negatively impacted than the conduct of strength and mobility activities. In areas where sustained running cannot be conducted, military movement drills, 30:60s, 60:120s, and 300-yd SR should be employed to maintain physical readiness. Commanders can also recommend the use of endurance training equipment (treadmills, elliptical trainers, steppers, and cycle ergometers) for individual and small unit training. Strength and mobility may be trained individually or collectively using the strength and mobility activities specified in Chapter 9. When training individually or in small groups, much benefit is gained by using strength STMs and equipment such as free weights. See the sample individual and collective deployment PRT schedule listed later in this chapter.

SUSTAINING PHASE PRT SCHEDULES

5-41. The following paragraphs discuss sustaining phase PRT schedules as they apply to operational units.

SCHEDULE OVERVIEW

5-42. PRT should be conducted four to five days per week according to AR 350-1. Unlike the toughening phase schedule, activities will vary from week to week in order to train more PRT activities and specifically train for the physical requirements in support of C- and/or D-METL performance. Not all sustaining phase activities are listed on the sample Sustaining Phase PRT Daily Session Overview (Table 5-9). This table is an example of activity sequencing and session purpose. Tables 5-10 through 5-14 describe the ordered sequence of training to be used for unit PRT during phases of ARFORGEN. The following special considerations apply to the sustaining phase schedules:

- PRT sessions should be 60 to 90 minutes to allow adequate conditioning for all components.
- Foot marching under fighting or approach march load may be substituted for sustained running.
- Speed running should be conducted at least one time per week. If conducted only once, preferably, it should be scheduled in the middle of the training week.
- The APFT is best conducted on Monday to ensure adequate recovery and performance.
- Post-deployment, a 1-mile unit run reveals detraining and allows reassignment by ability group.
- During FTXs, the FTX or deployment PRT schedules are used.
- Chapter 6 gives sample training schedules for Soldiers who fail standards or are on medical profile.
    - Fail AWCP standards.
    - Fail APFT standards or unit goals.
    - Temporary or permanent physical profile.

Table 5-9. Sustaining phase PRT daily session overview (ARFORGEN)

Monday	Tuesday	Wednesday	Thursday	Friday
Preparation: PD Activities: MMD 1&2, AGR or RR or HR or TR Recovery: RD	Preparation: PD Activities: GD, CD 1&2, CL 1 and PSD or STC Recovery: RD	Preparation: PD Activities: MMD 1&2, 60:120s and 300 yd SR Recovery: RD	Preparation: PD Activities: GD, CD 3 and CL 1&2 or STC and PSD Recovery: RD	Preparation: PD Activities: MMD 1&2, AGR or RR or HR or TR or 10K FM w/aml Recovery: RD

	Monday
Preparation: PD Activities: MMD 1&2, AGR or RR or HR or TR Recovery: RD	The purpose of this session is to improve the endurance and mobility needed for the successful performance of WTBDs. Preparation readies and conditions the body for a variety of body management competencies. The Military Movement Drills 1 and 2 help improve running form while preparing the Soldier for sustained and speed running. Ability Group, Release and Terrain Runs along with Hill Repeats improve aerobic and anaerobic endurance through sustained running. Recovery safely returns Soldiers to a pre-exercise state while improving mobility.

	Tuesday
Preparation: PD Activities: GD, CD 1&2 CL 1 and PSD or STC Recovery: RD	The purpose of this session is to improve the strength and mobility needed for the successful performance of WTBDs. Preparation readies and conditions the body for a variety of body management competencies. The Guerrilla Drill develops functional mobility for the performance of combatives and the ability to carry another Soldier. Conditioning Drills 1& 2 consist of intermediate and advanced exercises that are designed to functionally train upper body and trunk muscular strength, and endurance needed to successfully perform WTBDs. Climbing Drill 1 improves upper body and trunk strength needed for manipulating body weight. The Push-up and Sit-up Drill improves APFT performance. The Strength Training Circuit develops total body strength and movement proficiency. Recovery safely returns Soldiers to a pre-exercise state while improving mobility.

	Wednesday
Preparation: PD Activities: MMD 1&2, 60:120s and 300 yd SR Recovery: RD	The purpose of this session is to improve the conditioning required for successful performance WTBDs, such as individual movement techniques and move under direct and indirect fire. Preparation readies and conditions the body for a variety of body management competencies. Military Movement Drills 1 & 2 help improve running form while preparing the Soldier for speed running. 60:120s enhance speed and anaerobic power through sustained repeats of high intensity running with intermittent periods of recovery. The 300-yd Shuttle Run develops anaerobic endurance and functional mobility. Recovery safely returns Soldiers to a pre-exercise state while improving mobility.

	Thursday
Preparation: PD Activities: GD, CD 3 and CL 1&2 or STC and PSD Recovery: RD	The purpose of this session is to improve the strength and mobility needed for the successful performance of WTBDs. Preparation readies and conditions the body for a variety of body management competencies. The Guerrilla Drill develops functional mobility for the performance of combatives and the ability to carry another Soldier. Conditioning Drill 3 consists of advanced exercises that are designed to functionally train upper body and trunk muscular strength and endurance needed to successfully perform WTBDs. Climbing Drills 1&2 improve upper body and trunk strength needed for manipulating body weight with and without load. The Strength Training Circuit develops total body strength and movement proficiency. The Push-up and Sit-up Drill improves APFT performance. Recovery safely returns Soldiers to a pre-exercise state while improving mobility.

	Friday
Preparation: PD Activities: MMD 1&2, AGR or RR or HR or TR or 10K FM w/aml Recovery: RD	The purpose of this session is to improve the endurance and mobility needed for the successful performance of WTBDs. Preparation readies and conditions the body for a variety of body management competencies. The Military Movement Drills 1&2 help improve running form while preparing the Soldier for sustained and speed running. Ability Group, Release, and Terrain Runs along with Hill Repeats improve aerobic and anaerobic endurance through sustained running and improves the Soldier's ability to move quickly with agility over various terrains. The foot march improves the muscular and aerobic endurance needed for foot marching under load. Recovery safely returns Soldiers to a pre-exercise state while improving mobility.
Abbreviations	PD – Preparation Drill 4C – Four for the Core HSD – Hip Stability Drill RD – Recovery Drill CD – Conditioning Drill CL – Climbing Drill PSD – PU/SU Drill SR – Shuttle Run AGR – Ability Group Run FM – Foot March RR – Release Run TR – Terrain Run STC – Strength Training Circuit FM-aml (foot march approach march load)

Table 5-10. Unit PRT reset schedule, Month 1

OCTOBER (RESET MONTH 1)				
Monday	*Tuesday*	*Wednesday*	*Thursday*	*Friday*
1 Preparation: PD (5 reps) Activities: 1-mile run assessment Recovery: RD (30 sec)	**2** Preparation: PD (5 reps) Activities: 4C (60 sec). CD 1 & 2 (5 reps ea), CL1 (5 reps) Recovery: RD (30 sec)	**3** Preparation: PD (5 reps) Activities: HSD (5 reps), MMD 1 (1 rep), 30:60s (6 reps), 300-yd SR (1 rep) Recovery: RD (30 sec)	**4** Preparation: PD (5 reps) Activities: 4C (60 sec). CD 1 & 2 (5 reps ea), CL1 (5 reps) Recovery: RD (30 sec)	**5** COLUMBUS DAY TRAINING HOLIDAY
8 COLUMBUS DAY TRAINING HOLIDAY	**9** Preparation: PD (5 reps) Activities: HSD (5 reps), MMD 1 (2 reps), 30:60s (7 reps), 300-yd SR (1 rep) Recovery: RD (30 sec)	**10** Preparation: PD (5 reps) Activities: 4C (60 sec). CD 1 & 2 (5 reps ea), CL1 (5 reps) Recovery: RD (30 sec)	**11** Preparation: PD (5 reps) Activities: HSD (5 reps), MMD 1 (1 rep), AGR (20 min) Recovery: RD (30 sec)	**12** Preparation: PD (5 reps) Activities: 4C (60 sec). CD 1 & 2 (5 reps ea), CL1 (5 reps) Recovery: RD (30 sec)
15 Preparation: PD (5 reps) Activities: HSD (5 reps), MMD 1 (1 rep), AGR (20 min) Recovery: RD (30 sec)	**16** Preparation: PD (5 reps) Activities: 4C (60 sec). CD 1 & 2 (5 reps ea), CL1 (5 reps) Recovery: RD (30 sec)	**17** Preparation: PD (5 reps) Activities: HSD (5 reps), MMD 1 (2 reps), 30:60s (7 reps), 300-yd SR (1 rep) Recovery: RD (30 sec)	**18** Preparation: PD (5 reps) Activities: 4C (60 sec). CD 1 & 2 (5 reps ea), CL1 (5 reps) Recovery: RD (30 sec)	**19** Preparation: PD (5 reps) Activities: HSD (5 reps), MMD 1 (1 rep), AGR (20 min) Or FM w/fl Recovery: RD (30 sec)
22 Preparation: PD (5 reps) Activities: 4C (60 sec). CD 1 & 2 (5 reps ea), CL1 (5 reps) Recovery: RD (30 sec)	**23** Preparation: PD (5 reps) Activities: HSD (5 reps), MMD 1 (2 reps), 30:60s (9 reps), 300-yd SR (1 rep) Recovery: RD (30 sec)	**24** Preparation: PD (5 reps) Activities: 4C (60 sec). CD 1 & 2 (5 reps ea), CL1 (5 reps) Recovery: RD (30 sec)	**25** Preparation: PD (5 reps) Activities: HSD (5 reps), MMD 1 (1 rep), AGR (20 min) Recovery: RD (30 sec)	**26** Preparation: PD (5 reps) Activities: 4C (60 sec). CD 1 & 2 (5 reps ea), CL1 (5 reps) Recovery: RD (30 sec)
29 Preparation: PD (5 reps) Activities: HSD (5 reps), MMD 1 (1 rep), AGR (20 min) Recovery: RD (30 sec)	**30** Preparation: PD (5 reps) Activities: 4C (60 sec). CD 1 & 2 (5 reps ea), CL1 (5 reps) Recovery: RD (30 sec)	**31** Preparation: PD (5 reps) Activities: HSD (5 reps), MMD 1 (2 reps), 30:60s (9 reps), 300-yd SR (1 rep) Recovery: RD (30 sec)		

Table 5-10. Unit PRT reset schedule, Month 2 (continued)

NOVEMBER (RESET MONTH 2)				
Monday	Tuesday	Wednesday	Thursday	Friday
			1 Preparation: PD (5 reps) Activities: 4C (60 sec), CD 1 (5 reps), CL 1 (5 reps), PSD (2 x 30 sec) Recovery: RD (30 sec)	**2** Preparation: PD (5 reps) Activities: HSD (5 reps), MMD 1 (1 rep), AGR (20 min) or FM w/fl ($\leq$10 k) Recovery: RD (30 sec)
5 Preparation: PD (10 reps) Activities: 4C (60 sec), CD 2 (2 x 5 reps), CL 1 (2 x 5 reps), PSD (2 x 45 sec) Recovery: RD (30 sec)	**6** Preparation: PD (10 reps) Activities: HSD (10 reps), MMD 1 (1 rep), 30:60s (8 reps), 300-yd SR (1 rep) Recovery: RD (30 sec)	**7** Preparation: PD (10 reps) Activities: 4C (60 sec), CD 2 (2 x 5 reps), CL 1 (2 x 5 reps), PSD (2 x 45 sec) Recovery: RD (30 sec)	**8** Preparation: PD (10 reps) Activities: HSD (10 reps), MMD 1 (1 rep), AGR (25 min) Recovery: RD (30 sec)	**9** VETERAN'S DAY TRAINING HOLIDAY
12 VETERAN'S DAY TRAINING HOLIDAY	**13** Preparation: PD (10 reps) Activities: 4C (60 sec), CD 2 (2 x 5 reps), CL 1 (2 x 5 reps), PSD (2 x 45 sec) Recovery: RD (30 sec)	**14** Preparation: PD (10 reps) Activities: HSD (10 reps), MMD 1 (1 rep), 30:60s (8 reps), 300-yd SR (1 rep) Recovery: RD (30 sec)	**15** Preparation: PD (10 reps) Activities: 4C (60 sec), CD 2 (2 x 5 reps), CL 1 (2 x 5 reps), PSD (2 x 45 sec) Recovery: RD (30 sec)	**16** Preparation: PD (10 reps) Activities: HSD (10 reps), MMD 1 (1 rep), AGR (25 min) or FM w/fl ($\leq$10 k) Recovery: RD (30 sec)
19 Preparation: PD (10 reps) Activities: 4C (60 sec), CD1 (2 x 5 reps), CL 1 (2 x 5 reps), PSD (2 x 45 sec) Recovery: RD (30 sec)	**20** Preparation: PD (10 reps) Activities: HSD (10 reps), MMD 1 (1 rep), 30:60s (10 reps), 300-yd SR (1 rep) Recovery: RD (30 sec)	**21** Preparation: PD (10 reps) Activities: 4C (60 sec), CD1 (2 x 5 reps), CL 1 (2 x 5 reps), PSD (2 x 45 sec) Recovery: RD (30 sec)	**22** THANKSGIVING TRAINING HOLIDAY	**23** THANKSGIVING TRAINING HOLIDAY
26 Preparation: PD (10 reps) Activities: HSD (10 reps), MMD 1&2 (1 rep ea), AGR (25 min) or RR (25 min) Recovery: RD (30 sec)	**27** Preparation: PD (10 reps) Activities: 4C (60 sec), CD1 & 2 (5 reps ea), CL 1 (2 x 5-10 reps), PSD (2 x 45 sec) Recovery: RD (30 sec)	**28** Preparation: PD (10 reps) Activities: HSD (10 reps), MMD 1&2 (1 rep ea), 60:120s (6 reps), 300-yd SR (1 rep) Recovery: RD (30 sec)	**29** Preparation: PD (10 reps) Activities: 4C (60 sec), CD1 & 2 (5 reps ea), CL 1 (2 x 5-10 reps), PSD (2 x 45 sec) Recovery: RD (30 sec)	**30** Preparation: PD (10 reps) Activities: HSD (10 reps), MMD 1&2 (1 rep ea.), AGR (25 min) or FM w/fl ($\leq$10 k) Recovery: RD (30 sec)

Chapter 5

Table 5-10. Unit PRT reset schedule, Month 3 (continued)

DECEMBER (RESET MONTH 3)					
Monday	*Tuesday*	*Wednesday*	*Thursday*	*Friday*
3 Preparation: PD (10 reps) Activities: 4C (60 sec), CD1 (10 reps), CL 1 (10 reps), PSD (2 x 60 sec) Recovery: RD (30 sec)	**4** Preparation: PD (10 reps) Activities: HSD (10 reps), MMD 1&2 (1 rep ea), AGR (25 min) or RR (25 min) Recovery: RD (30 sec)	**5** Preparation: PD (10 reps) Activities: 4C (60 sec), CD1 (10 reps), CL 1 (10 reps), PSD (2 x 60 sec) Recovery: RD (30 sec)	**6** Preparation: PD (10 reps) Activities: HSD (10 reps), MMD 1&2 (1 rep ea), 60:120s (8 reps) Recovery: RD (30 sec)	**7** Preparation: PD (10 reps) Activities: 4C (60 sec), CD1 (10 reps), CL 1 (10 reps), PSD (2 x 60 sec) Recovery: RD (30 sec)
10 Preparation: : PD (10 reps) Activities: APFT Recovery: RD (30 sec)	**11** Preparation: PD (10 reps) Activities: 4C (60 sec), CD2 (10 reps), CL 1 (10 reps), PSD (2 x 60 sec) Recovery: RD (30 sec)	**12** Preparation: PD (10 reps) Activities: ACUs/Boots, HSD (10 reps), MMD 1&2 (1 rep ea) 300-yd SR (1 rep), 30:60s (10 reps) Recovery: RD (30 sec)	**13** Preparation: PD (10 reps) Activities: 4C (60 sec), CD2 (10 reps), CL 1 (10 reps), PSD (2 x 60 sec) Recovery: RD (30 sec)	**14** Preparation: PD (10 reps) Activities: HSD (10 reps), MMD 1&2 (1 rep ea), AGR (30 min) or FM w/fl ($\leq$10 k) Recovery: RD (30 sec)
17 Preparation: PD (10 reps) Activities: 4C (60 sec), CD1 (10 reps), CL 1 (10 reps), PSD (2 x 60 sec) or STC (3 rotations x 60 sec) or STM1 (3 rotations x 60 sec) Recovery: RD (30 sec)	**18** Preparation: PD (10 reps) Activities: ACUs/Boots, HSD (10 reps), MMD 1&2 (1 rep ea), 300-yd SR (1 rep), 60:120s (8 reps) Recovery: RD (30 sec)	**19** Preparation: PD (10 reps) Activities: 4C (60 sec), CD2 (10 reps), CL 1 (10 reps), PSD (2 x 60 sec) or STC (3 rotations x 60 sec) or STM1 (3 rotations x 60 sec) Recovery: RD (30 sec)	**20** Preparation: PD (10 reps) Activities: HSD (10 reps), MMD 1&2 (1 rep ea), AGR (25 min) or RR (25 min) Recovery: RD (30 sec)	**21** Preparation: PD (10 reps) Activities: 4C (60 sec), CD1 (10 reps), CL 1 (10 reps), PSD (2 x 60 sec) or STC (3 rotations x 60 sec) or STM1 (3 rotations x 60 sec) Recovery: RD (30 sec)
24 CHRISTMAS TRAINING HOLIDAY	**25** CHRISTMAS TRAINING HOLIDAY	**26** BLOCK LEAVE	**27** BLOCK LEAVE	**28** BLOCK LEAVE
31 NEW YEARS TRAINING HOLIDAY				

Table 5-10. Unit PRT reset schedule, Month 4 (continued)

JANUARY (RESET MONTH 4)

Monday	Tuesday	Wednesday	Thursday	Friday
	1 NEW YEAR'S DAY TRAINING HOLIDAY	**2** BLOCK LEAVE	**3** BLOCK LEAVE	**4** BLOCK LEAVE
7 Preparation: PD (10 reps) Activities: HSD (10 reps), MMD 1&2 (1 rep ea), AGR (25 min) or RR (25 min) Recovery: RD (30 sec)	**8** Preparation: PD (10 reps) Activities: 4C (60 sec), CD1 (10 reps), CL 1 (10 reps), PSD (2 x 60 sec) or STC (3 rotations x 60 sec) or STM1 (3 rotations x 60 sec) Recovery: RD (30 sec)	**9** Preparation: PD (10 reps) Activities: ACUs/Boots, HSD (10 reps), MMD 1&2 (1 rep ea), 300-yd SR (1 rep), 30:60s (10 reps) Recovery: RD (30 sec)	**10** Preparation: PD (10 reps) Activities: 4C (60 sec), CD2 (10 reps), CL 1 (10 reps), PSD (2 x 60 sec) or STC (3 rotations x 60 sec) or STM1 (3 rotations x 60 sec) Recovery: RD (30 sec)	**11** MLK TRAINING HOLIDAY
14 MLK TRAINING HOLIDAY	**15** Preparation: PD (10 reps) Activities: HSD (10 reps), MMD 1&2 (1 rep ea), AGR (30 min) or RR (30 min) Recovery: RD (30 sec)	**16** Preparation: PD (10 reps) Activities: 4C (60 sec), CD1 (10 reps), CL 1 (10 reps), PSD (2 x 60 sec) or STC (3 rotations x 60 sec) or STM1 (3 rotations x 60 sec) Recovery: RD (30 sec)	**17** Preparation: PD (10 reps) Activities: ACUs/Boots, HSD (10 reps), MMD 1&2 (1 rep ea), 300-yd SR (1 rep), 30:60s (10 reps) Recovery: RD (30 sec)	**18** Preparation: PD (10 reps) Activities: 4C (60 sec), CD2 (10 reps), CL 1 (10 reps), PSD (2 x 60 sec) or STC (3 rotations x 60 sec) or STM1 (3 rotations x 60 sec) Recovery: RD (30 sec)
21 Preparation: PD (10 reps) Activities: HSD (10 reps), MMD 1&2 (1 rep ea), AGR (30 min) or RR (30 min) Recovery: RD (30 sec)	**22** Preparation: PD (10 reps) Activities: 4C (60 sec), CD1 (10 reps), CL 1 (10 reps), PSD (2 x 60 sec) or STC (3 rotations x 60 sec) or STM1 (3 rotations x 60 sec) Recovery: RD (30 sec)	**23** Preparation: PD (10 reps) Activities: HSD (10 reps), MMD 1&2 (1 rep ea), 60:120s (10 reps) Recovery: RD (30 sec)	**24** Preparation: PD (10 reps) Activities: 4C (60 sec), CD2 (10 reps), CL 1 (10 reps), PSD (2 x 60 sec) or STC (3 rotations x 60 sec) or STM1 (3 rotations x 60 sec) Recovery: RD (30 sec)	**25** Preparation: PD (10 reps) Activities: HSD (10 reps), MMD 1&2 (1 rep ea), AGR (30 min) or FM w/fl ($\leq$10 k) Recovery: RD (30 sec)
28 Preparation: PD (10 reps) Activities: 4C (60 sec), CD2 (10 reps), CL 1 (10 reps), PSD (2 x 60 sec) or STC (3 rotations x 60 sec) or STM1 (3 rotations x 60 sec) Recovery: RD (30 sec)	**29** Preparation: PD (10 reps) Activities: HSD (10 reps), MMD 1&2 (1 rep ea), 60:120s (10 reps) Recovery: RD (30 sec)	**30** Preparation: PD (10 reps) Activities: 4C (60 sec), CD2 (10 reps), CL 1 (10 reps), PSD (2 x 60 sec) or STC (3 rotations x 60 sec) or STM1 (3 rotations x 60 sec) Recovery: RD (30 sec)	**31** Preparation: PD (10 reps) Activities: HSD (10 reps), MMD 1&2 (1 rep ea), AGR (30 min) or RR (30 min) Recovery: RD (30 sec)	

Chapter 5

Table 5-10. Unit PRT reset schedule, Month 5 (continued)

\multicolumn{5}{c	}{FEBRUARY (RESET MONTH 5)}
Monday	Tuesday	Wednesday	Thursday	Friday
				1 Preparation: PD (10 reps) Activities: 4C (60 sec), CD2 (10 reps), CL 1 (10 reps), PSD (2 x 60 sec) or STC (3 rotations x 60 sec) or STM1 (3 rotations x 60 sec) Recovery: RD (30 sec)
4 Preparation: PD (10 reps) Activities: HSD (10 reps), MMD 1&2 (1 rep ea), AGR (30 min) or RR (30 min) Recovery: RD (30 sec)	**5** Preparation: PD (10 reps) Activities: ACUs/Boots, 4C (60 sec), GD (1 rep), CD 1&2 (10 reps ea), CL 2 (2x5 reps), PSD (2 x 60 sec) or STC (3 rotations x 60 sec) or STM1 (3 rotations x 60 sec) Recovery: RD (30 sec)	**6** Preparation: PD (10 reps) Activities: HSD (10 reps) MMD 1&2 (1 rep ea) 60:120s (10 reps) or300 yd-SR (2 reps) & Hill Repeats (6-8 reps) Recovery: RD (30 sec)	**7** Preparation: PD (10 reps) Activities: ACUs/Boots, 4C (60 sec), GD (1 rep), CD 1&2 (10 reps ea), CL 2 (2x5 reps), PSD (2 x 60 sec) or STC (3 rotations x 60 sec) or STM1 (3 rotations x 60 sec) Recovery: RD (30 sec)	**8** Preparation: PD (10 reps) Activities: HSD (10 reps), MMD 1&2 (1 rep ea), AGR (30 min) or FM w/fl (TBD) Recovery: RD (30 sec)
11 Preparation: PD (10 reps) Activities: 4C (60 sec), CD 1&2 (10 reps ea), CL 1 (2x5 reps), PSD (2 x 60 sec) or STC (3 rotations x 60 sec) or STM1 (3 rotations x 60 sec) Recovery: RD (30 sec)	**12** Preparation: PD (10 reps) Activities: ACUs/Boots, HSD (10 reps), MMD 1&2 (1 rep ea), TR (20 min) Recovery: RD (30 sec)	**13** Preparation: PD (10 reps) Activities: 4C (60 sec), CD 3 (5-10 reps), CL 1 (2x5 reps), PSD (2 x 60 sec) or STC (3 rotations x 60 sec) or STM1 (3 rotations x 60 sec) Recovery: RD (30 sec)	**14** Preparation: PD (10 reps) Activities: ACUs/Boots, HSD (10 reps) MMD 1&2 (1 rep ea), 30:60s (6-8 reps) or 300 yd-SR (2 reps) & Hill Repeats (6-8 reps) Recovery: RD (30 sec)	**15** PRESIDENT'S DAY TRAINING HOLIDAY
18 PRESIDENT'S DAY TRAINING HOLIDAY	**19** Preparation: PD (10 reps) Activities: ACUs/Boots, 4C (60 sec), GD (1 rep), CD 3 (5 reps), CL 2 (2x5 reps), PSD (2 x 60 sec) or STC (3 rotations x 60 sec) or STM1 (3 rotations x 60 sec) Recovery: RD (30 sec)	**20** Preparation: PD (10 reps) Activities: HSD (10 reps), MMD 1&2 (1 rep ea), 60:120s (10 reps) or 300 yd-SR (2 reps) & Hill Repeats (6-8 reps) Recovery: RD (30 sec)	**21** Preparation: PD (10 reps) Activities: ACUs/Boots, 4C (60 sec), GD (1 rep), CD 3 (5 reps), CL 2 (2x5 reps), PSD (2 x 60 sec) or STC (3 rotations x 60 sec) or STM1 (3 rotations x 60 sec). Recovery: RD (30 sec)	**22** Preparation: PD (10 reps) Activities: HSD (10 reps), MMD 1&2 (1 rep ea), AGR (30 min) or FM w/fl (TBD) Recovery: RD (30 sec)
25 Preparation: PD (10 reps) Activities: 4C (60 sec), CD 3 (5-10 reps), CL 1 (2x5 reps), PSD (2 x 60 sec) or STC (3 rotations x 60 sec) or STM1 (3 rotations x 60 sec) Recovery: RD (30 sec)	**26** Preparation: PD (10 reps) Activities: HSD (10 reps), MMD 1&2 (1 rep ea), AGR (30 min) or RR (30 min) Recovery: RD (30 sec)	**27** Preparation: PD (10 reps) Activities: 4C (60 sec), CD 1&2 (10 reps ea), CL 1 (2x5 reps), PSD (2 x 60 sec) or STC (3 rotations x 60 sec) or STM1 (3 rotations x 60 sec) Recovery: RD (30 sec)	**28** Preparation: PD (10 reps) Activities: HSD (10 reps) MMD 1&2 (1 rep ea) 60:120s (10 reps) or 300 yd-SR (2 reps) & Hill Repeats (6-8 reps) Recovery: RD (30 sec)	**29** Preparation: PD (10 reps) Activities: 4C (60 sec), CD 1&2 (10 reps ea), CL 1 (2x5 reps), PSD (2 x 60 sec) or STC (3 rotations x 60 sec) or STM1 (3 rotations x 60 sec) Recovery: RD (30 sec)

Table 5-10. Unit PRT reset schedule, Month 6 (continued)

\multicolumn{5}{c}{MONTH MARCH (RESET MONTH 6)}
Monday	*Tuesday*	*Wednesday*	*Thursday*	*Friday*
3 Preparation: PD (10 reps) Activities: MMD 1 (1 rep), 1-mile run assessment Recovery: RD (30 sec)	**4** Preparation: PD (10 reps) Activities: ACUs/Boots, GD (1 rep), CD 3 (5-10 reps), CL 2 (2x5 reps), PSD (2 x 60 sec) or STC (3 rotations x 60 sec) or other strength training modalities Recovery: RD (30 sec)	**5** Preparation: PD (10 reps) Activities: MMD 1&2 (1 rep ea) 60:120s (10 reps) or 300 yd-SR (2 reps) & Hill Repeats (6-8 reps) Recovery: RD (30 sec)	**6** Preparation: PD (10 reps) Activities: ACUs/Boots, GD (1 rep), CD 3 (5-10 reps), CL 2 (2x5 reps), PSD (2 x 60 sec) or STC (3 rotations x 60 sec) or other strength training modalities Recovery: RD (30 sec)	**7** Preparation: PD (10 reps) Activities: MMD 1&2 (1 rep ea), AGR (30 min) or FM w/fl or aml (TBD) Recovery: RD (30 sec)
10 Preparation: PD (10 reps) Activities: CD 1&2 (10 reps ea), CL 1 (2x5 reps), PSD (2 x 60 sec) or STC (3 rotations x 60 sec) or other strength training modalities Recovery: RD (30 sec)	**11** Preparation: PD (10 reps) Activities: ACUs/Boots, MMD 1&2 (1 rep ea), TR (20 min) Recovery: RD (30 sec)	**12** Preparation: PD (10 reps) Activities: CD 1&2 (10 reps ea), CL 1 (2x5 reps), PSD (2 x 60 sec) or STC (3 rotations x 60 sec) or other strength training modalities Recovery: RD (30 sec)	**13** Preparation: PD (10 reps) Activities: ACUs/Boots, MMD 1&2 (1 rep ea), 30:60s (10 reps) or 300 yd-SR (2 reps) & Hill Repeats (8-10 reps) Recovery: RD (30 sec)	**14** Preparation: PD (10 reps) Activities: CD 1&2 (10 reps ea), CL 1 (2x5 reps), PSD (2 x 60 sec) or STC (3 rotations x 60 sec) or other strength training modalities Recovery: RD (30 sec)
17 Preparation: PD (10 reps) Activities: MMD 1&2 (1 rep ea), AGR (30 min) or RR (30 min) Recovery: RD (30 sec)	**18** Preparation: PD (10 reps) Activities: ACUs/Boots, GD (1 rep), CD 3 (5-10 reps), CL 2 (2x5 reps), PSD (2 x 60 sec) or STC (3 rotations x 60 sec) or other strength training modalities Recovery: RD (30 sec)	**19** Preparation: PD (10 reps) Activities: MMD 1&2 (1 rep ea) 60:120s (10 reps) or 300 yd-SR (2 reps) & Hill Repeats (6-8 reps) Recovery: RD (30 sec)	**20** Preparation: PD (10 reps) Activities: ACUs/Boots, GD (1 rep), CD 3 (5-10 reps), CL 2 (2x5 reps), PSD (2 x 60 sec) or STC (3 rotations x 60 sec) or other strength training modalities Recovery: RD (30 sec)	**21** Preparation: PD (10 reps) Activities: MMD 1&2 (1 rep ea), AGR (30 min) or FM w/fl or aml (TBD) Recovery: RD (30 sec)
24 Preparation: PD (10 reps) Activities: ACUs/Boots, GD (1 rep), CD 3 (5-10 reps), CL 2 (2x5 reps), PSD (2 x 60 sec) or STC (3 rotations x 60 sec) or other strength training modalities Recovery: RD (30 sec)	**25** Preparation: PD (10 reps) Activities: MMD 1&2 (1 rep ea), 60:120s (10 reps) or 300 yd-SR (2 reps) & Hill Repeats (8-10 reps) Recovery: RD (30 sec)	**26** Preparation: PD (10 reps) Activities: ACUs/Boots, GD (1 rep), CD 3 (5-10 reps), CL 2 (2x5 reps), PSD (2 x 60 sec) or STC (3 rotations x 60 sec) or other strength training modalities Recovery: RD (30 sec)	**27** Preparation: PD (10 reps) Activities: MMD 1&2 (1 rep ea), AGR (30 min) or RR (30 min) Recovery: RD (30 sec)	**28** Preparation: PD (10 reps) Activities: ACUs/Boots, GD (1 rep), CD 3 (5-10 reps), CL 2 (2x5 reps), PSD (2 x 60 sec) or STC (3 rotations x 60 sec) or other strength training modalities Recovery: RD (30 sec)
31 Preparation: PD (10 reps) Activities: APFT Recovery: RD (30 sec)				

Table 5-11. Unit PRT train/ready schedule, Month 1

\multicolumn{5}{c}{APRIL (TRAIN/READY MONTH 1)}
Monday	**Tuesday**	**Wednesday**	**Thursday**	**Friday**
	1 Preparation: PD (10 reps) Activities: CD 1&2 (10 reps ea), CL 1 (10 reps), PSD (2 x 60 sec) or STC (3 rotations x 60 sec) or other strength training modalities Recovery: RD (30 sec)	**2** Preparation: PD (10 reps) Activities: MMD 1&2 (1 rep ea), 60:120s (10 reps) or 300 yd-SR (2 reps) & Hill Repeats (8-10 reps) Recovery: RD (30 sec)	**3** Preparation: PD (10 reps) Activities: CD 1&2 (10 reps ea), CL 1 (10 reps), PSD (2 x 60 sec) or STC (3 rotations x 60 sec) or other strength training modalities Recovery: RD (30 sec)	**4** Preparation: PD (10 reps) Activities: MMD 1&2 (1 rep ea), AGR (30 min) or RR (30 min) Recovery: RD (30 sec)
7 Preparation: PD (10 reps) Activities: CD 1&2 (10 reps ea), CL 1 (10 reps), PSD (2 x 60 sec) or STC (3 rotations x 60 sec) or other strength training modalities Recovery: RD (30 sec)	**8** Preparation: PD (10 reps) Activities: MMD 1&2 (1 rep ea), 30:60s (15 reps) or 300 yd-SR (3 reps) or Hill Repeats (10 reps) Recovery: RD (30 sec)	**9** Preparation: PD (10 reps) Activities: ACUs/Boots, GD (1 rep), CD 3 (5-10 reps), CL 2 (2x5 reps), PSD (2 x 60 sec) or STC (3 rotations x 60 sec) or other strength training modalities Recovery: RD (30 sec)	**10** Preparation: PD (10 reps) Activities: MMD 1&2 (1 rep ea), 30:60s (10 reps) & 300 yd-SR (1 rep) or Hill Repeats (10 reps) Recovery: RD (30 sec)	**11** Preparation: PD (10 reps) Activities: CD 1&2 (10 reps ea), CL 1 (10 reps), PSD (2 x 60 sec) or STC (3 rotations x 60 sec) or other strength training modalities Recovery: RD (30 sec)
14 Preparation: PD (10 reps) Activities: MMD 1&2 (1 rep ea), AGR (30 min) or RR (30 min) or APFT Recovery: RD (30 sec)	**15** Preparation: PD (10 reps) Activities: ACUs/Boots, GD (1 rep), CD 3 (5-10 reps), CL 2 (2x5 reps), PSD (2 x 60 sec) or STC (3 rotations x 60 sec) or other strength training modalities Recovery: RD (30 sec)	**16** Preparation: PD (10 reps) Activities: MMD 1&2 (1 rep ea), 30:60s (15 reps) or 300 yd-SR (3 reps) or Hill Repeats (10 reps) Recovery: RD (30 sec)	**17** Preparation: PD (10 reps) Activities: ACUs/Boots, GD (1 rep), CD 3 (5-10 reps), CL 2 (2x5 reps), PSD (2 x 60 sec) or STC (3 rotations x 60 sec) or other strength training modalities Recovery: RD (30 sec)	**18** Preparation: PD (10 reps) Activities: MMD 1&2 (1 rep ea), AGR (30 min) or Unit Run (30 min) or FM w/fl or aml (TBD) Recovery: RD (30 sec)
21 Preparation: PD (10 reps) Activities: CD 1&2 (10 reps ea), CL 1 (10 reps), PSD (2 x 60 sec) or STC (3 rotations x 60 sec) or other strength training modalities Recovery: RD (30 sec)	**22** Preparation: PD (10 reps) Activities: MMD 1&2 (1 rep ea), 60:120s (10 reps) or 300 yd-SR (2 reps) & Hill Repeats (8-10 reps) Recovery: RD (30 sec)	**23** Preparation: PD (10 reps) Activities: CD 1&2 (10 reps ea), CL 1 (10 reps), PSD (2 x 60 sec) or STC (3 rotations x 60 sec) or other strength training modalities Recovery: RD (30 sec)	**24** Preparation: PD (10 reps) Activities: MMD 1&2 (1 rep ea), AGR (30 min) or RR (30 min) Recovery: RD (30 sec)	**25** Preparation: PD (10 reps) Activities: CD 1&2 (10 reps ea), CL 1 (10 reps), PSD (2 x 60 sec) or STC (3 rotations x 60 sec) or other strength training modalities Recovery: RD (30 sec)
28 Preparation: PD (10 reps) Activities: ACUs/Boots, MMD 1&2 (1 rep ea), TR (20 min) Recovery: RD (30 sec)	**29** Preparation: PD (10 reps) Activities: CD 1&2 (10 reps ea), CL 1 (10 reps), PSD (2 x 60 sec) or STC (3 rotations x 60 sec) or other strength training modalities Recovery: RD (30 sec)	**30** Preparation: PD (10 reps) Activities: ACUs/Boots, MMD 1&2 (1 rep ea), 30:60s (10-15 reps) or 300 yd-SR (2 reps) & Hill Repeats (8-10 reps) Recovery: RD (30 sec)		

Table 5-11. Unit PRT train/ready schedule, Months 2 through 6 (continued)

MAY-SEPTEMBER (TRAIN/READY MONTHS 2-6)				
Monday	Tuesday	Wednesday	Thursday	Friday
			1 Preparation: PD (10 reps) Activities: CD 3 (10 reps), CL 1 (10 reps), PSD (2 x 60 sec) or STC (3 rotations x 60 sec) or other strength training modalities Recovery: RD (30 sec)	**2** Preparation: PD (10 reps) Activities: MMD 1&2 (1 rep ea), AGR (30 min) or Unit Run (30 min) or FM w/fl or aml (TBD) Recovery: RD (30 sec)
5 Preparation: PD (10 reps) Activities: CD 1&2 (10 reps ea), CL 1 (10 reps), PSD (2 x 60 sec) or STC (3 rotations x 60 sec) or other strength training modalities Recovery: RD (30 sec)	**6** Preparation: PD (10 reps) Activities: MMD 1&2 (1 rep ea), 60:120s (10 reps) or 300 yd-SR (2 reps) & Hill Repeats (8-10 reps) Recovery: RD (30 sec)	**7** Preparation: PD (10 reps) Activities: CD 1&2 (10 reps ea), CL 1 (10 reps), PSD (2 x 60 sec) or STC (3 rotations x 60 sec) or other strength training modalities Recovery: RD (30 sec)	**8** Preparation: PD (10 reps) Activities: MMD 1&2 (1 rep ea), 30:60s (10-15 reps) or 300 yd-SR (2 reps) & Hill Repeats (8-10 reps) Recovery: RD (30 sec)	**9** Preparation: PD (10 reps) Activities: CD 1&2 (10 reps ea), CL 1 (10 reps), PSD (2 x 60 sec) or STC (3 rotations x 60 sec) or other strength training modalities Recovery: RD (30 sec)
12 Preparation: PD (10 reps) Activities: ACUs/Boots, MMD 1&2 (1 rep ea), TR (20 min) Recovery: RD (30 sec)	**13** Preparation: PD (10 reps) Activities: CD 1&2 (10 reps ea), CL 1 (10 reps), PSD (2 x 60 sec) or STC (3 rotations x 60 sec) or other strength training modalities Recovery: RD (30 sec)	**14** Preparation: PD (10 reps) Activities: ACUs/Boots, MMD 1&2 (1 rep ea), 30:60s (10-15 reps) or 300 yd-SR (2 reps) & Hill Repeats (8-10 reps) Recovery: RD (30 sec)	**15** Preparation: PD (10 reps) Activities: CD 1&2 (10 reps ea), CL 1 (10 reps), PSD (2 x 60 sec) or STC (3 rotations x 60 sec) or other strength training modalities Recovery: RD (30 sec)	**16** Preparation: PD (10 reps) Activities: MMD 1&2 (1 rep ea), AGR (30 min) or Unit Run (30 min) or FM w/fl or aml (TBD) Recovery: RD (30 sec)
19 Preparation: PD (10 reps) Activities: CD 1&2 (10 reps ea), CL 1 (10 reps), PSD (2 x 60 sec) or STC (3 rotations x 60 sec) or other strength training modalities. Recovery: RD (30 sec)	**20** Preparation: PD (10 reps) Activities: MMD 1&2 (1 rep ea), 60:120s (10 reps) or 300 yd-SR (2 reps) & Hill Repeats (8-10 reps) Recovery: RD (30 sec)	**21** Preparation: PD (10 reps) Activities: CD 1&2 (10 reps ea), CL 1 (10 reps), PSD (2 x 60 sec) or STC (3 rotations x 60 sec) or other strength training modalities Recovery: RD (30 sec)	**22** Preparation: PD (10 reps) Activities: MMD 1&2 (1 rep ea), 30:60s (10-15 reps) or 300 yd-SR (2 reps) & Hill Repeats (8-10 reps) Recovery: RD (30 sec)	**23** Preparation: PD (10 reps) Activities: CD 1&2 (10 reps ea), CL 1 (10 reps), PSD (2 x 60 sec) or STC (3 rotations x 60 sec) or other strength training modalities Recovery: RD (30 sec)
26 Preparation: PD (10 reps) Activities: ACUs/Boots, MMD 1&2 (1 rep ea), TR (20 min) Recovery: RD (30 sec)	**27** Preparation: PD (10 reps) Activities: CD 1&2 (10 reps ea), CL 1 (10 reps), PSD (2 x 60 sec) or STC (3 rotations x 60 sec) or other strength training modalities Recovery: RD (30 sec)	**28** Preparation: PD (10 reps) Activities: ACUs/Boots, MMD 1&2 (1 rep ea), 30:60s (10-15 reps) or 300 yd-SR (2 reps) & Hill Repeats (8-10 reps) Recovery: RD (30 sec)	**29** Preparation: PD (10 reps) Activities: CD 1&2 (10 reps ea), CL 1 (10 reps), PSD (2 x 60 sec) or STC (3 rotations x 60 sec) or other strength training modalities Recovery: RD (30 sec)	**30** Preparation: PD (10 reps) Activities: MMD 1&2 (1 rep ea), AGR (30 min) or Unit Run (30 min) or FM w/fl or aml (TBD) Recovery: RD (30 sec)

Table 5-12. Unit PRT, available schedule

OCTOBER or until DEPLOYED (AVAILABLE)				
Monday	*Tuesday*	*Wednesday*	*Thursday*	*Friday*
		1 Preparation: PD (10 reps) Activities: CD 1&2 (10 reps ea), CL 1 (10 reps), PSD (2 x 60 sec) or STC (3 rotations x 60 sec) or other strength training modalities Recovery: RD (30 sec)	**2** Preparation: PD (10 reps) Activities: MMD 1&2 (1 rep ea), 60:120s (10 reps) or 300 yd-SR (2 reps) & Hill Repeats (8-10 reps) Recovery: RD (30 sec)	**3** Preparation: PD (10 reps) Activities: CD 1&2 (10 reps ea), CL 1 (10 reps), PSD (2 x 60 sec) or STC (3 rotations x 60 sec) or other strength training modalities Recovery: RD (30 sec)
6 Preparation: PD (10 reps) Activities: ACUs/Boots, MMD 1&2 (1 rep ea), TR (30 min) Recovery: RD (30 sec)	**7** Preparation: PD (10 reps) Activities: CD 3 (10 reps), CL 1 (10 reps), PSD (2 x 60 sec) or STC (3 rotations x 60 sec) or other strength training modalities Recovery: RD (30 sec)	**8** Preparation: PD (10 reps) Activities: ACUs/Boots, MMD 1&2 (1 rep ea), 30:60s (10-15 reps) or 300 yd-SR (2 reps) & Hill Repeats (8-10 reps) Recovery: RD (30 sec)	**9** Preparation: PD (10 reps) Activities: CD 3 (10 reps), CL 1 (10 reps), PSD (2 x 60 sec) or STC (3 rotations x 60 sec) or other strength training modalities Recovery: RD (30 sec)	**10** Preparation: PD (10 reps) Activities: MMD 1&2 (1 rep ea), AGR (30 min) or Unit Run (30 min) or FM w/fl or aml (TBD) Recovery: RD (30 sec)
13 Preparation: PD (10 reps) Activities: ACUs/Boots, GD (1 rep), CD 1&2 (10 reps ea), CL 2 (2x5 reps), PSD (2 x 60 sec) or STC (3 rotations x 60 sec) or other strength training modalities Recovery: RD (30 sec)	**14** Preparation: PD (10 reps) Activities: MMD 1&2 (1 rep ea), 60:120s (10 reps) or 300 yd-SR (2 reps) & Hill Repeats (8-10 reps) Recovery: RD (30 sec)	**15** Preparation: PD (10 reps) Activities: ACUs/Boots, GD (1 rep), CD 3 (5-10 reps), CL 2 (2x5 reps), PSD (2 x 60 sec) or STC (3 rotations x 60 sec) or other strength training modalities Recovery: RD (30 sec)	**16** Preparation: PD (10 reps) Activities: MMD 1&2 (1 rep ea), AGR (30 min) or RR (30 min) Recovery: RD (30 sec)	**17** Preparation: PD (10 reps) Activities: ACUs/Boots, GD (1 rep), CD 1&2 (10 reps ea), CL 2 (2x5 reps), PSD (2 x 60 sec) or STC (3 rotations x 60 sec) or other strength training modalities Recovery: RD (30 sec)
20 Preparation: PD (10 reps) Activities: MMD 1&2 (1 rep ea), AGR (30 min) or RR (30 min) Recovery: RD (30 sec)	**21** Preparation: PD (10 reps) Activities: ACUs/Boots, GD (1 rep), CD 3 (5-10 reps), CL 2 (2x5 reps), PSD (2 x 60 sec) or STC (3 rotations x 60 sec) or other strength training modalities Recovery: RD (30 sec)	**22** Preparation: PD (10 reps) Activities: MMD 1&2 (1 rep ea), 60:120s (10 reps) or 300 yd-SR (2 reps) & Hill Repeats (8-10 reps) Recovery: RD (30 sec)	**23** Preparation: PD (10 reps) Activities: ACUs/Boots, GD (1 rep), CD 3 (5-10 reps), CL 2 (2x5 reps), PSD (2 x 60 sec) or STC (3 rotations x 60 sec) or other strength training modalities Recovery: RD (30 sec)	**24** Preparation: PD (10 reps) Activities: MMD 1&2 (1 rep ea), AGR (30 min) or Unit Run (30 min) or FM w/fl or aml (TBD) Recovery: RD (30 sec)
27 Preparation: PD (10 reps) Activities: APFT Recovery: RD (30 sec)	**28** Preparation: PD (10 reps) Activities: CD 1 (10 reps), CL 1 (10 reps), PSD (2 x 60 sec) or STC (3 rotations x 60 sec) or other strength training modalities Recovery: RD (30 sec)	**29** Preparation: PD (10 reps) Activities: MMD 1&2 (1 rep ea), 60:120s (10 reps) or 300 yd-SR (2 reps) & Hill Repeats (8-10 reps) Recovery: RD (30 sec)	**30** Preparation: PD (10 reps) Activities: CD 2 (10 reps), CL 1 (10 reps), PSD (2 x 60 sec) or STC (3 rotations x 60 sec) or other strength training modalities Recovery: RD (30 sec)	**31** Preparation: PD (10 reps) Activities: MMD 1&2 (1 rep ea), AGR (30 min) or RR (30 min) Recovery: RD (30 sec)

Planning Considerations

Table 5-13. Deployed PRT, collective schedule

		DEPLOYED PRT-COLLECTIVE		
Monday	*Tuesday*	*Wednesday*	*Thursday*	*Friday*
Preparation: PD (10 reps) Activities: (ACUs, Boots, IOTV) MMD 1&2 (1 rep ea) 300-yd SR (3 reps) Recovery: RD (30 sec)	Preparation: PD (10 reps) Activities: (ACUs, Boots, IOTV) GD (1 rep) CD 1&2 (10 reps ea) CL 1 (10 reps) or other strength training modalities Recovery: RD (30 sec)	Preparation: PD (10 reps) Activities: (ACUs, Boots, IOTV) MMD 1&2 (1 rep ea) 30:60s (10 reps) Recovery: RD (30 sec)	Preparation: PD (10 reps) Activities: (ACUs, Boots, IOTV) GD (1 rep) CD 1&2 (10 reps ea) CL 1 (10reps) or other strength training modalities Recovery: RD (30 sec)	Preparation: PD (10 reps) Activities: (ACUs, Boots, IOTV) MMD 1&2 (1 rep ea) 300-yd SR (3 reps) Recovery: RD (30 sec)
Preparation: PD (10 reps) Activities: (ACUs, Boots, IOTV) GD (1 rep) CD 3 (10 reps), PSD (4 x 30 sec) CL 2 (5-10 reps) or other strength training modalities Recovery: RD (30 sec)	Preparation: PD (10 reps) Activities: (ACUs, Boots, IOTV) MMD 1&2 (1 rep ea) 300-yd SR (3 reps) Recovery: RD (30 sec)	Preparation: PD (10 reps) Activities: (ACUs, Boots, IOTV) GD (1 rep) CD 1&2 (10 reps ea) CL 1 (10 reps) or other strength training modalities Recovery: RD (30 sec)	Preparation: PD (10 reps) Activities: (ACUs, Boots, IOTV) MMD 1&2 (1 rep ea) 30:60s (10 reps) Recovery: RD (30 sec)	Preparation: PD (10 reps) Activities: (ACUs, Boots, IOTV) GD (1 rep) CD 3 (10 reps) PSD (4 x 30 sec) CL 2 (5-10 reps) or other strength training modalities Recovery: RD (30 sec)

Table 5-14. Deployed PRT, individual schedule

		DEPLOYED PRT-INDIVIDUAL		
Monday	*Tuesday*	*Wednesday*	*Thursday*	*Friday*
Preparation: PD (10 reps) Activities: (ACUs, Boots, IOTV) MMD 1&2 (1 rep ea), 300-yd SR (3 reps) Recovery: RD (30 sec)	Preparation: PD (10 reps) Activities: (ACUs, Boots, IOTV) CD 1&2 (10 reps ea), PSD (4 x 30 sec), CL 1 (5 reps) or other strength training modalities Recovery: RD (30 sec)	Preparation: PD (10 reps) Activities: (ACUs, Boots, IOTV) MMD 1&2 (1 rep ea), 300-yd SR (1 rep), 30:60s (10 reps) Recovery: RD (30 sec)	Preparation: PD (10 reps) Activities: (ACUs, Boots, IOTV) STM (3 sets x 10 reps) or other strength training modalities Recovery: RD (30 sec)	Preparation: PD (10 reps) Activities: (ACUs, Boots, IOTV) MMD 1&2 (1 rep ea), 300-yd SR (3 reps), or ETM (30 min) Recovery: RD (30 sec)
Preparation: PD (10 reps) Activities: (ACUs, Boots, IOTV) CD 1&2 (10 reps ea), PSD (2 x 60 sec), CL 1 (5 reps) or other strength training modalities Recovery: RD (30 sec)	Preparation: PD (10 reps) Activities: (ACUs, Boots, IOTV) MMD 1&2 (1 rep ea), 300-yd SR (3 reps) Recovery: RD (30 sec)	Preparation: PD (10 reps) Activities: (ACUs, Boots, IOTV) CD 3 (10 reps), PSD (4 x 60 sec), CL 1 (5 reps) or other strength training modalities Recovery: RD (30 sec)	Preparation: PD (10 reps) Activities: (ACUs, Boots, IOTV) MMD 1&2 (1 rep ea), 300-yd SR (1 rep), 30:60s (10 reps) Recovery: RD (30 sec)	Preparation: PD (10 reps) Activities: (ACUs, Boots, IOTV) STM Drill 2 (3 x 10 reps) or other strength training modalities Recovery: RD (30 sec)

Chapter 5

RESERVE COMPONENT

5-43. The sample schedules shown in Table 5-15 (quarterly collective training) and Table 5-16 (annual collective training) show PRT instruction and training for an RC unit. Table 5-15 shows individual training.

Table 5-15. RC quarterly and annual PRT schedule

	1st Main Support Battalion					
Requirements	1ST QTR			2ND QTR		
	7-9 OCT	18-19 NOV	1-2 DEC	6-7 JAN	17-18 FEB	3-4 MAR
Higher HQ	FTX	CMD Inspection	MOBEX	IG	Mandatory Briefings	CPX
Unit Training	Company Collective Tasks	SM Tasks	NBC Proficiency Preparation	NBC Proficiency Test	SM Tasks NBC Collective Tasks	Plan Bn Defense
Sustaining Phase PRT Assessment & Instruction	Preparation: PD Activities: MMD 1&2 Activities: 1-mile Run Assessment Recovery: RD	Preparation: PD Activities: ACU/Boots GD, CD 1&2, CL 1, PSD Recovery: RD	Preparation: PD Activities: 30:60s, 60:120s, 300-yd SR Recovery: RD	Preparation: PD Activities: ACU/Boots CD 3, CL 2, Recovery: RD	Preparation: PD Activities AGR, TR, HR, RR Recovery: RD	Preparation: PD Activities: ACU/Boots STM, and other strength training modalities Recovery: RD

	1st Main Support Battalion					
Requirements	3RD QTR			4TH QTR		
	15-16 APR	12-14 MAY	10-11 JUN	9-10 JUL	11-13 AUG	9-10 SEP
Higher HQ	CMD Inspection	FTX	CMD Inspection	Prep for AT AT 11-25 JUL	Weapons Qualification	Civil Disturbance Training
Unit Training	Plan, Prepare Battalion Operations order	CSS Operations	ARTEP Mission Preparation	Load Vehicles, Equipment ARTEP Missions Download	PMI DECONEX Weapons Fire in MOPP 4	SM Tasks
Sustaining Phase PRT, Assessment & Instruction	Preparation: PD Activities: APFT Recovery: RD	Preparation: PD Activities: FM-FL/AML (Instruction) Recovery: RD	Preparation: PD Activities: ACU/Boots GD, CD 3, CL 1&2, PSD Recovery: RD	Refer to AT Collective PRT Schedule (Fig. 5-22)	Preparation: PD Activities: STM 1or other strength training modalities (Instruction) Recovery: RD	Preparation: PD Activities: APFT Recovery: RD

Table 5-16. RC annual collective PRT schedule

		JULY – Annual Training		
Monday	*Tuesday*	*Wednesday*	*Thursday*	*Friday*
11 Preparation: PD (5 reps) Activities: CD 1&2 (5 reps ea), CL 1 (5 reps), PSD (4x30 sec) Recovery: RD (30 sec)	**12** Preparation: PD (5 reps) Activities: MMD 1&2 (1 rep ea), AGR (A/B 30 min C/D 20 min) Recovery: RD (30 sec)	**13** Preparation: PD (5 reps) Activities: ACU/Boots GD (1 rep), CD 3 (5 reps), CL 2 (5 reps) Recovery: RD (30 sec)	**14** Preparation: PD (5 reps) Activities: MMD 1&2 (1 rep ea), 300-yd SR (1 rep), 60:120s (6 reps) Recovery: RD (30 sec)	**15** Preparation: PD (5 reps) Activities: CD 1&2 (5 reps ea), CL 1 (5 reps), PSD (4x30 sec) Recovery: RD (30 sec)
18 Preparation: PD (5 reps) Activities: FM-fl (10 K) Recovery: RD (30 sec)	**19** Preparation: PD (5 reps) Activities: GD (1 rep), CD 1&2 (5 reps ea), CL 1 (5 reps) Recovery: RD (30 sec)	**20** Preparation: PD (5 reps) Activities: ACUs/Boots/IOTV MMD 1&2 (1 rep ea), 300-yd SR (1 rep), 30:60s (6 reps) Recovery: RD (30 sec)	**21** Preparation: PD (5 reps) Activities: CD 1&2 (5 reps ea), CL 1 (5 reps), PSD (4x30 sec) Recovery: RD (30 sec)	**22** Preparation: PD (5 reps) Activities: MMD 1 & 2 (1 rep ea), Unit Run (30 min) Recovery: RD (30 sec)

Chapter 5

INDIVIDUAL

5-44. Table 5-17 shows the individual schedule. Figure 5-2 shows an example of a commander's policy letter.

Table 5-17. RC individual PRT schedule

5 Days @ 60 Minutes per Day				
Monday	**Tuesday**	**Wednesday**	**Thursday**	**Friday**
1 Preparation: PD (5 reps) Activities: Sustained Running **Or** ETM (20-30 min) Recovery: RD (30 sec)	**2** Preparation: PD (5 reps) Activities: STM (2x10 reps) & PSD (4x30 sec) or other strength training modalities Recovery: RD (30 sec)	**3** Preparation: PD (5 reps) Activities: MD 1&2 (1 rep ea) & 60:120s (8 reps) Recovery: RD (30 sec)	**4** Preparation: PD (5 reps) Activities: STM (2x10reps) & PSD (4x30 sec) or other strength training modalities Recovery: RD (30 sec)	**5** Preparation: PD (5 reps) Activities: MMD 1&2 (1 rep ea) & Sustained Running or ETM (20-30 min) Recovery: RD (30 sec)
8 Preparation: PD (5 reps) Activities: STM (2x10 reps) & PSD (4x30 sec) or other strength training modalities Recovery: RD (30 sec)	**9** Preparation: PD (5 reps) Activities: Sustained Running or ETM (20-30 min) Recovery: RD (30 sec)	**10** Preparation: PD (5 reps) Activities: STM (2x10 reps) & PSD (4x30 sec) or other strength training modalities Recovery: RD (30 sec)	**11** Preparation: PD (5 reps) Activities: MMD 1&2 (1 rep ea) & 60:120s (8 reps) Recovery: RD (30 sec)	**12** Preparation: PD (5 reps) Activities: STM (2x10 reps) & PSD (4x30 sec) or other strength training modalities Recovery: RD (30 sec)

SAMPLE COMMANDER'S POLICY LETTER

5-45. Refer to Figure 5-2. For an example of a unit PRT policy letter.

Physical Readiness Training

DEPARTMENT OF THE ARMY
Headquarters, XX Battalion, XX Infantry
XX Brigade, XX Division
Fort XXXX, XX State XX zip code
Office Symbol

Date

MEMORANDUM FOR All XX Infantry Leaders and Soldiers

SUBJECT: Commander's Policy Letter # X: Physical Readiness Training

1. **References.**
    FM 7-22, *Army Physical Readiness Training*, XX Date
    AR 350-1, *Army Training and Leader Development*
    Installation AR 350-1, XX Date
2. **General.** In accordance with AR 350-1, all Soldiers will participate in either collective or individual 60 to 90 minute daily PRT sessions four or five times per week.
    a. All sessions will include the exercises, drills, and activities listed in the *sustaining* phase from TC 7-22.
    b. Normal PRT time is 0630 to 0800. Commanders will protect PRT time. Combatives training will be scheduled separately from unit PRT.
    c. Special Conditioning programs will be conducted according to Chapter 6, FM 7-22. Soldiers on temporary or permanent physical profile will be evaluated and assigned to the battalion reconditioning program. 1SG will coordinate with the reconditioning program leader (RPL) for all matters concerning Soldiers in the reconditioning program.
    d. The APFT will be conducted according to FM 7-22, Appendix A.
    e. AR 600-9 is the standard for conduct of the Army Weight Control Program (AWCP).
3. **Schedules.** Commanders will follow Chapter 5, FM 7-22, which contains a doctrinal template for the conduct of collective and individual PRT exercises, drills, and activities.
    a. The *sustaining* phase of PRT supports the Army Force Generation (ARFORGEN) model using reset, train/ready, and available phases. The model is designed using a "structured progression of increased unit readiness over time."
    b. This results in recurring periods of availability of trained, ready, and cohesive units prepared for operational deployment as specified in the Army Campaign Plan. The recurring structured progression of increasing unit readiness focuses on reset, train/ready, and available phases according to the operational readiness cycles. The result is full mission readiness.
    c. Structured progression also allows units time to prepare for an operational deployment cycle and surge capability.
    d. As shown in AR 350-1, active Army forces plan for one deployment every three years, while reserve forces plan for one deployment every six years. Active and reserve forces can be called from the ready force pool for a surge to meet strategic requirements.
    e. Commanders must be prepared to move to any position along the ARFORGEN cycle. Core-METL (C-METL) or Directed-METL (D-METL) paths for units must prepare them to operate anywhere or anytime within the spectrum of conflict. With the potential to have shortened ARFORGEN cycles, commanders need to stay vigilant in planning and programming PRT.

Figure 5-2. Sample, commander's policy letter

4. **Uniforms.** All Soldiers in a unit PRT formation will wear the same uniform. The Army physical fitness uniform (PFU) and ACUs (as described in AR 670-1) with boots, ACH, IOTV, and weapon are appropriate uniforms for the conduct of PRT. The high visibility reflective belt or reflective vest will be worn by all Soldiers during the conduct of unit or individual PRT. The reflective belt will be worn diagonally over the right shoulder to the hip. During the conduct of unit foot marches the reflective belt or reflective vest will be placed horizontally around the rucksack.

5. **Execution.** Commanders will develop their PRT programs around mission and METL requirements.
   a. PRT sessions are not solely devoted to preparation for the APFT.
   b. Company, platoon, and squad level PRT is authorized. Individual PRT should be the exception and not the rule.
   c. The PRT formation is for accountability and the execution of PRT, not a platform for administrative announcements.
   d. Commanders must ensure that all PRT follows the Task, Conditions, Standards format.
   e. Commanders will annotate "Preparation," "PRT drills and activities," and "Recovery" on their unit training schedules.
   f. Organized athletics are not to be executed during PRT time.
   g. Foot marching will be conducted 2 to 3 times per month, replacing sustained running on the PRT schedule. Speed running will always be conducted at least one time per week.
   h. To ensure leaders conduct PRT training to standard, commanders require them to receive the training themselves before they may conduct PRT sessions. (This training is described in Chapters 7 through 10, FM 7-22.) This helps ensure that the PRT program is conducted to standard.

6. **APFT.** In accordance with AR 350-1, the APFT will be administered for record at least twice a year.
   a. Ideally, testing dates should fall within the months of April and October.
   b. Record testing, to include make-up testing, is annotated on the unit training schedule. Soldiers who score 270 points or higher with at least 90 points in each APFT event are awarded the Army Physical Fitness Excellence badge.
   c. Height and weight screening to meet AR 600-9 standards may be conducted on the day of the record APFT or up to 30 days before or after the record test.

7. **Unit Goals.** Commanders should establish goals based on the physical requirements of the unit's mission/METL. The following is an example battery of unit-level assessments used to measure individual and collective unit readiness:
   a. Collectively foot march 20 km with fighting load, under 5 hours (Chapter 10, FM 7-22, and FM 21-18).
   b. Perform five unassisted pull-ups using overhand grip.
   c. Complete a 300-yard shuttle run in under 90 seconds.
   d. Perform individual Soldier carry with a Soldier of equal weight for 50 yards.
   e. Soldiers score 270 or higher.
   f. Soldiers meet AR 600-9 standards.

Figure 5-2. Sample, commander's policy letter (continued)

Summary

The PRT schedules prescribed in this chapter are adaptable to unit missions, individual capabilities, and unit OPTEMPO. The principles of *train as you will fight, train to standard,* and *train to develop agile leaders and organizations* are fundamental in the construction of PRT program development. Commanders must understand and apply the doctrinal templates in this chapter to chart a clear and achievable direction for the physical readiness of their units.

Chapter 6
Special Conditioning Programs

"When Soldiers become ill, injured, or have other medical conditions, leaders have the responsibility to recondition these Soldiers and safely return them to duty at an equal or higher physical fitness level."

COL William R. Rieger, Commandant, U.S. Army Physical Fitness School, 1999 to 2006

AR 350-1 states special conditioning programs are appropriate for Soldiers who have difficulty meeting unit goals or Army standards. These programs are not punitive; their purpose is to improve the physical readiness of Soldiers. Special conditioning programs designed to accommodate these needs will be conducted during normal duty hours. Special conditioning programs include:

- APFT or unit PRT goal failure.
- Soldiers on the AWCP.
- Reconditioning.

APFT OR UNIT PRT GOAL FAILURE

6-1. When Soldiers fail to meet APFT standards or unit goals, leaders should consider many factors that may contribute to these failures, including:

- Time in training.
- Regular PRT participation.
- Prolonged deployment.
- Recovery from injury, illness or medical condition (physical profile).

TIME IN TRAINING

6-2. The Soldier who is fresh out of IMT may have a level of physical performance below the minimum threshold of his gaining unit. He may be a borderline APFT performer or borderline overweight. Regardless of the situation, he will not be accustomed to the demands placed on the lower extremities during a normal duty day. These Soldiers will face new conditions relating to physical performance such as acclimatization to altitude, temperature, and humidity. It can take up to four weeks to adapt to these unfamiliar conditions. Although Soldiers leave IMT prepared for the transition to the sustaining phase, they may detrain due to leave, transit, and in-processing at their new duty assignments. The same holds true for Soldiers reassigned to different units throughout the Army.

REGULAR PARTICIPATION

6-3. Many factors may influence regular participation in PRT sessions. The most common factors include OPTEMPO and related mission requirements. Leaders must anticipate and plan for these, and must make PRT as important as any other programmed training. In accordance with AR 350-1, Soldiers are required to participate in collective or individual PRT activities at least three times per week. Optimal participation in PRT may be achieved through conducting training sessions anytime during the duty day; not necessarily only in the early morning. Leaders must understand this and make it known. Soldiers should only be excused from regular unit PRT when they have performed exhaustive duties with little or no rest, or have a temporary or permanent physical profile according to AR 40-501, *Standards of Medical Fitness*.

6-4. All Soldiers must understand that it is their personal responsibility to achieve and sustain a high level of physical readiness. Many Soldiers are assigned to duty positions that restrict participation in collective unit PRT programs. Commanders must therefore develop leadership environments that encourage and motivate Soldiers to accept individual responsibility for their own physical readiness. Leaders and individual Soldiers need to use the PRT system outlined in this FM to help achieve and sustain high levels of physical readiness.

PROLONGED DEPLOYMENT

6-5. It is well documented that detraining may occur during prolonged deployments. Significant losses in strength, endurance, and mobility occur after a period of 14 days when little or no PRT is conducted. Every effort should be made by leaders to conduct PRT activities as often as mission requirements allow during deployment. Chapter 5, Planning Considerations, provides sample schedules of PRT activities that may be conducted during deployment when both time and space are limited. During post-deployment, when fitness levels may have declined, special considerations must be taken to ensure Soldiers meet or exceed their pre-deployment physical readiness levels. Adequate rest and recovery are especially important to successfully bring Soldiers back to a high level of readiness. Leaders must recognize the amount of time that is required to condition these Soldiers. Furthermore, Soldiers need at least 90 days post-deployment to retrain and prepare for the APFT or unit physical readiness goal.

RECOVERY FROM INJURY, ILLNESS, OR MEDICAL CONDITION

6-6. Soldiers recovering from injury, illness, or other medical conditions must train within the limits of their medical profiles (DA Form 3349 [Physical Profile]) and be afforded a minimum train-up period of twice the length of the profile. Prescribed train-up periods must not exceed 90 days before APFT administration or other unit physical readiness goal requirements according to AR 350-1.

ARMY WEIGHT CONTROL PROGRAM

6-7. See AR 600-9 for the policy and procedures that apply to screening and enrollment in the AWCP. AR 350-1 specifies that the AWCP will be kept separate and distinct from other special conditioning programs. Soldiers recovering from injury, illness or other medical conditions will be in reconditioning. Soldiers who fail the APFT or other unit physical readiness goals will continue participation in regular PRT sessions with the unit. Soldiers who fail to meet AR 600-9 standards will be enrolled in the AWCP and continue participation in regular unit PRT sessions. They should also participate in additional low impact, caloric expenditure activities.

RECONDITIONING

6-8. Injuries, illness, and other medical conditions impact readiness. Commanders are faced with the daily challenge of controlling injuries in the conduct of rigorous military training. Leaders must be familiar with the factors that influence injury risk. Adherence to the fundamental principles of PRT allows the commander to manage injury risk effectively. When injuries, illness, or other medical conditions limit the Soldier's ability to participate in PRT, units should offer organized and effective reconditioning programs that expedite his return to unit PRT.

INJURIES

6-9. Injuries are defined as any intentional or unintentional damage to the body resulting from acute or chronic exposure to mechanical, thermal, electrical, or chemical energy, and from the absence of such essentials as heat or oxygen. The information in this section will focus specifically on musculoskeletal (orthopedic) conditions, since they represent the type of injury risk most responsive to sound PRT practices. Among the other conclusions from the DoD Injury Work Group: In the Army alone, musculoskeletal conditions account for over half of all disabilities creating compensation of about $125 million per year. Knee and back injuries constitute a significant proportion of disability and limited duty. Training injuries treated on an outpatient basis and sports injuries may have the biggest impact on readiness.

6-10. According to the Atlas of Injuries in the Armed Forces:

Special Conditioning Programs

"...injuries pose the single most significant medical impediment to readiness in the military. Not only do injuries impact the strength and ability of our Armed Forces to effectively respond to their mission, they levy staggering annual costs in the hundreds of millions of dollars against the operating budgets of all the services."

DoD Injury Surveillance and Prevention Work Group (Injury Work Group)

PREVENTION

6-11. The reconditioning program described in this FM responds to the DoD Injury Work Group recommendation to *"...implement programs designed to enhance fitness and reduce training injury rates."* By enhancing the fitness level of Soldiers during the profile and post-profile recovery period, this program is expected to reduce training injury rates. The Army Physical Readiness Training System, shown in Figure 6-1, was developed with Soldier performance and injury control as its two primary objectives. Though these objectives may seem to oppose one another at first glance, the principles of PRT that improve Soldier performance also contribute to reducing injury risk. The DoD Injury Work Group recommends the following measures for injury prevention:

- Implement programs designed to enhance fitness and reduce training injury rates.
- Target knee and back injuries for additional efforts toward prevention.
- Place greater emphasis on prevention of training and sports injuries.

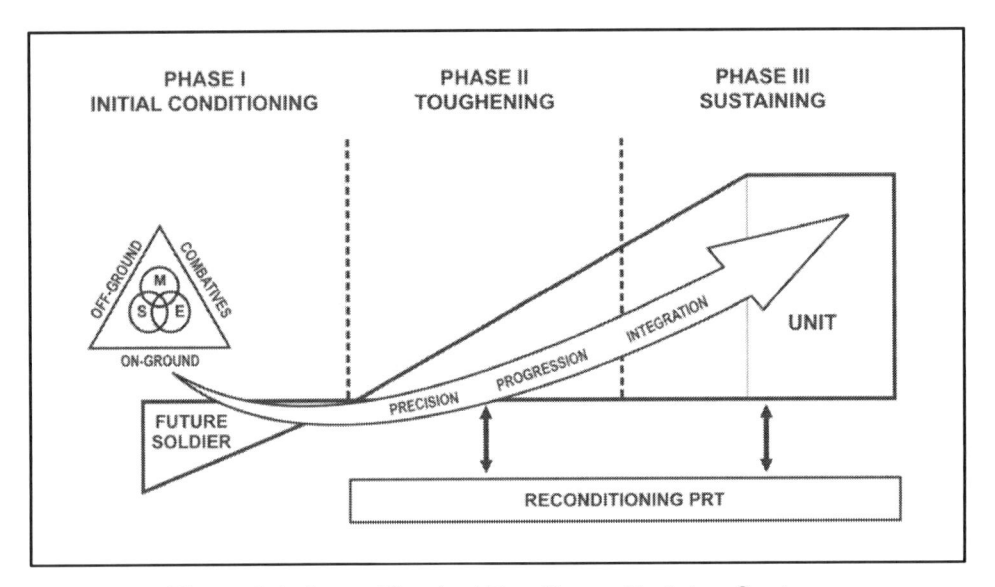

Figure 6-1. Army Physical Readiness Training System

6-12. The Army PRT System shown in Figure 6-1 includes reconditioning as part of the toughening and sustaining phases for Soldiers to facilitate recovery from illness, injury, or other medical conditions. Soldiers in need of recovery should return to unit PRT at a level equal to or higher than their physical state previous to the condition that brought them to reconditioning. Commanders and NCOs must take an active role to control avoidable injuries; however, in spite of every effort to limit injuries in the Army, Soldiers and situations will continue to produce overuse, accidental, and/or traumatic injuries. Keeping this in mind, a plan to bridge the gap between injury and physical readiness is essential. Reconditioning bridges this gap.

"Injuries are not random events; they are the predictable result of a complex set of risk factors, many of which can and should be controlled."

MG Patrick Scully, Deputy Surgeon General, U.S. Army (1998-2002)

Commander's Role in Injury Control

6-13. Precise execution of all PRT activities is essential to the injury control effort. Commanders must allow trained PRT leaders and AIs the time to teach proper execution of PRT activities. PRT leaders and AIs must be able to recognize and offer corrective guidance to Soldiers who are not executing drills to the standards described in this FM. It is especially important for PRT leaders and AIs to maintain the standard since transition from the toughening to the sustaining phase of training depends on execution of the drills to standard. For example, to control back injuries, postural awareness should be stressed during execution of all drills and activities. This is evident when the PRT leader or the AI prompts Soldiers to "set the hips and tighten the abs" while performing the exercises.

6-14. Both military and civilian research has shown that reduced running volume is associated with lower injury rates. Accordingly, PRT schedules prescribed in this FM involve less sustained running than is currently performed in Army units. Several studies of military units have shown that reduced running volume does not hinder performance on two- or three-mile run assessments as long as the quality (intensity) of running is maintained.

6-15. In addition to using appropriate PRT schedules, units must also look for conflicts between the PRT schedule and the unit training schedule. By considering the physical demands of tasks on the unit training schedule, PRT leaders are better prepared to plan appropriate PRT sessions. For example, if a 10-km foot march to a range is scheduled for Friday, speed work should not be scheduled for PRT on Thursday. Time should be allotted for leg recovery. Monday and Wednesday's PRT should not involve CLs 1 and 2 or the strength training circuit if Tuesday's unit training schedule takes the unit to an obstacle course where upper body strength is heavily challenged.

Executing Unit Reconditioning Programs

6-16. The following paragraphs assist leaders as they plan and execute a reconditioning program within their units. Army Reserve and National Guard units may tailor this program to meet their specific requirements. The purpose of a reconditioning program is to safely restore a level of physical readiness that enables Soldiers to successfully re-enter unit PRT after injury, illness or other medical condition. A physical profile defines, in writing, limitations to physical activity due to injury, illness or medical condition. The authorized forms for written profiles in the Army are the DD Form 689 and DA Form 3349. DA Form 3349 is better than DD Form 689, because it requires a much more detailed description of the Soldier's injury and the activities and exercises that the Soldier can perform with the injury. Soldiers assigned to the reconditioning program include:

- Soldiers on temporary medical profile.
- Soldiers in the recovery period after a temporary profile expires.
- Soldiers on permanent medical profile with specific limitations and special fitness requirements.

Level 1

6-17. To address the needs of Soldiers who are on profile and those recovering from profile, reconditioning employs a two-level system. Level I is a gym-based program designed to maximize the potential of a profiled Soldier while protecting the injured area. Soldiers enter level I once cleared to begin limited activity by the profiling health care provider. Activities in level I include the use of STMs and ETMs. Functional criteria are used to determine whether a Soldier is able to begin reconditioning, at level I or level II.

> **Example**
> A Soldier with a permanent profile that prohibits sustained or speed running may be assigned to the level I program. This allows him the use of aerobic training equipment on unit endurance and mobility training days.

Level II

6-18. To begin at level II, the profile or recovery reconditioning program, Soldiers must meet the level II reconditioning entry criteria requirements shown in Figure 6-2. Upon entering level II, Soldiers will begin to perform the PRT program. In this level the Soldier is on profile, just off of profile, or cleared to begin level II reconditioning. Preparation will be exactly the same as for unit PRT. The activity may be modified to follow a safe exercise progression. Recovery will be exactly the same as unit PRT.

PARTIAL SQUATS WITHOUT PAIN	5 REPETITIONS IN 5 SECONDS
PUSH-UPS	10 REPETITIONS TO STANDARD
SIT-UPS	10 REPETITIONS TO STANDARD
HANG FROM PULL-UP BAR	15 SECONDS
WALK	30 MINUTES UNASSISTED, AT NORMAL GAIT WITOUT PAIN

Figure 6-2. Level II reconditioning entry criteria

6-19. Before being discharged from level II and returning to unit PRT, Soldiers must meet the level II exit criteria requirements shown in Figure 6-3.

PREPARATION	5 REPETITIONS TO STANDARD
MILITARY MOVEMENT DRILL 1	1 REPETITION TO STANDARD
CONDITIONING DRILL 1	5 REPETITIONS TO STANDARD
CLIMBING DRILL 1	5 REPETITIONS TO STANDARD
CONTINUOUS RUNNING	30 MINUTES AT SLOWEST AGR PACE IN THE UNIT
RECOVERY	HOLD EACH STRETCH FOR 20 SECONDS TO STANDARD

Figure 6-3. Level II exit criteria

TOUGHENING PHASE RECONDITIONING

6-20. Rehabilitation and reconditioning programs within IMT are currently conducted at all Army Training Centers as a part of the physical training and rehabilitation program (PTRP). The purpose of the PTRP is to provide physical rehabilitation and physical conditioning for Soldiers who are injured during BCT or OSUT. These programs usually fall under the training command and act independently under the supervision of a physical therapist. Soldiers remain in the PTRP until they are capable of returning to the same phase of BCT/OSUT that they left or as a "new start" at day one of IMT. If an injury is minor and only requires short-term limitations (with minimal impact to training); it may not require assignment to the PTRP.

SUSTAINING PHASE RECONDITIONING

6-21. Reconditioning in the sustaining phase includes AIT and operational units. Consolidation of reconditioning programs at the battalion (or equivalent) level minimizes the administrative and logistical strain on operational unit assets. The brigade surgeon should have medical oversight of the unit reconditioning program. Battalion medical officers are the liaisons between reconditioning program leaders (RPLs) and the brigade surgeon. The first local military treatment facility with rehabilitation services may provide a physical therapist and a physical therapy assistant as consultants to oversee the gym-based reconditioning program level I. The physical therapist can assist/coordinate training efforts with the RPL.

6-22. The medical platoon leader is the RPL, and the medical platoon sergeant is the assistant RPL or assistant reconditioning program leader (ARPL). If this is not possible, the RPL and the ARPL should be chosen based on the following criteria:
- Thorough understanding of the Army's PRT program.
- Ability to instruct all activities.
- Understanding of regulations that govern profiling (AR 40-501, *Standards of Medical Fitness*).
- Ability to adapt activities to profiled Soldiers.
- Ability to effectively interact with medical personnel to ensure that Soldiers are fully capable of returning to the unit PRT program.

6-23. It is recommended that each company in the battalion should provide an NCO to assist the RPL on a daily basis. These NCOs should meet criteria mentioned above for the ARPL. In addition, training sessions should be provided on a quarterly basis by the physical therapist and/or physical therapy assistant to ensure proper supervision and optimal safety practices are observed. Trained NCOs will provide supervision and group instruction to Soldiers in the reconditioning program. To meet supervision requirements, at least two NCOs per company should be trained in the conduct and supervision of the reconditioning program.

6-24. Units should ensure adequate space and equipment are provided for the reconditioning program to accommodate STM and ETM drills. The reconditioning program is best executed at the brigade or installation fitness facilities. Because lower extremity injuries prevent many Soldiers from running activities, it is essential to have an adequate number of ETMs that offer cardio-respiratory conditioning while limiting weight-bearing stress to the body. Examples are cycle ergometers, steppers, elliptical machines, rowing machines, and treadmills. Treadmills are full weight bearing machines and are most appropriate for Soldiers cleared by medical personnel to begin a walk-to-run progression. Of these machines, cycle ergometers offer the most body weight support.

6-25. Pool activities such as swimming or deep-water running can eliminate weight-bearing stress. All Soldiers who are recovering from surgery or have open wounds will receive a physician's clearance before entering the swimming pool. Swimming laps, aqua-jogging, and aquatic exercises are excellent ways to maintain or improve cardio-respiratory fitness without putting undue stress on joints and bones. Limitations to one leg or one arm are minimal deficits in a pool environment. Kick board workouts or upper body workouts allow for strenuous activity with minimal risk of re-injury to an affected limb. If staffing is adequate, specialized aquatics programs may be implemented to work on water aerobics or deep-water running programs for non-swimmers. It is important to plan activities that keep everyone active during group pool sessions. Even if a regular pool program is not practical, an occasional trip to the pool may be scheduled to break up the routine and provide cross-training.

6-26. Units that must rely on installation or shared facilities should make arrangements to ensure that space and STM/ETM equipment are available during the time dedicated to the reconditioning program. This may require policies that restrict the use of these facilities to only reconditioning programs. Leaders might need to schedule reconditioning outside typical PRT times such as after 0800 or before 1600 to best achieve dedicated access to gym space and equipment.

Command Responsibilities

6-27. The reconditioning program is the battalion commander's and command sergeant's major program. A well-run program will assist force reconstitution efforts. The success of the program is dependent on the priority placed on it from the top down. Company commanders and first sergeants must care enough about the program to ensure NCO support.

6-28. The brigade surgeon and battalion medical officers should maintain constant awareness of the program. A medical officer with a background in rehabilitation should act as the installation medical consultant for reconditioning programs. The primary responsibility of the medical consultant is to act as a liaison or advocate for RPLs. The medical consultant should also provide training for the RPLs, ARPLs, and unit reconditioning NCOs. Figure 6-4 shows rehabilitation and reconditioning responsibilities.

Special Conditioning Programs

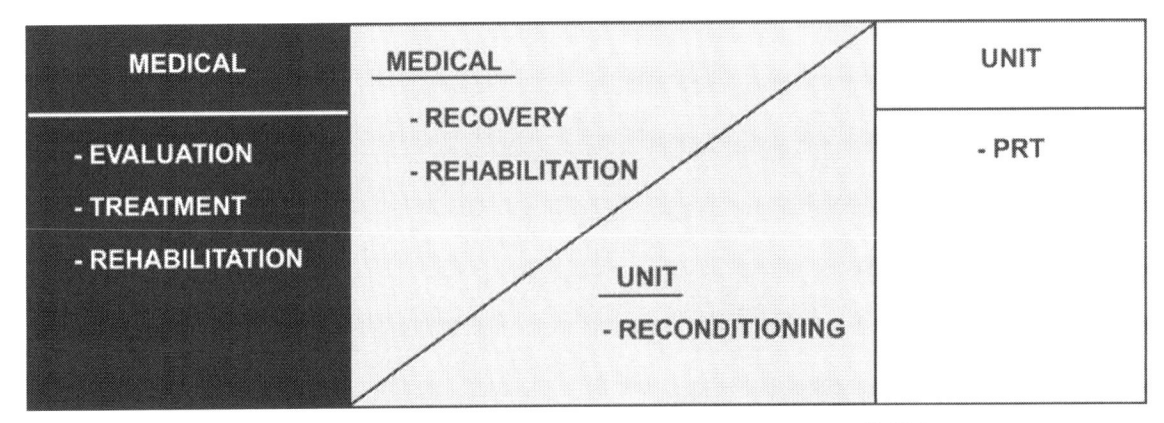

MEDICAL	MEDICAL	UNIT
- EVALUATION - TREATMENT - REHABILITATION	- RECOVERY - REHABILITATION UNIT - RECONDITIONING	- PRT

Figure 6-4. Rehabilitation and reconditioning responsibilities

6-29. Trainers for the reconditioning program must possess the same knowledge of the program that the RPL have and must have additional education in exercise science. For this reason a physical therapist or a physical therapy assistant is well suited for the role. The following outline should be used when developing training for this program:

- STM Orientation
  - Equipment familiarization: purpose, technique, safety.
  - Etiquette: observe posted rules, replace all weights and equipment to original positions, and wipe down all surfaces after use.
- ETM Orientation
  - Equipment familiarization: purpose, technique, safety.
  - Etiquette: observe posted rules, replace equipment to original position, and wipe down all surfaces after use.
- Reconditioning Session Orientation
  - Preparation: increase heart rate, muscle temperature to prepare the body for more vigorous activity.
  - Activity: provide neural adaptation and improve strength, endurance, and mobility.
  - Recovery: gradually return to resting heart rate (below 100 beats per minute) and bring body safely back to pre-exercise state.
- Level I (Gym-Based) Reconditioning Objectives
  - Prevent de-conditioning.
  - Work within profile limitations.
  - Restore functional strength, endurance, and mobility.
  - Avoid injury or re-injury.
  - Transition to level II reconditioning.
- Level II Reconditioning Objectives
  - Progress to pre-injury level of fitness.
  - Avoid injury or re-injury.
  - Transition to unit PRT.

PROFILES AND RECOVERY PERIODS

6-30. Soldiers in the reconditioning program will be on a physical profile or in the authorized recovery period from a temporary profile. Commanders may assign Soldiers with a permanent profile to the reconditioning program or allow them to remain in unit PRT. Soldiers on convalescence leave may be exempted from reconditioning at the discretion of the profiling medical officer. In no case can a Soldier carry a temporary profile that has been extended for more than 12 months without positive action taken to correct the problem or effect other appropriate disposition according to a military medical review board. <u>Once a profile is lifted, the Soldier must be given twice the time of the temporary profile (but not more than 90 days) to train for the APFT.</u>

Chapter 6

It is not a requirement to take an APFT after the recovery period if a Soldier is not due to take the semi-annual test. Refer to AR 350-1 and Appendix A of this FM for APFT policy and procedures. The RPL follows the medical guidance on the profile for Soldiers on profile. If there are any questions about the limitations of the profile, the RPL will contact the medical officer for clarification. Once a profile has expired, Soldiers will remain in the reconditioning program until they have met transition criteria to return to unit PRT activities. During this period, the RPL/ARPL, and unit reconditioning NCOs will reinforce the precise execution of PRT activities with each Soldier in small groups or individually. See Figures 6-2 and 6-3 for transition criteria to move from level I to level II or return to unit PRT.

6-31. Soldiers with permanent profiles that do not allow them to meet all reconditioning exit criteria may return to unit PRT once they demonstrate proficiency at all non-profiled activities. For example, a Soldier whose permanent profile only prohibits running would not be in the reconditioning program. Rather, he would do PRT with the unit and perform all activities except running. The Soldier in this example would walk or use ETMs when PRT activities call for sustained or speed running. When a permanent profile is so restrictive that the Soldier is unable to perform several PRT activities, the commander may direct the Soldier to the reconditioning program. This scenario is more likely to occur with Soldiers who are awaiting medical boarding procedures. For less clearly defined cases, the commander can solicit input from the battalion medical officer or brigade surgeon.

EXERCISE PROGRESSION

6-32. Progressing injured Soldiers to a "return-to-duty" level of fitness is the goal of any reconditioning program. There are two possible pitfalls to exercise progression. First, if the exercise progression is too rapid it may aggravate the injury, resulting in a further delay to recovery. Second, if the exercise progression is too slow it risks general deconditioning and a loss of effectiveness when returned to duty. A gap between recovery fitness and unit expectations may also cause undue physical and psychological stress. To assist the RPL/ARPL in decisionmaking regarding exercise progression, the following recommendations are made:

- Soldiers on profile will have specific limitations as defined by their DD Form 689 or DA Form 3349. These limits will be strictly adhered to.
- Communication with the profile writer is encouraged if a Soldier is clearly improving faster than written limits allow. There may be a reason that is not obvious for the slow progression. If there is no clear reason to limit the progression, instruct the Soldier to get a new profile that reflects communication with the health care provider. A written request is preferable to relying on the individual's memory for this.
- Limitations that are in place for a given injury may not affect other areas. A case of tendonitis in the right shoulder should not affect the ability to do leg presses or ride a stationary cycle. Get a clear understanding from the Soldier of what they can and cannot do. Do not read between the lines of the profile. Once again, contact the profile writer if clarification is needed.
- Maintain an exercise workout log to track progress of each individual who will require more than two weeks of gym reconditioning. When a profile expires, work with unit leaders to ensure the recovery period is used for reconditioning until the Soldier can meet the criteria to re-enter unit PRT.

LEVEL I RECONDITIONING DRILLS AND ACTIVITIES

6-33. The exercise schedule shown in Table 6-1 provides guidance for conducting level I reconditioning. This schedule of activities will ensure safe reconditioning of Soldiers during the profile period. The physical profile of a medical officer supersedes the following:

- The RPL briefs the profiled Soldier concerning which exercises are restricted and which they are to perform. The Soldier is also briefed on the use of ETMs (walking and swimming may also be appropriate).
- As the Soldier improves and profiling limitations are removed, the Soldier may be transitioned into level II of the reconditioning program when transition criteria is met.

Special Conditioning Programs

Table 6-1. Reconditioning Level I training schedule

MON	TUE	WED	THU	FRI
PREPARATION: PD ETM 5 MIN ACTIVITIES: HIP STABILITY DRILL 4 FOR THE CORE ETM 20-30 MIN RECOVERY: RD HOLD EACH STRETCH FOR 20-30 SECONDS	PREPARATION: PD ETM 5 MIN ACTIVITIES: HIP STABILITY DRILL 4 FOR THE CORE STM 1 (1-3 SETS@10 REPS) RECOVERY:RD HOLD EACH STRETCH FOR 20-30 SECONDS	PREPARATION: PD ETM 5 MIN ACTIVITIES: HIP STABILITY DRILL 4 FOR THE CORE ETM 20-30 MIN RECOVERY: RD HOLD EACH STRETCH FOR 20-30 SECONDS	PREPARATION: PD ETM 5 MIN ACTIVITIES HIP STABILITY DRILL 4 FOR THE CORE STM 1 (1-3 SETS@10 REPS) RECOVERY: RD HOLD EACH STRETCH FOR 20-30 SECONDS	PREPARATION: PD ETM 5 MIN ACTIVITIES: HIP STABILITY DRILL 4 FOR THE CORE ETM 20-30 MIN RECOVERY: RD HOLD EACH STRETCH FOR 20-30 SECONDS

6-34. Before transition to level II, the RPL/ARPL ensures that the Soldier meets the criteria in Figure 6-2. If the Soldier cannot meet the transition criteria, he should be directed to the medical officer for re-evaluation.

6-35. Before releasing the Soldier back to unit PRT, the RPL/ARPL ensures the Soldier meets the criteria in Figure 6-3. If the Soldier does not meet these criteria before the recovery period ends, the RPL/ARPL will consult with the battalion medical officer to determine a proper disposition.

EQUIPMENT

6-36. When using equipment, endurance training includes four primary variables: exercise mode, training frequency, exercise duration, and training intensity. Exercise prescription specifies training frequency, exercise duration, and training intensity. The mode of exercise (type of equipment) is determined by environmental constraints and training according to physical profile limitations (temporary/permanent). Each ETM and STM contains specific instructions for proper use and adjustments to obtain optimal posture and technique during exercise (seat position on cycle ergometers, rowing machines, and STMs). If a piece of training equipment has no visible list of operating instructions, the RPL, ARPL, or gym personnel should be consulted for assistance.

EXERCISE MODE

6-37. Exercise mode refers to the specific activity performed by a Soldier: running, cycling, swimming, strength training, and endurance training equipment. Environmental constraints, safety for Soldiers on physical profile, and isolation of specific muscle groups to be trained during rehabilitation and reconditioning are some of the advantages of using STMs and ETMs. Consideration for use of specific types of equipment may be based on a Soldier's range of movement, limb limitation and/or the ability to participate in weight-bearing or non-weight-bearing activities. Weight-bearing activities include walking or running on a treadmill and climbing on a stair climbing or stepping machine. Non-weight bearing and limited weight-bearing activities include use of cycle ergometers (upright/recumbent), elliptical trainers, rowers, climbing machines, and cross-country ski machines. Use of limited or non-weight-bearing endurance training equipment is desirable for obtaining higher caloric expenditure through additional training sessions by overweight Soldiers. Each of these modes typically provide the Soldier with a variety of individual exercise routines that monitor and display exercise duration, training intensity (heart rate/pace/watts, caloric expenditure, and distance completed miles/km). See Figure 6-5 for examples of various types of endurance training equipment. Use of STMs not only improves strength, but also builds muscle mass for higher caloric expenditure and stability for rehabilitation and reconditioning of the injured body part.

Chapter 6

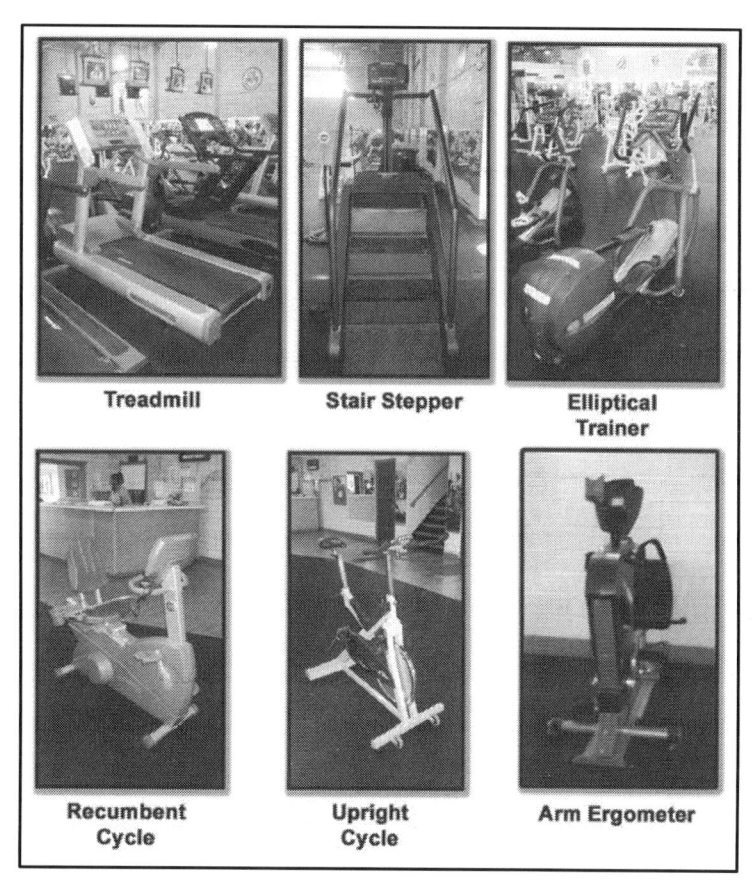

Figure 6-5. Endurance training equipment

TRAINING FREQUENCY

6-38. Training frequency refers to the number of training sessions conducted per day or week. Training frequency is determined by exercise duration and training intensity. Training sessions that involve high intensity or longer duration may necessitate less frequent training to allow for adequate recovery. Endurance and mobility, as well as strength and mobility training frequency, is three exercise sessions per week for each, for a total of six reconditioning PRT sessions. If five days of training occur, then three days are dedicated to endurance and mobility and two days are dedicated to strength and mobility for one week. The following week will consist of three days of strength and mobility and two days of endurance and mobility training.

EXERCISE DURATION

6-39. Exercise duration is 20 minutes or longer and varies from machine to machine, depending on the intensity of the exercise routine being performed (hill profile, speed, degree of incline, resistance). Most exercise sessions of high or moderate intensity should last 20 to 30 minutes. Endurance exercise sessions that address additional caloric expenditure for body fat reduction should be of low intensity and may last up to 60 minutes. The duration for STM exercise is 1-3 sets of 10 repetitions of each exercise for each major muscle group. Refer to the STM drill later in this chapter for specific instructions on the conduct of each exercise.

TRAINING INTENSITY

6-40. Training intensity is typically monitored and displayed on the exercise equipment control panel in terms of heart rate, pace (mph/kph, step rate), watts, kiloponds, caloric expenditure (kcals), or resistance for ETMs and weight lifted (number of plates, pin placement, pounds, or kilograms) for STMs.

Stability Training

6-41. Stability is dependent upon structural strength and body management. Regular precise performance of 4C and the HSD form a foundation of good stability for physical performance. These drills are listed in detail throughout the following pages in this chapter.

4 FOR THE CORE

6-42. The abdomen, lower spine, and pelvis comprise the trunk (core) of the body. This area must be stable so the limbs have a fixed base from which to create powerful movements. The abdominal and back muscles form a supportive ring around the spine. Soldiers are only as strong as their weakest link; so all these muscles must be trained in a manner that mimics their function. In reconditioning, 4C and HSD are performed daily before engaging in other PRT activities. During the toughening phase, 4C is performed after preparation and prior to strength and mobility activities. Four for the core may also be performed outside regular PRT sessions as supplemental training. Do not exceed 60 seconds for each 4C exercise. The following commands are used for 4C exercises.

6-43. Exercises 1 and 3 (bent leg raise and back bridge):
- Starting Position, MOVE.
- Ready, EXERCISE.
- Starting Position, MOVE.
- Position of Attention, MOVE.

6-44. Exercises 2 and 4 (side bridge and quadraplex) are both performed on the right and left sides. The commands for execution for this exercise and changing sides are as follows:
- Starting Position, MOVE.
- Ready, EXERCISE.
- Starting Position, MOVE.
- Change Position, Ready EXERCISE.
- Starting Position, MOVE.
- Position of Attention, MOVE.

6-45. The goal is to hold each exercise position for 60 seconds. If the Soldier is unable to do this, he will follow the instructions for each exercise to momentarily change position and return to the prescribed exercise position. Detailed descriptions of each 4C exercise follows.

4 FOR THE CORE

EXERCISE 1: BENT-LEG RAISE

6-46. Lying in the starting position for the sit-up, place the fingers of both hands underneath the small of the back. Raise the feet off of the ground until both the hips and knees flex to 90 degrees. Holding the head two or three inches off the ground, contract the abdominals as if preparing for a blow to the stomach. Another way to perform this drawing in maneuver is to imagine pulling the navel toward the spine. Think about the amount of pressure on the fingers created by the contraction of the abdominals. Maintain the same degree of pressure while slowly straightening the legs. As soon as the Soldier can no longer maintain the same degree of pressure on his fingers, he brings his legs back to the 90-degree position for three to five seconds, and repeats until one minute has elapsed (Figure 6-6).

Figure 6-6. Bent-leg raise (4 for the core)

4 FOR THE CORE

EXERCISE 2: SIDE BRIDGE

6-47. Lay on either side with the upper body off the ground, supported by the elbow, forearm, and fist. Cross the bottom leg in front of the top leg, keeping the feet together. The legs may also be positioned with the knees together and bent 90 degrees. Firmly press into the ground with the supporting arm, and then raise the trunk and pelvis straight upward until they form a straight line with the legs and knees. Hold this position while continuing to breathe. Switch to the other side after one minute. If he cannot hold for one minute, lower, rest briefly, then repeat until one minute has elapsed (Figure 6-7).

Figure 6-7. Side bridge

4 FOR THE CORE

EXERCISE 3: BACK BRIDGE

6-48. Lying on the back with knees bent at 90 degrees, arms extended sideward at 45 degrees, with head and feet on the marching surface, perform the drawing-in maneuver. Once the abdominal contraction is established, raise the hips off of the ground until the trunk and thighs form a generally straight line. The spine must not arch to achieve this position. With the buttocks still up, straighten the left leg until it aligns with the trunk and thigh. Don't let the trunk and pelvis sag on the unsupported side. Hold five seconds, and then switch to the other leg. Repeat for one minute. If the spine begins to sag, arch, or tilt, lower to the starting position, rest for 3 to 5 seconds, then, try again (Figure 6-8). The goal is to maintain the back bridge position for 60 seconds alternating leg raises as needed.

Figure 6-8. Back bridge

4 FOR THE CORE

EXERCISE 4: QUADRAPLEX

6-49. The starting position is on the hands and knees with the back flat. Contract the abdominal muscles as described in the bent-leg raise. Without rotating the trunk or sagging or arching the spine, straighten the left leg to the rear and the right arm to the front. Hold for at least 5 seconds, recover to the starting position if needed, then return to the quadraplex. The goal is to hold each quadraplex position (left and right) for 60 seconds each. Alternate the arm and leg movements on subsequent repetitions, repeating for one minute. The key to this exercise is controlled lowering and raising of the opposite arm and leg while keeping the rest of the body aligned and still (Figure 6-9).

Figure 6-9. Quadraplex

HIP STABILITY DRILL

6-50. The HSD, like 4C, trains the hip and upper thigh areas three-dimensionally, developing the basic strength and mobility needed for stability to perform functional movements. In reconditioning, the HSD is performed daily immediately after 4C and before engaging in other PRT activities. During the toughening phase, the HSD is performed after preparation and prior to endurance and mobility activities. The HSD may also be performed outside regular PRT sessions as supplemental training. In the HSD, perform no more than 10 repetitions of exercises 1 through 4 and do not exceed 30 seconds for each exercise position in exercise 5. If more repetitions are desired, repeat the entire drill.

HIP STABILITY DRILL

EXERCISE 1: LATERAL LEG RAISE

(5 repetitions on each side)

Purpose: This exercise strengthens lateral hip and upper leg muscles (Figure 6-10).

Starting Position 1: Lay on the right side with the legs extended straight to the side and feet together with toes pointing straight ahead. Support the upper body with the right elbow. The elbow is bent at 90 degrees, the upper arm is perpendicular to the ground and the right hand makes a fist vertical to the ground.

Starting Position 2: Lay on the left side with the legs extended straight to the side and feet together with toes pointing straight ahead. Support the upper body with the left elbow. The elbow is bent at 90 degrees, the upper arm is perpendicular to the ground and the left hand makes a fist vertical to the ground.

Commands: The commands for the lateral leg raise are as follows:
- Starting Position, MOVE.
- In Cadence, EXERCISE (count as a 4-count exercise according to Chapter 7, Execution of Training, paragraph 7-29).
- Change Position, MOVE.
- In Cadence, EXERCISE (count as a 4-count exercise according to Chapter 7, Execution of Training, paragraph 7-29).
- Position of Attention, MOVE.

Cadence: SLOW

Count:
1. Raise the top leg so the top foot is 6 to 8 inches above the ground.
2. Return to the starting position.
3. Raise the top leg so the top foot is 6 to 8 inches above the ground.
4. Return to the starting position.

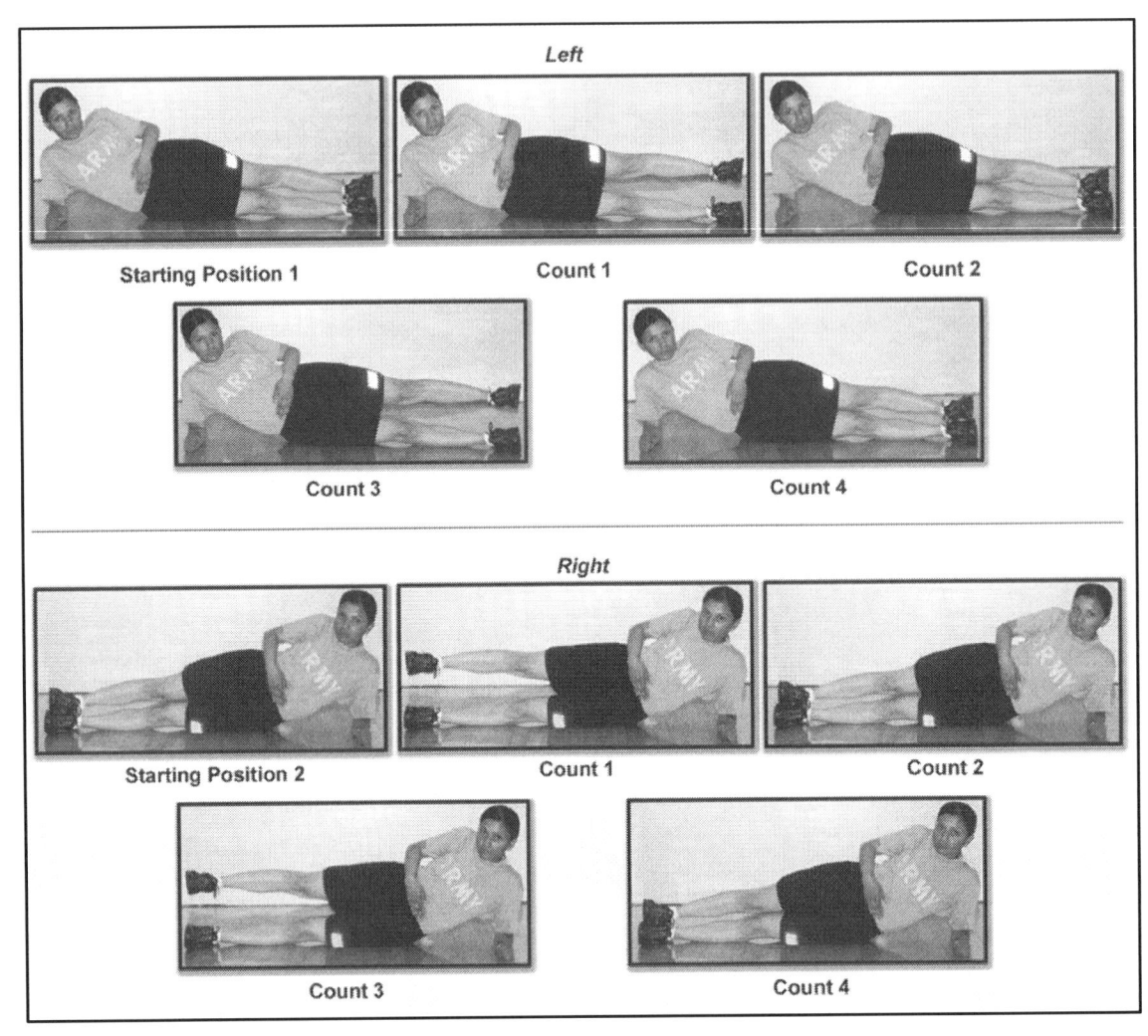

Figure 6-10. Lateral leg raise

Check Points:
- Face to the front of the formation, maintaining a generally straight line with the body.
- On counts 1 and 3, keep the knee of the raised leg straight and the foot pointing forward. The top leg raises no more than 6-8 inches above the ground.
- Place the top hand over the stomach throughout the exercise.

Precaution: N/A

HIP STABILITY DRILL

EXERCISE 2: MEDIAL LEG RAISE

(5 repetitions on each side)

Purpose: This exercise strengthens the inner thigh and hip muscles (Figure 6-11).

Starting Position 1: Lay on the left side with the left leg extended straight to the side and the right leg bent at 90 degrees with the right foot flat on the ground behind the left leg. Support the upper body with the left elbow. The elbow is bent at 90 degrees, the upper arm is perpendicular to the ground and the left hand makes a fist vertical to the ground.

Starting Position 2: Lay on the right side with the right leg extended straight to the side and the left leg bent at 90 degrees with the left foot flat on the ground behind the right leg. Support the upper body with the right elbow. The elbow is bent at 90 degrees, the upper arm is perpendicular to the ground and the right hand makes a fist vertical to the ground.

Commands: The commands for the lateral leg raise are as follows:
- Starting Position, MOVE.
- In Cadence, EXERCISE (count as a 4-count exercise according to Chapter 7, Execution of Training, paragraph 7-29).
- Change Position, MOVE.
- In Cadence, EXERCISE (count as a 4-count exercise according to Chapter 7, Execution of Training, paragraph 7-29).
- Position of Attention, MOVE.

Cadence: SLOW

Count:
1. Raise the bottom leg so the bottom foot is 6-8 inches above the ground.
2. Return to the starting position.
3. Raise the bottom leg so the bottom foot is 6-8 inches above the ground.
4. Return to the starting position.

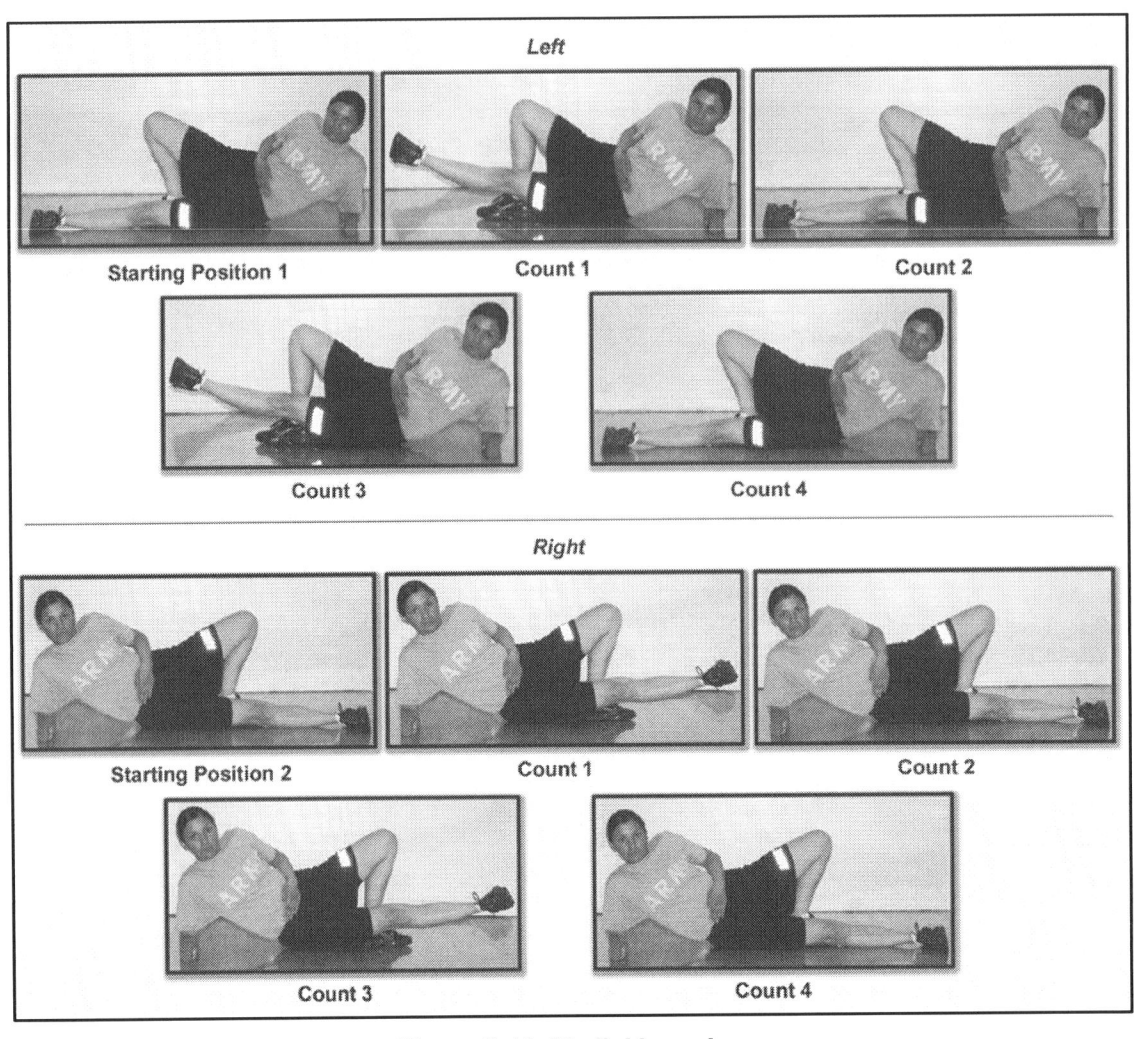

Figure 6-11. Medial leg raise

Check Points:
- Keep the hips facing forward and the body in a generally straight line.
- Keep the toes facing forward on the bottom leg.
- Place the top hand over the stomach throughout the exercise.
- Do not raise the bottom foot higher than 6-8 inches above the ground.

Precaution: N/A

HIP STABILITY DRILL

EXERCISE 3: BENT-LEG LATERAL RAISE

(5 repetitions on each side)

Purpose: This exercise strengthens hip rotator muscles (Figure 6-12).

Starting Position 1: Lay on the right side with the legs bent at 90 degrees and feet together with toes pointing straight ahead. Support the upper body with the right elbow. The elbow is bent at 90 degrees, the upper arm is perpendicular to the ground and the right hand makes a fist vertical to the ground.

Starting Position 2: Lay on the left side with the legs bent at 90 degrees and feet together with toes pointing straight ahead. Support the upper body with the left elbow. The elbow is bent at 90 degrees, the upper arm is perpendicular to the ground, and the left hand makes a fist vertical to the ground.

Commands: The commands for the lateral leg raise are as follows:
- Starting Position, MOVE.
- In Cadence, EXERCISE (count as a 4-count exercise according to Chapter 7, Execution of Training, paragraph 7-29).
- Change Position, MOVE.
- In Cadence, EXERCISE (count as a 4-count exercise according to Chapter 7, Execution of Training, paragraph 7-29).
- Position of Attention, MOVE.

Cadence: SLOW

Count:
1. Raise the top leg about 12 inches above the ground, keeping the feet together.
2. Return to the starting position.
3. Raise the top leg about 12 inches above the ground, keeping the feet together.
4. Return to the starting position.

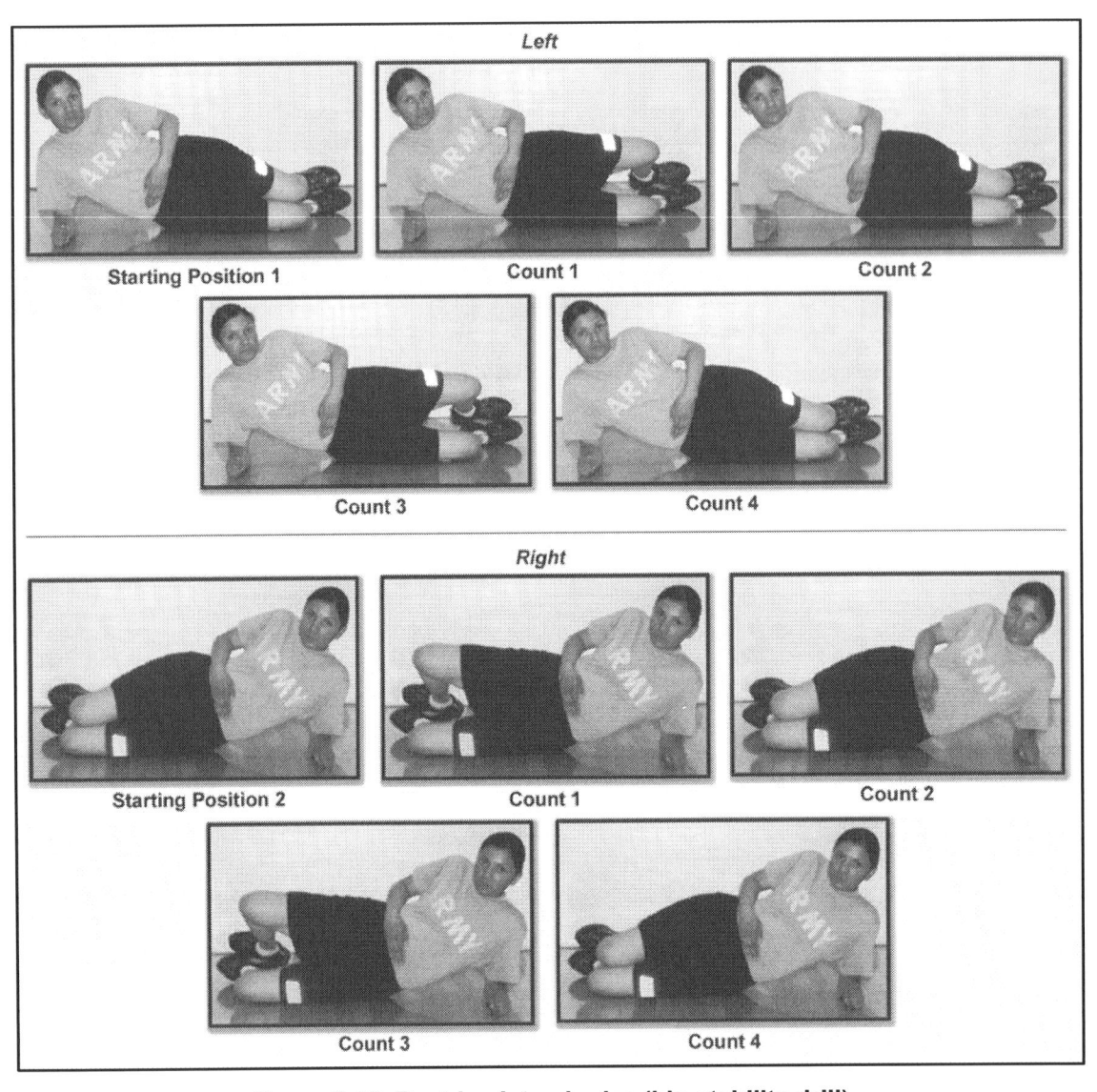

Figure 6-12. Bent-leg lateral raise (hip stability drill)

Check Points:
- Face to the front of the formation, maintaining a generally straight line with the body, from the knees to the torso.
- Keep the feet together throughout the exercise.
- Place the top hand over the stomach throughout the exercise.

Precaution: N/A

HIP STABILITY DRILL

EXERCISE 4: SINGLE-LEG TUCK

(5 repetitions on each side)

Purpose: This exercise strengthens the hip flexors, lateral hip, and upper leg muscles (Figure 6-13).

Starting Position 1: Lay on the right side with the legs extended straight to the side, with the left leg 6 to 8 inches above the ground, and toes pointing straight ahead. Support the upper body with the right elbow. The elbow is bent at 90 degrees, the upper arm is perpendicular to the ground, and the right hand makes a fist vertical to the ground.

Starting Position 2: Lay on the left side with the legs extended straight to the side with the right leg 6 to 8 inches above the ground and toes pointing straight ahead. Support the upper body with the left elbow. The elbow is bent at 90 degrees, the upper arm is perpendicular to the ground, and the left hand makes a fist vertical to the ground.

Commands: The commands for the lateral leg raise are as follows:
- Starting Position, MOVE.
- In Cadence, EXERCISE (count as a 4-count exercise according to Chapter 7, Execution of Training, paragraph 7-29).
- Change Position, MOVE.
- In Cadence, EXERCISE (count as a 4-count exercise according to Chapter 7, Execution of Training, paragraph 7-29).
- Position of Attention, MOVE.

Cadence: SLOW

Count:
1. Bring the thigh of the top leg toward the chest, bending the knee at 90-degrees.
2. Return to the starting position.
3. Bring the thigh of the top leg toward the chest, bending the knee at 90-degrees.
4. Return to the starting position.

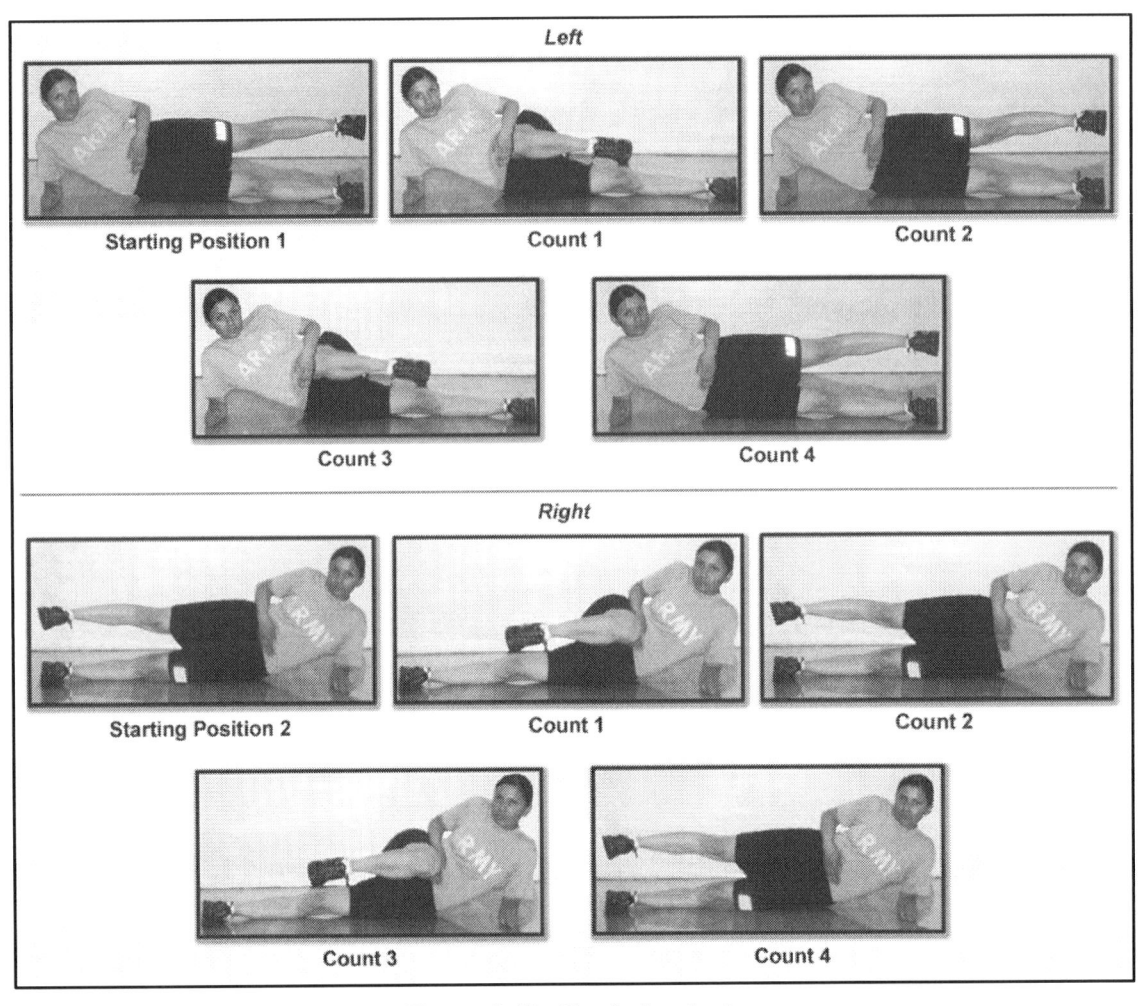

Figure 6-13. Single-leg tuck

Check Points:
- Face to the front of the formation, maintaining a generally straight line with the body.
- The top foot remains 6-8 inches above the ground throughout the exercise.
- Place the top hand over the stomach throughout the exercise.

Precaution: N/A

HIP STABILITY DRILL

EXERCISE 5: SINGLE-LEG OVER

(20-30 seconds on each side)

Purpose: This exercise develops flexibility of the hips and lower back muscles (Figure 6-14).

Starting Position 1: Supine position with arms sideward, palms down, and head on the ground.

Movement: On the command, "Ready, STRETCH," turn the body to right, bend the left knee to 90 degrees over the right leg, grasp the outside of the left knee with the right hand, and pull toward the right. Hold this position for 20-30 seconds. On the command, "Starting Position, MOVE," assume the starting position. On the command, "Change Position, Ready, STRETCH," turn the body to left, bend the right knee to 90-degrees over the left leg and grasp the outside of the right knee with the left hand and pull toward the left. Hold this position for 20-30 seconds. On the command, "Starting Position, MOVE," assume the starting position.

Figure 6-14. Single-leg over

Check Points:
- At the starting position, the arms are directed to the sides at 90-degrees to the trunk; the fingers and thumbs are extended and joined.
- In Exercise Position 1, keep the left shoulder, arm, and hand on the ground.
- In Exercise Position 2, keep the right shoulder, arm, and hand on the ground.
- Head remains on the ground throughout the exercise.

Precaution: N/A

SHOULDER STABILITY DRILL

6-51. The shoulder stability drill (SSD), Figure 6-2, is designed to develop strength and stability of the shoulders. This drill consists of five, 4-count exercises performed at a SLOW cadence for five repetitions each. The SSD may be performed between preparation, strength, and mobility activities along with 4C and the HSD to better prepare Soldiers in the toughening phase for the rigors of conditioning, climbing, push-up and sit-up drills, and the STC. Soldiers recovering from shoulder injuries may perform exercises in this drill as part of rehabilitation and reconditioning according to their medical profile.

Table 6-2. Shoulder stability drill (SSD)

1. "I" raise	5 repetitions	SLOW cadence
2. "T" raise	5 repetitions	SLOW cadence
3. "Y" raise	5 repetitions	SLOW cadence
4. "L" raise	5 repetitions	SLOW cadence
5. "W" raise	5 repetitions	SLOW cadence

SHOULDER STABILITY DRILL

EXERCISE 1: "I" RAISE

Purpose: This exercise develops shoulder strength and stability (Figure 6-15).

Starting Position: Prone position with the head slightly elevated and aligned with the spine. Feet are together and toes are pointed to the rear. The arms remain on the ground and are extended overhead, forming an "I" straight in line with the body. The hands are in a neutral position (perpendicular to the ground) with the thumbs and fingers extended and joined.

Cadence: SLOW

Count:
1. Raise both arms 3-6 inches off the ground.
2. Return to the starting position.
3. Repeat count 1.
4. Return to the starting position.

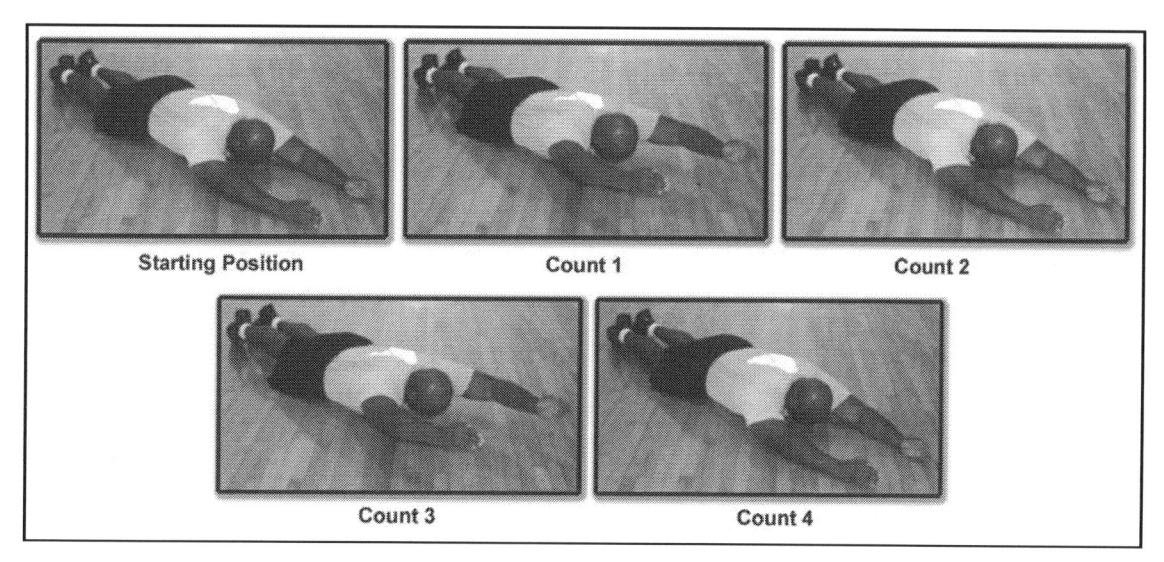

Figure 6-15. "I" raise

Check Points:
- At the starting position, tighten the abdominals to stabilize the trunk. The head is slightly elevated and aligned with the spine.
- On counts 1 and 3, keep the back generally straight with the head up.
- Throughout the exercise, the arms should be fully extended and the trunk and legs should also be aligned.

Precaution: Keep the head slightly elevated throughout the exercise and do not jerk the body into the up positions on counts 1 and 3.

SHOULDER STABILITY DRILL

EXERCISE 2: "T" RAISE

Purpose: This exercise develops shoulder strength and stability (Figure 6-16).

Starting Position: Prone position with the head slightly elevated and aligned with the spine. Feet are together and toes are pointed to the rear. The arms remain on the ground and are extended sideward at 90 degrees to the trunk, forming a "T." The hands are in a neutral position (perpendicular to the ground) with the thumbs and fingers extended and joined.

Cadence: SLOW

Count:

1. Raise both arms 3-6 inches off the ground.
2. Return to the starting position.
3. Repeat count 1.
4. Return to the starting position.

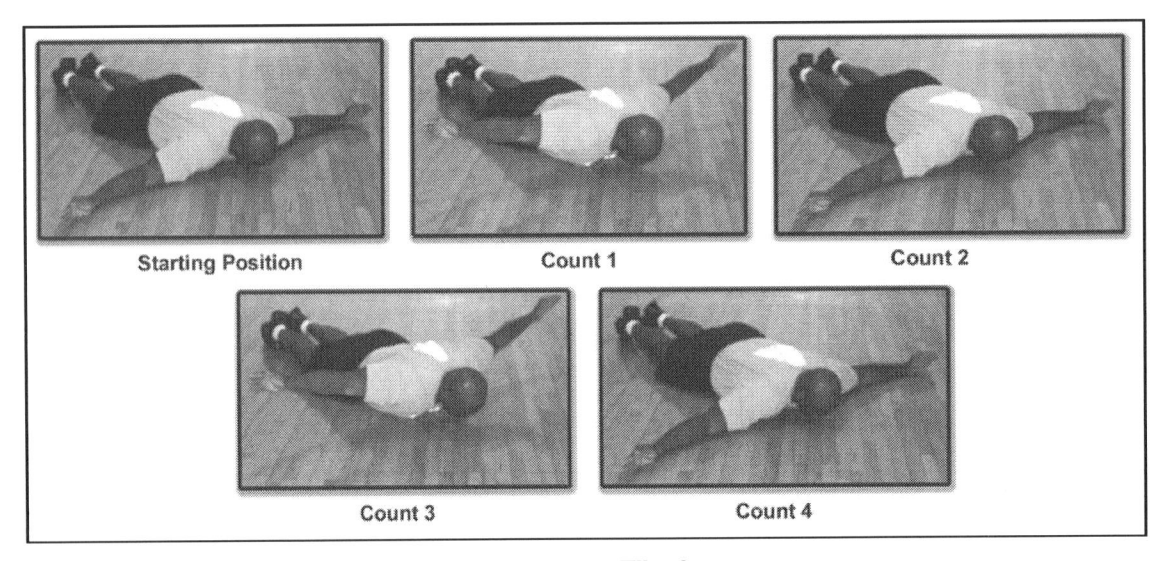

Figure 6-16. "T" raise

Check Points:

- At the starting position, tighten the abdominals to stabilize the trunk. The head is slightly elevated and aligned with the spine.
- On counts 1 and 3, keep the back generally straight with the head up.
- Throughout the exercise, the arms should be fully extended and the trunk and legs should also be aligned.

Precaution: Keep the head slightly elevated throughout the exercise and do not jerk the body into the up positions on counts 1 and 3.

SHOULDER STABILITY DRILL

EXERCISE 3: "Y" RAISE

Purpose: This exercise develops shoulder strength and stability (Figure 6-17).

Starting Position: Prone position with the head slightly elevated and aligned with the spine. Feet are together and toes are pointed to the rear. The arms remain on the ground and are extended overhead at 45 degrees to the trunk, forming a "Y." The hands are in a neutral position (perpendicular to the ground) with the thumbs and fingers extended and joined.

Cadence: SLOW

Count:

1. Raise both arms 3-6 inches off the ground.
2. Return to the starting position.
3. Repeat count 1.
4. Return to the starting position.

Figure 6-17. "Y" raise

Check Points:

- At the starting position, tighten the abdominals to stabilize the trunk. The head is slightly elevated and aligned with the spine.
- On counts 1 and 3, keep the back generally straight with the head up.
- Throughout the exercise, the arms should be fully extended and the trunk and legs should also be aligned.

Precaution: Keep the head slightly elevated throughout the exercise and do not jerk the body into the up positions on counts 1 and 3.

SHOULDER STABILITY DRILL

EXERCISE 4: "L" RAISE

Purpose: This exercise develops shoulder strength and stability (Figure 6-18).

Starting Position: Prone position with the head slightly elevated and aligned with the spine. Feet are together and toes are pointed to the rear. The arms remain on the ground and are extended sideward and the elbows are bent at 90 degrees, forming an "L." The hands are in a neutral position (perpendicular to the ground) with the thumbs and fingers extended and joined.

Cadence: SLOW

Count:
1. Raise both arms 3-6 inches off the ground.
2. Return to the starting position.
3. Repeat count 1.
4. Return to the starting position.

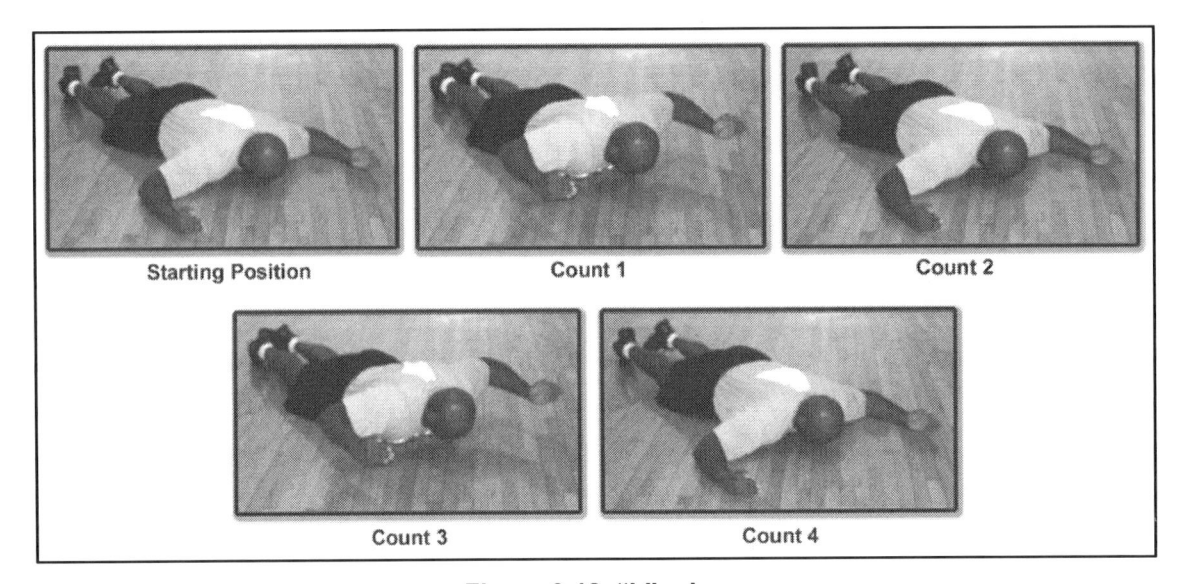

Figure 6-18. "L" raise

Check Points:
- At the starting position, tighten the abdominals to stabilize the trunk. The head is slightly elevated and aligned with the spine.
- On counts 1 and 3, keep the back generally straight with the head up.
- Throughout the exercise, the arms maintain an "L" and the trunk and legs should also be aligned.

Precaution: Keep the head slightly elevated throughout the exercise and do not jerk the body into the up positions on counts 1 and 3.

SHOULDER STABILITY DRILL

EXERCISE 5: "W" RAISE

Purpose: This exercise develops shoulder strength and stability (Figure 6-19).

Starting Position: Prone position with the head slightly elevated and aligned with the spine. Feet are together and toes are pointed to the rear. The arms remain on the ground and are extended downward at 45 degrees to the trunk and the elbow bent also at 45 degrees, forming a "W." The hands are in a neutral position (perpendicular to the ground) with the thumbs and fingers extended and joined.

Cadence: SLOW

Count:

1. Raise both arms 3-6 inches off the ground.
2. Return to the starting position.
3. Repeat count 1.
4. Return to the starting position.

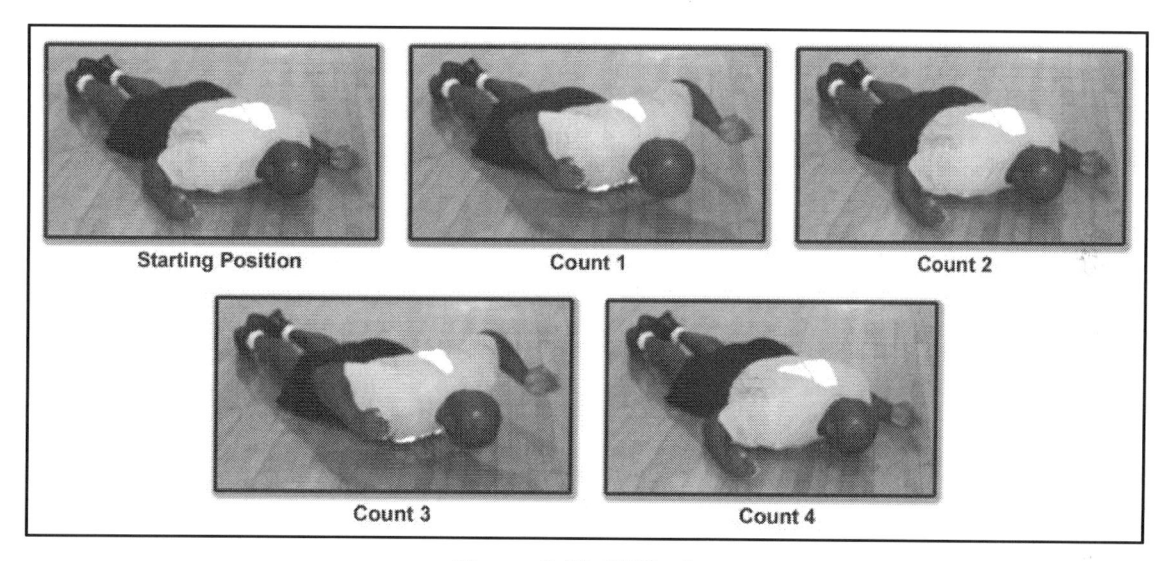

Figure 6-19. "W" raise

Check Points:

- At the starting position, tighten the abdominals to stabilize the trunk. The head is slightly elevated and aligned with the spine.
- On counts 1 and 3, keep the back generally straight with the head up.
- Throughout the exercise, the arms maintain a "W" and the trunk and legs should also be aligned.

Precaution: Keep the head slightly elevated throughout the exercise and do not jerk the body into the up positions on counts 1 and 3.

STRENGTH AND MOBILITY TRAINING

6-52. Strength and mobility training in reconditioning consists of the STM drill for level I, CD 1 and 2, and the PSD with modifications for level II. The following pages in this chapter describe in detail the conduct of these drills and modifications.

STRENGTH TRAINING MACHINE DRILL

6-53. The STM drill is conducted on strength and mobility training days according to the Soldiers' physical profile. The exercises may be modified to meet the Soldiers' capabilities. The following exercises are examples of each exercise in the STM and modifications of these exercises that may be employed to accommodate Soldiers' specific profiles.

Chapter 6

STRENGTH TRAINING MACHINE DRILL

EXERCISE 1: LEG PRESS

Purpose: This exercise develops strength in the hip and thigh muscles (Figure 6-20).

Starting Position: Seated position with the knees bent at 90-degrees and feet flat on the foot platform. The hips, low back, shoulders, and head are firmly against the seat back with the eyes looking straight ahead. A natural arch is maintained in the lower back. Select the appropriate weight and ensure the pin is secure in the weight stack. Hands are relaxed and placed on the handgrips.

Cadence: SLOW

Count:

1. Straighten the legs slowly until they are fully extended, not locked.
2. Return to the starting position in a slow, controlled motion.

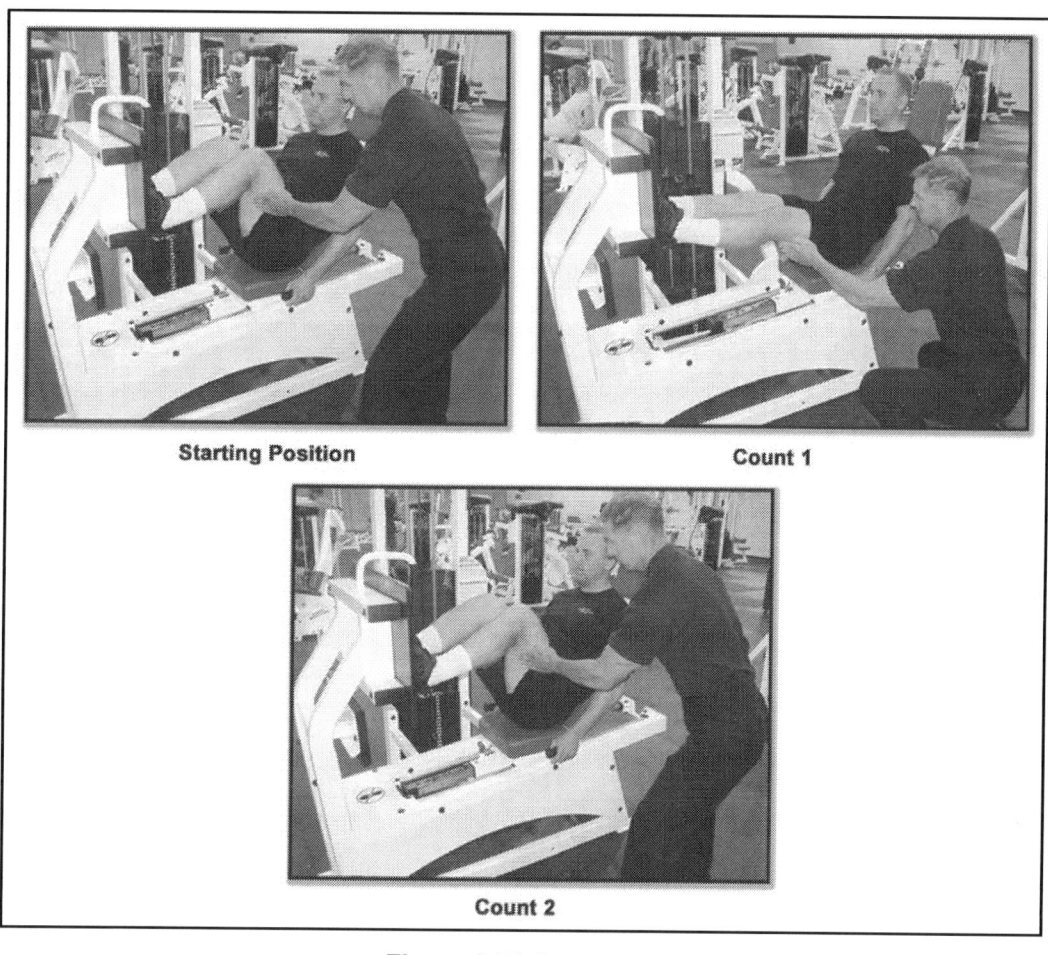

Figure 6-20. Leg press

Check Points:

- The hips, low back, shoulders, and head are firmly against the seat back.
- Maintain a natural arch in the lower back.
- Exhale on count 1 and inhale on count 2.

Precautions: Do not arch the back or allow the hips to rise off the seat. Do not grip the handgrips tightly.

STRENGTH TRAINING MACHINE DRILL

MODIFIED EXERCISE 1A: MODIFIED LEG PRESS

6-54. This exercise (Figure 6-21) is performed the same as the leg press. However, the range of motion is much less. As the Soldier's condition improves, the range of motion may gradually increase until the exercise is performed to standard. The resistance should not be increased until the Soldier can move through the full range of motion and perform the exercise to standard. The Soldier may also employ the single-leg press to maintain a heavy resistance on the good leg and/or to reduce the resistance on the injured leg.

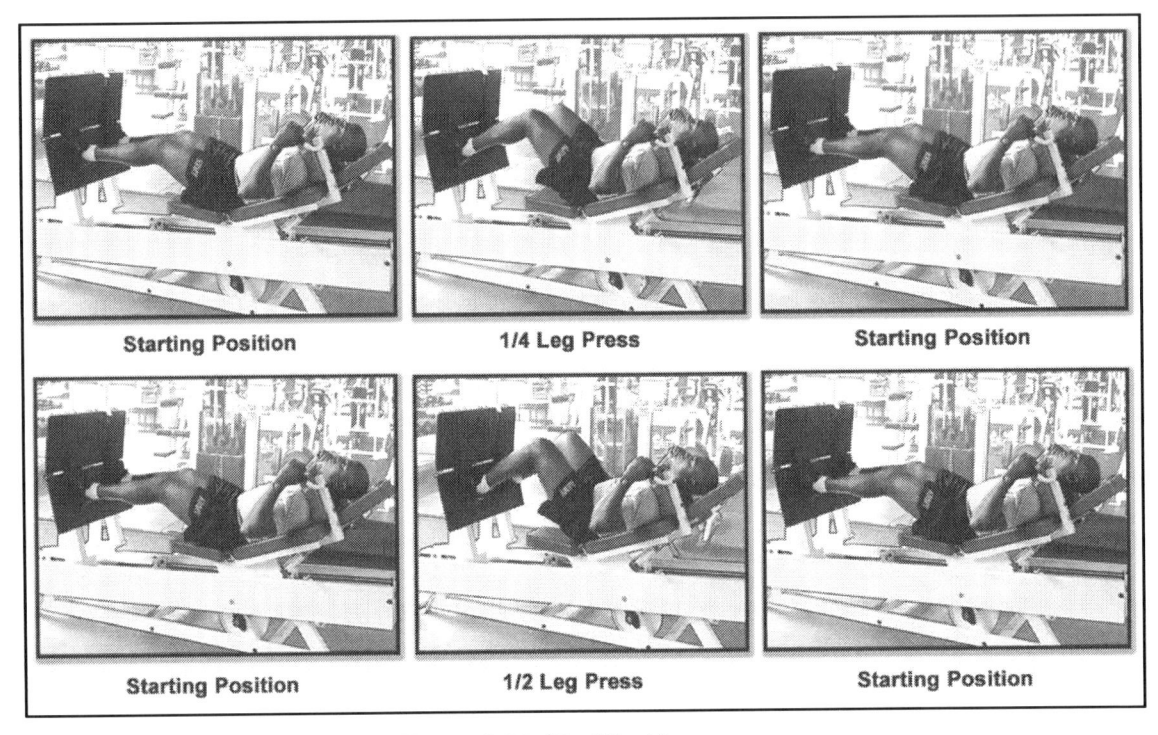

Figure 6-21. Modified leg press

MODIFIED EXERCISE 1B: SINGLE-LEG PRESS

6-55. This exercise (Figure 6-22) is performed much like the leg press, using only one leg at a time. The range of motion and resistance is decreased for the injured leg. As the Soldier's condition improves, the range of motion may gradually increase until the exercise is performed to standard. The resistance should not be increased until the Soldier can move through the full range of motion. The single leg press is used to maintain a heavy resistance on the good leg and/or to reduce the resistance on the injured leg.

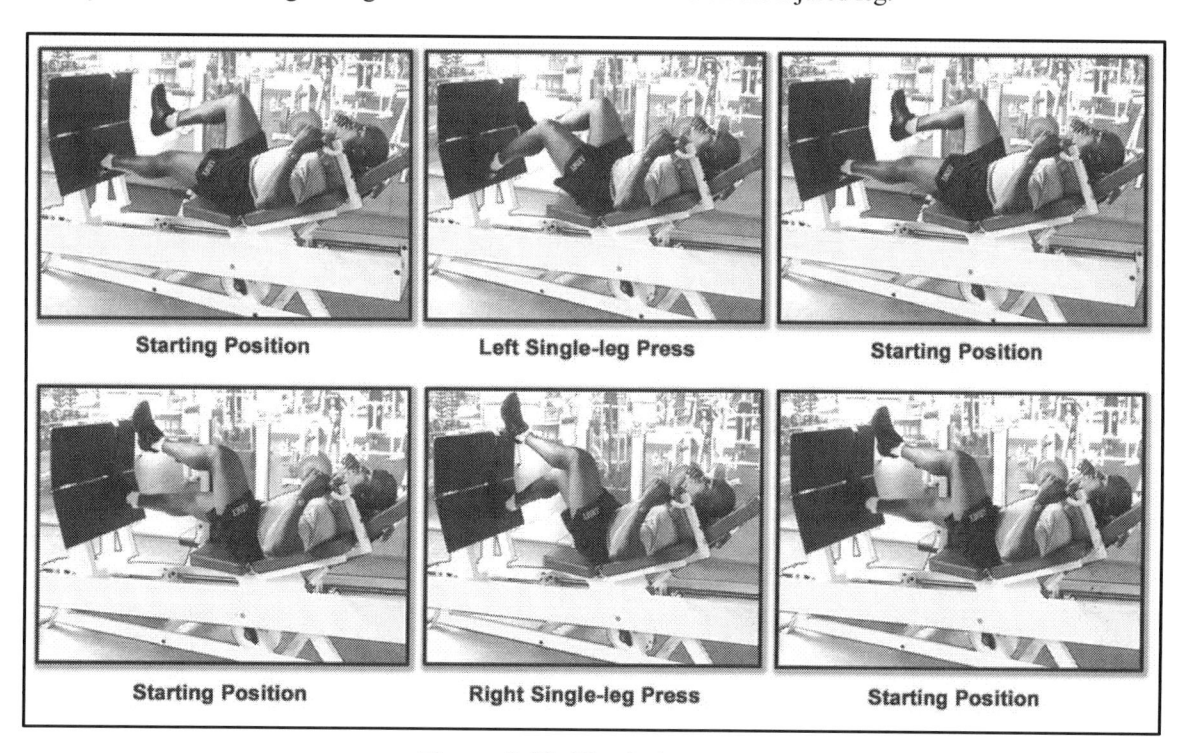

Figure 6-22. Single-leg press

STRENGTH TRAINING MACHINE DRILL

EXERCISE 2: LEG CURL

Purpose: This exercise develops strength in the back of the upper leg muscles (Figure 6-23).

Starting Position: Seated position, knees aligned with the center axis of the machine. The lower leg pad is adjusted to contact the lower legs just above the ankle, allowing the lower leg to be fully extended, but not locked. The lower legs and feet are relaxed. The thigh pad is positioned just above the knees. The hips, low back, shoulders, and head are firmly against the seat back with the eyes looking straight ahead. A natural arch is maintained in the lower back. Select the appropriate weight and ensure the pin is secure in the weight stack. Hands are relaxed and placed on the handgrips on the top of the thigh pad.

Cadence: SLOW

Count:

1. Pull the lower legs to the rear slowly until the lower legs are flexed, forming a 90-degree angle between the upper and lower legs.
2. Return to the starting position by slowly raising the lower legs.

Figure 6-23. Leg curl

Check Points:

- Knees are aligned with the center axis of the machine.
- The leg pad contacts the lower legs just behind the ankles.
- The hips, low back, shoulders, and head are firmly against the seat back.
- Maintain a natural arch in the lower back.
- Exhale on count 1 and inhale on count 2.

Precautions: Do not arch the back or allow the hips to rise off the seat. Do not grip the handgrips tightly.

STRENGTH TRAINING MACHINE DRILL

MODIFIED EXERCISE 2A: MODIFIED LEG CURL (SEATED)

6-56. This exercise (Figure 6-24) is performed the same as the leg curl; however, the range of motion is much less. As the Soldier's condition improves, the range of motion may gradually increase until the exercise is performed to standard. The resistance should not be increased until the Soldier can move through the full range of motion and perform the exercise to standard.

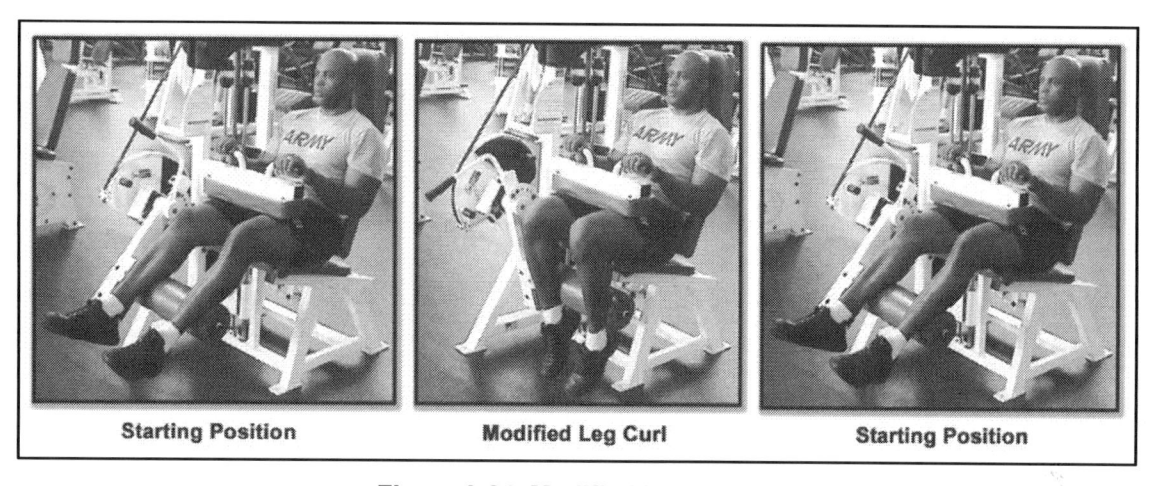

Figure 6-24. Modified leg curl

MODIFIED EXERCISE 2B: SINGLE-LEG CURL (SEATED)

6-57. This exercise (Figure 6-25) is performed much like the leg curl, using only one leg at a time. The range of motion and resistance is decreased for the injured leg. As the Soldier's condition improves, the range of motion may gradually increase until the exercise is performed to standard. The resistance should not be increased until the Soldier can move through the full range of motion. The single-leg curl is used to maintain a heavy resistance on the good leg and/or to reduce the resistance on the injured leg.

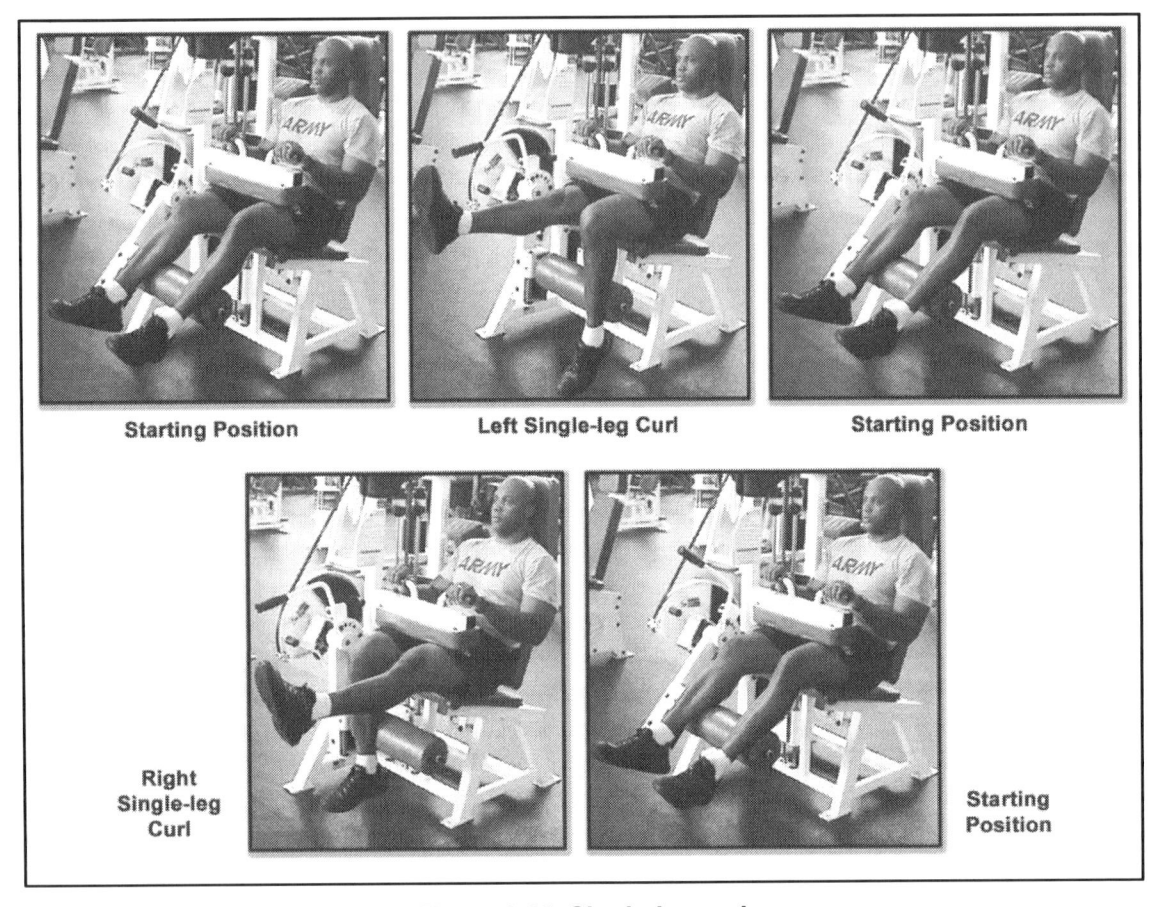

Figure 6-25. Single-leg curl

STRENGTH TRAINING MACHINE DRILL

MODIFIED EXERCISE 2C: MODIFIED LEG CURL (PRONE)

6-58. This exercise (Figure 6-26) is performed in the prone position through a limited range of motion. Soldiers with low back or hip injuries may prefer to use the seated leg curl if it is available. As the Soldier's condition improves, the range of motion may gradually increase until the exercise is performed through a full range of motion (heels to the buttocks). The resistance should not be increased until the Soldier can move through the full range of motion.

Figure 6-26. Modified leg curl (prone)

MODIFIED EXERCISE 2D: SINGLE-LEG CURL (PRONE)

6-59. This exercise (Figure 6-27) is performed using only one leg at a time. Soldiers with low back or hip injuries may prefer to use the seated leg curl if it is available. The range of motion and resistance is decreased for the injured leg. As the Soldier's condition improves, the range of motion may gradually increase until the exercise is performed to standard (heel to the buttocks). The resistance should not be increased until the Soldier can move through the full range of motion. The single-leg curl is used to maintain a heavy resistance on the good leg and to reduce the resistance on the injured leg.

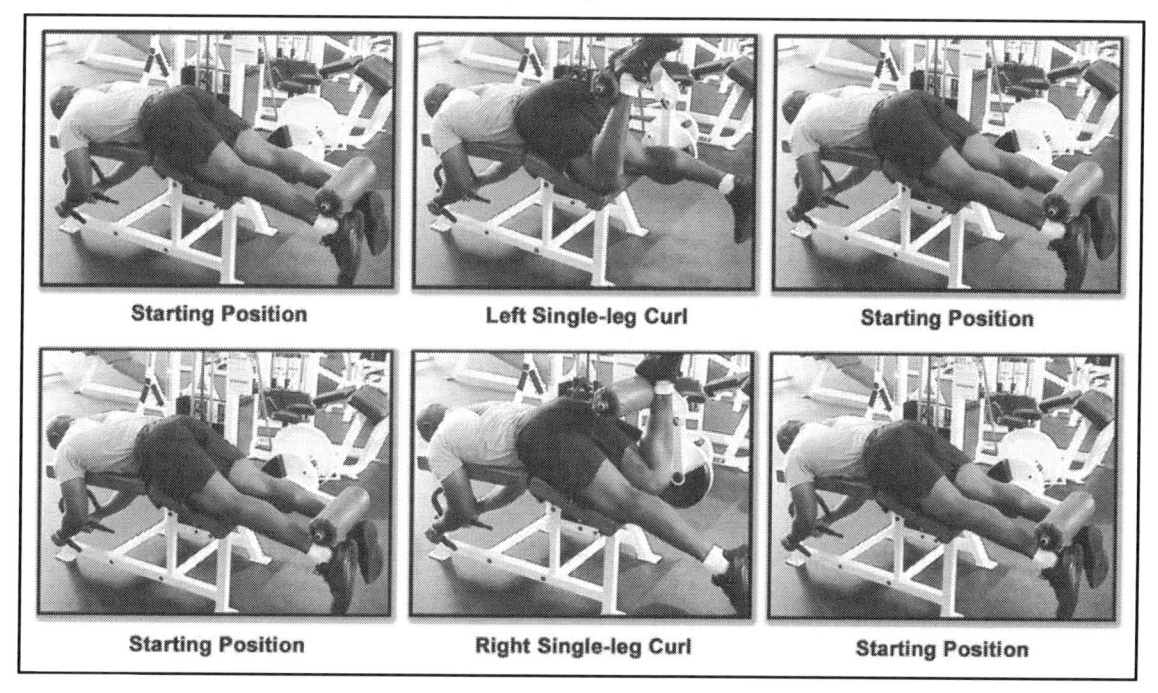

Figure 6-27. Single-leg curl (prone)

STRENGTH TRAINING MACHINE DRILL

EXERCISE 3: HEEL RAISE

Purpose: This exercise develops strength in the back of the lower leg muscles (Figure 6-28).

Starting Position: Stand with the balls of the feet on the elevated platform, toes pointing straight ahead, feet aligned directly below the hips, and the knees slightly flexed.

Cadence: SLOW

Count:
1. Raise the entire body slowly by pulling the heels up, maintaining a slight bend in the knees, and a natural arch in the low back.
2. Return to the starting position.

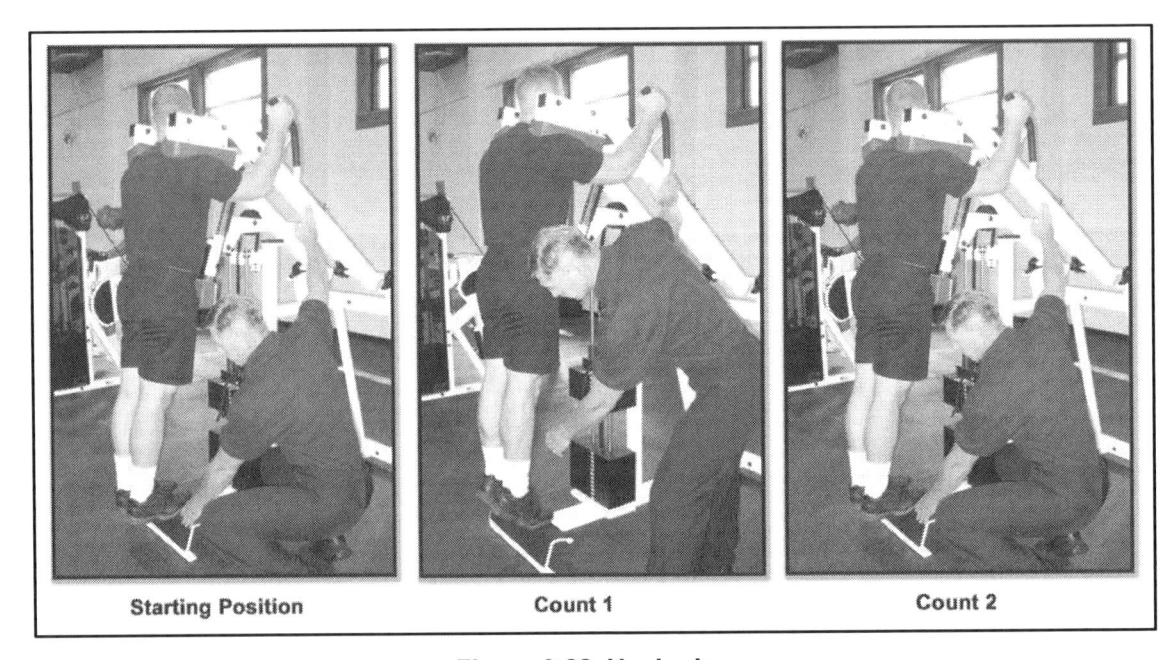

Figure 6-28. Heel raise

Check Points:
- Maintain a natural arch in the lower back.
- Keep the knees slightly flexed throughout the exercise.
- Keep the head and neck in a neutral position, looking straight ahead.
- Keep the knees aligned over the feet.
- Exhale on count 1 and inhale on count 2.

Precautions: Avoid flexing or extending the trunk. Do not allow the ankles to turn in or out.

STRENGTH TRAINING MACHINE DRILL

MODIFIED EXERCISE 3A: SINGLE-LEG HEEL RAISE

6-60. This exercise (Figure 6-29) is performed much like the heel raise, using only one leg at a time. The range of motion and resistance is decreased for the injured leg. As the Soldier's condition improves, the range of motion may gradually increase until the exercise is performed to standard. The resistance should not be increased until the Soldier can move through the full range of motion. The single leg is used to maintain a heavy resistance on the good leg and/or to reduce the resistance on the injured leg.

Figure 6-29. Single-leg heel raise

STRENGTH TRAINING MACHINE DRILL

EXERCISE 4: CHEST PRESS

Purpose: This exercise develops strength in the arms, shoulders, and chest muscles (Figure 6-30).

Starting Position: Seated position with the feet firmly on the ground. The seat is adjusted so a 90-degree angle is formed between the upper and lower arms with the shoulders directly below the handgrips. The hips, low back, shoulders, and head are firmly against the seat back with the eyes looking straight ahead. A natural arch is maintained in the lower back. Select the appropriate weight and ensure the pin is secure in the weight stack.

Cadence: SLOW

Count:
1. Push upward until both arms are fully extended, but not locked.
2. Return to the starting position.

Figure 6-30. Chest press

Check Points:
- Feet remain on the ground, with hips, back, shoulders, and head firmly on the bench.
- Keep the head and neck in a neutral position, looking straight ahead.
- Exhale on count 1 and inhale on count 2.

Precaution: Do not arch the back or allow the hips to rise off the bench.

STRENGTH TRAINING MACHINE DRILL

MODIFIED EXERCISE 4A: MODIFIED CHEST PRESS

6-61. This exercise (Figure 6-31) is performed the same as the chest press, but with much less range of motion. The elbows will not flex below 90 degrees as the resistance is lowered, nor will they fully straighten when the resistance is raised. As the Soldier's condition improves, the range of motion may gradually increase until the exercise is performed to standard. The resistance should not be increased until the Soldier can move through the full range of motion and perform the exercise to standard.

Figure 6-31. Modified chest press

MODIFIED EXERCISE 4B: SINGLE-ARM CHEST PRESS

6-62. This exercise (Figure 6-32) is performed much like the chest press, using only one arm at a time. The range of motion and resistance is decreased for the injured side. As the Soldier's condition improves, the range of motion may gradually increase until the exercise is performed to standard. The resistance should not be increased until the Soldier can move through the full range of motion. The single-arm chest press is used to maintain a heavy resistance on the good side and/or to reduce the resistance on the injured side.

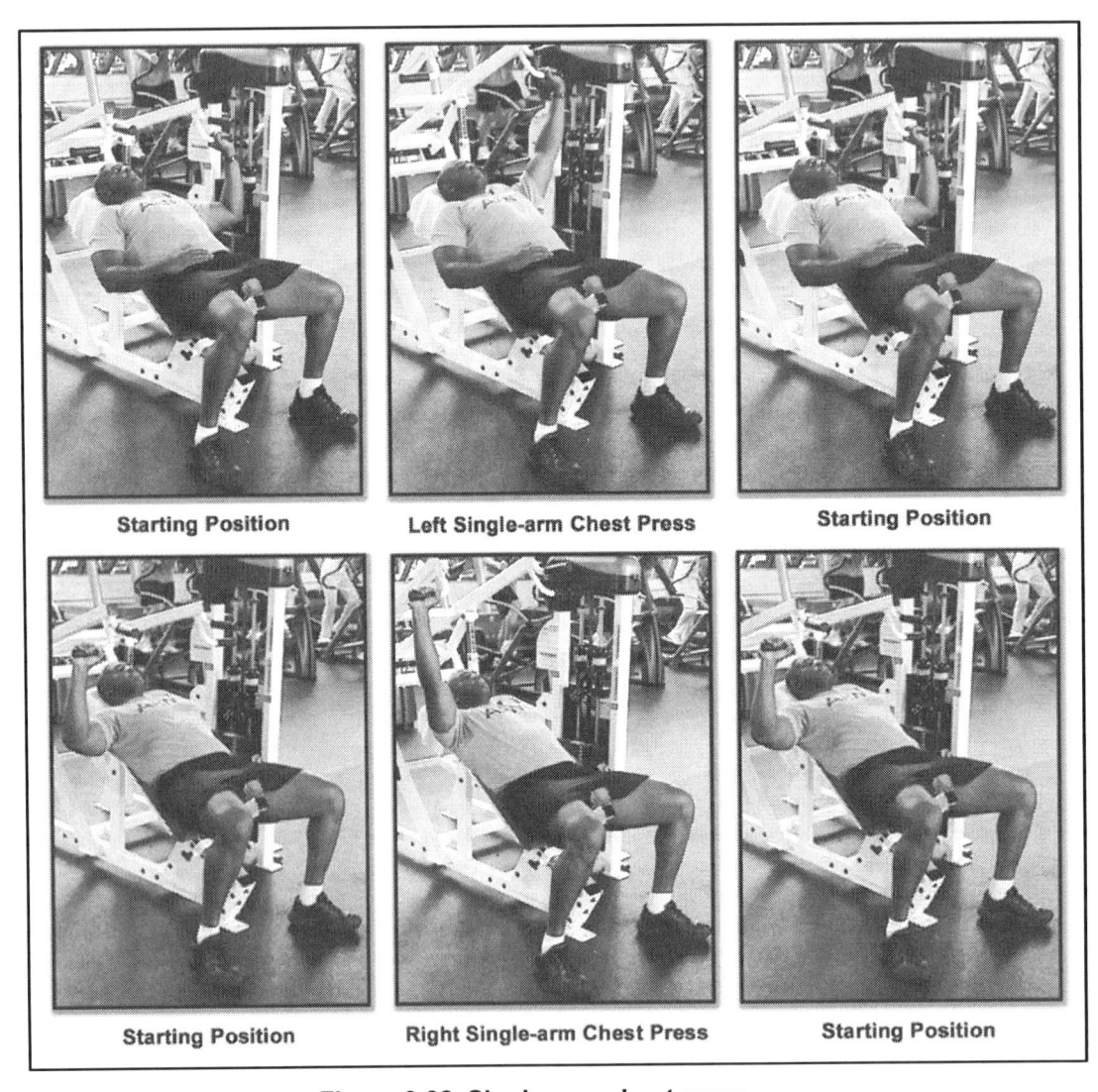

Figure 6-32. Single-arm chest press

STRENGTH TRAINING MACHINE DRILL

EXERCISE 5: SEATED ROW

Purpose: This exercise develops strength in the arm and back muscles (Figure 6-33).

Starting Position: Seated position with the feet firmly planted on the foot supports. Lean forward and grasp the handgrips with the hands in a neutral closed grip. Sit erect so the upper body is perpendicular to the floor. Select the appropriate weight and ensure the pin is secure in the weight stack.

Cadence: SLOW

Count:
1. Simultaneously, bend the elbows and pull the handgrips to the chest or upper abdomen while keeping the trunk rigid and the back flat.
2. Return to the starting position by slowly extending the elbows.

Figure 6-33. Seated row

Check Points:
- Feet remain flat on the ground or foot supports.
- The trunk is erect and the back is flat.
- Keep the head and neck in a neutral position, looking straight ahead or slightly downward.
- The arms are about parallel to the ground.
- On count 1 ensure the elbows point up and to the rear.
- Exhale on count 1 and inhale on count 2.

Precautions: Do not jerk the trunk to move the handgrips towards the chest. Maintain a flat back.

STRENGTH TRAINING MACHINE DRILL

MODIFIED EXERCISE 5A: STRAIGHT-ARM SEATED ROW

6-63. This exercise (Figure 6-34) is performed the same as the seated row, however, the range of motion is much less. The elbows remain fully extended and the arms straight, as the resistance is lowered and when the resistance is raised. As the Soldier's range of motion improves, he may employ the single-arm seated row to maintain a heavy resistance on the good side and/or to reduce the resistance on the injured side.

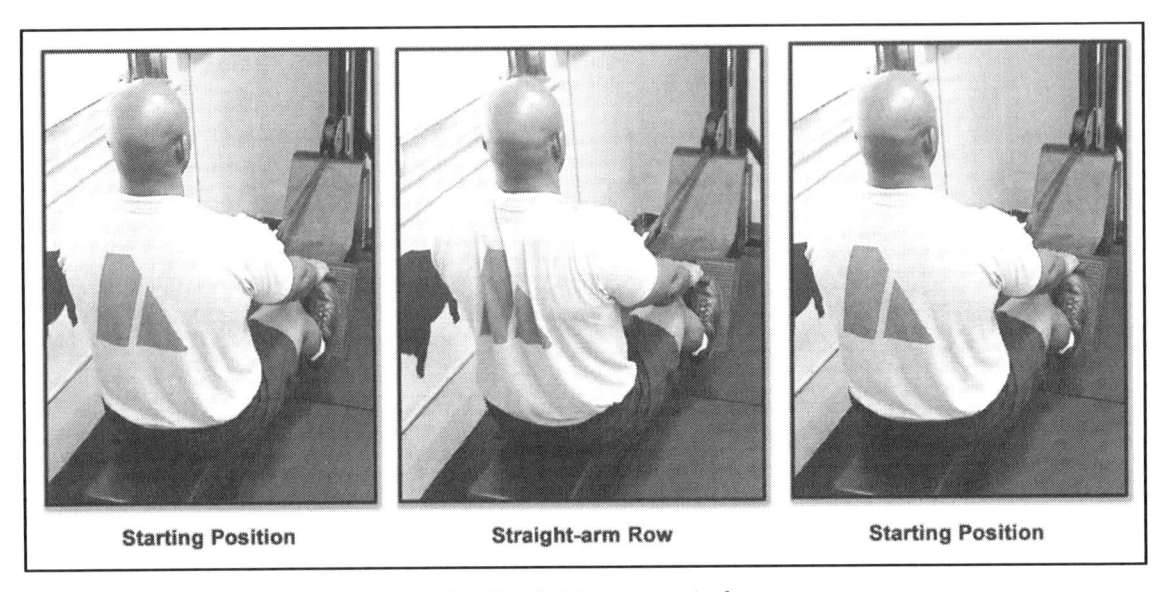

Figure 6-34. Straight-arm seated row

MODIFIED EXERCISE 5B: SINGLE-ARM SEATED ROW

This exercise (Figure 6-35) is performed much like the seated row, using only one arm at a time. The range of motion and resistance is decreased for the injured side. As the Soldier's condition improves, the range of motion may gradually increase until the exercise is performed to standard. The resistance should not be increased until the Soldier can move through the full range of motion. The single-arm seated row is used to maintain a heavy resistance on the good side and/or to reduce the resistance on the injured side.

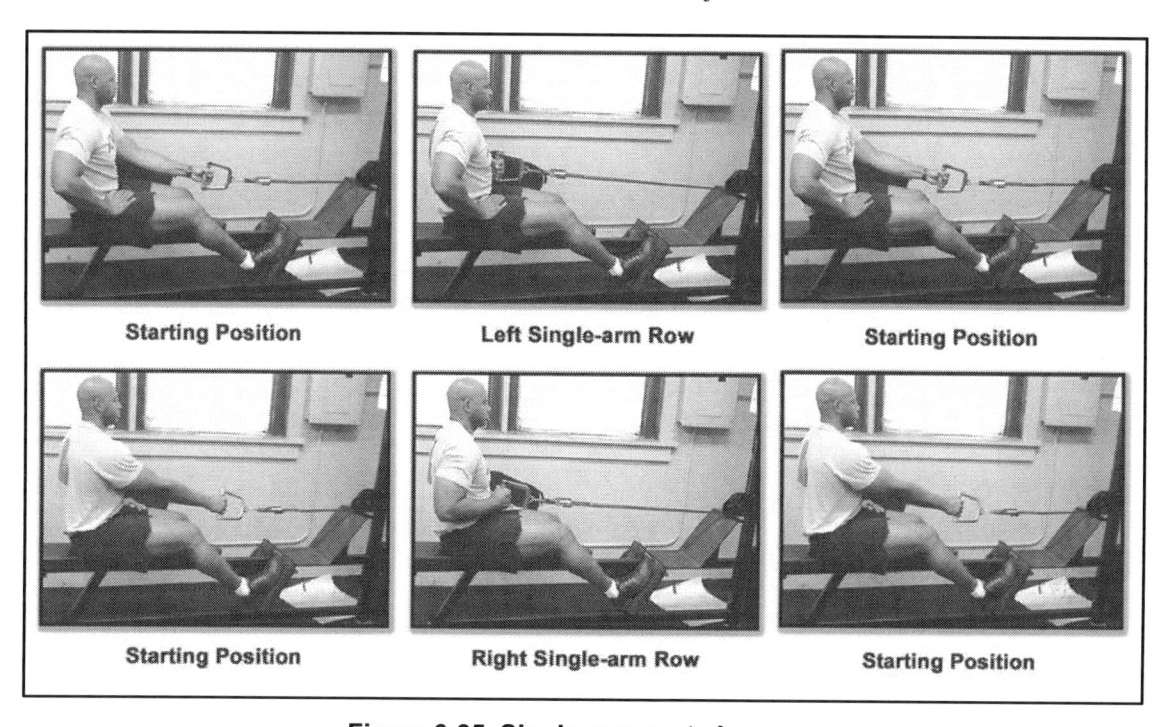

Figure 6-35. Single-arm seated row

STRENGTH TRAINING MACHINE DRILL

EXERCISE 6: OVERHEAD PRESS

Purpose: This exercise develops strength in the arm and shoulder muscles (Figure 6-36).

Starting Position: The Soldier assumes a seated position with the feet firmly on the ground. The Soldier adjusts the seat to achieve a 90-degree angle between the Soldier's upper and lower arms, with the shoulders directly below the handgrips. The hips, low back, shoulders, and head rest firmly against the seat back. The Soldier looks straight ahead, maintaining a natural arch in the lower back. The Soldier selects the appropriate weight and ensures the pin is secure in the weight stack.

Cadence: SLOW

Count:

1. Push upward until both arms are fully extended, but not locked.
2. Return to the starting position.

Figure 6-36. Overhead press

Check Points:

- Feet remain on the ground, with hips, back, shoulders, and head firmly on the bench.
- Keep the head and neck in a neutral position, looking straight ahead.
- Exhale on count 1 and inhale on count 2.

Precaution: Do not arch the back or allow the hips to rise off the bench.

STRENGTH TRAINING MACHINE DRILL

MODIFIED EXERCISE 6A: MODIFIED OVERHEAD PRESS

6-64. This exercise (Figure 6-37) is performed the same as the overhead press, but with much less range of motion. The elbows will not flex below 90 degrees as the resistance is lowered, nor will they fully straighten when the resistance is raised. As the Soldier's condition improves, the range of motion may gradually increase until the exercise is performed to standard. The resistance should not be increased until the Soldier can move through the full range of motion and perform the exercise to standard.

Figure 6-37. Modified overhead press

MODIFIED EXERCISE 6B: SINGLE-ARM OVERHEAD PRESS

6-65. This exercise (Figure 6-38) is performed much like the overhead press, using only one arm at a time. The range of motion and resistance is decreased for the injured side. As the Soldier's condition improves, the range of motion may gradually increase until the exercise is performed to standard. The resistance should not be increased until the Soldier can move through the full range of motion. The single-arm overhead press is used to maintain a heavy resistance on the good side and/or to reduce the resistance on the injured side.

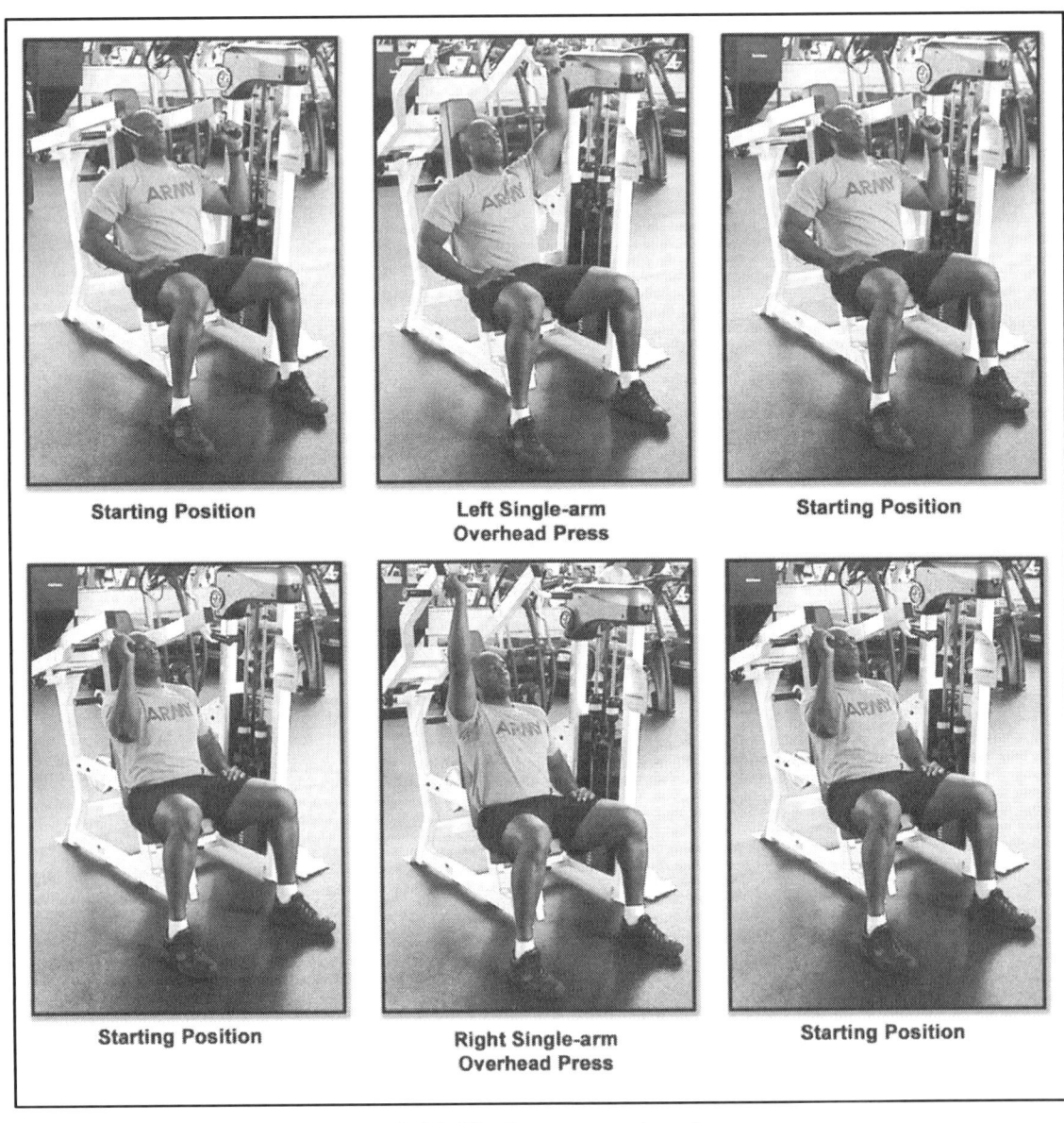

Figure 6-38. Single-arm overhead press

STRENGTH TRAINING MACHINE DRILL

EXERCISE 7: LAT PULL-DOWN

Purpose: This exercise develops strength in the arm and back muscles (Figure 6-39).

Starting Position: Select the appropriate weight and ensure the pin is secure in the weight stack before assuming the starting position. Sit erect and adjust the roller pad so it is firm against the upper thigh and hip. Grasp the bar with a closed, pronated grip and assume a seated position with the hips against the roller pad and the feet flat on the ground. The upper body is perpendicular to the floor.

Cadence: SLOW

Count:
1. Keeping the arms straight and elbows rotated out to the side and slightly flexed, simultaneously bend the elbows and pull bar toward the shoulders until the upper arms are parallel to the ground.
2. Return to the starting position by slowly extending the elbows.

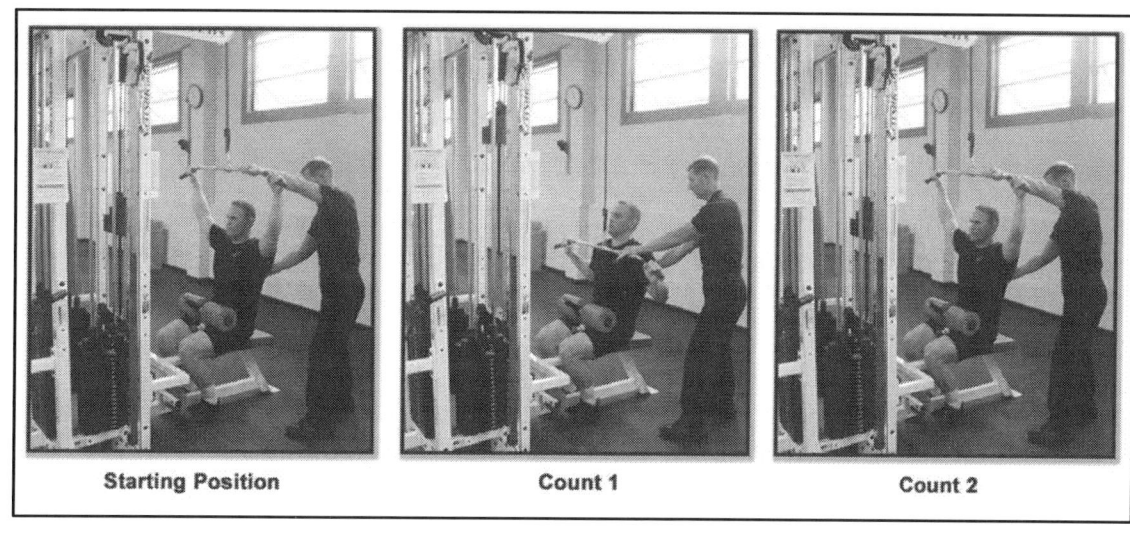

Figure 6-39. Lat pull-down

Check Points:
- Feet remain flat on the ground and the trunk is erect.
- Maintain a natural arch in the lower back.
- Keep the head and neck in a neutral position, looking straight ahead or slightly upward.
- Arms are straight and elbows rotated out to the side and slightly flexed and in direct line with the cable.
- Exhale on count 1 and inhale on count 2.

Precaution: Do not jerk the trunk or lean back to move the bar toward the shoulders.

STRENGTH TRAINING MACHINE DRILL

MODIFIED EXERCISE 7A: STRAIGHT-ARM LAT PULL-DOWN

6-66. This exercise (Figure 6-40) is performed the same as the lat pull-down, however, the range of motion is much less. The elbows remain fully extended and the arms straight, as the resistance is lowered and when the resistance is raised. As the Soldier's range of motion improves, he may employ the single-arm lat pull-down to maintain a heavy resistance on the good side and/or to reduce the resistance on the injured side.

Figure 6-40. Straight-arm lat pull-down

STRENGTH TRAINING MACHINE DRILL

MODIFIED EXERCISE 7B: SINGLE-ARM LAT PULL-DOWN

6-67. This exercise (Figure 6-41) is performed much like the lat pull-down, using only one arm at a time. The range of motion and resistance is decreased for the injured side. As the Soldier's condition improves, the range of motion may gradually increase until the exercise is performed to standard. The resistance should not be increased until the Soldier can move through the full range of motion. The single-arm lat pull-down is used to maintain a heavy resistance on the good side and/or to reduce the resistance on the injured side.

Figure 6-41. Single-arm lat pull-down

STRENGTH TRAINING MACHINE DRILL

EXERCISE 8: LATERAL RAISE

Purpose: This exercise develops strength in the shoulder and neck muscles (Figure 6-42).

Starting Position: Seated position with the feet firmly on the ground. The seat is adjusted so a 90-degree angle is formed between the upper and lower arms. The hips, lower back, shoulders, and head are firmly against the seat back with the eyes looking straight ahead. A natural arch is maintained in the lower back. Select the appropriate weight and ensure the pin is secure in the weight stack.

Cadence: SLOW

Count:
1. Raise both arms upward until they are parallel to the ground.
2. Return to the starting position.

Figure 6-42. Lateral raise

Check Points:
- Feet remain on the ground, with hips, back, shoulders, and head firmly on the bench.
- Keep the head and neck in a neutral position, looking straight ahead.
- Exhale on count 1 and inhale on count 2.

Precautions: Do not arch the back or allow the hips to rise off the bench. Do not raise arms above parallel to the ground.

STRENGTH TRAINING MACHINE DRILL

MODIFIED EXERCISE 8A: SINGLE-ARM LATERAL RAISE

6-68. This exercise (Figure 6-43) is performed much like the lateral raise, using only one arm at a time. The range of motion and resistance is decreased for the injured side. As the Soldier's condition improves, the range of motion may gradually increase until the exercise is performed to standard. The resistance should not be increased until the Soldier can move through the full range of motion. The single-arm lateral raise is used to maintain a heavy resistance on the good side and/or to reduce the resistance on the injured side.

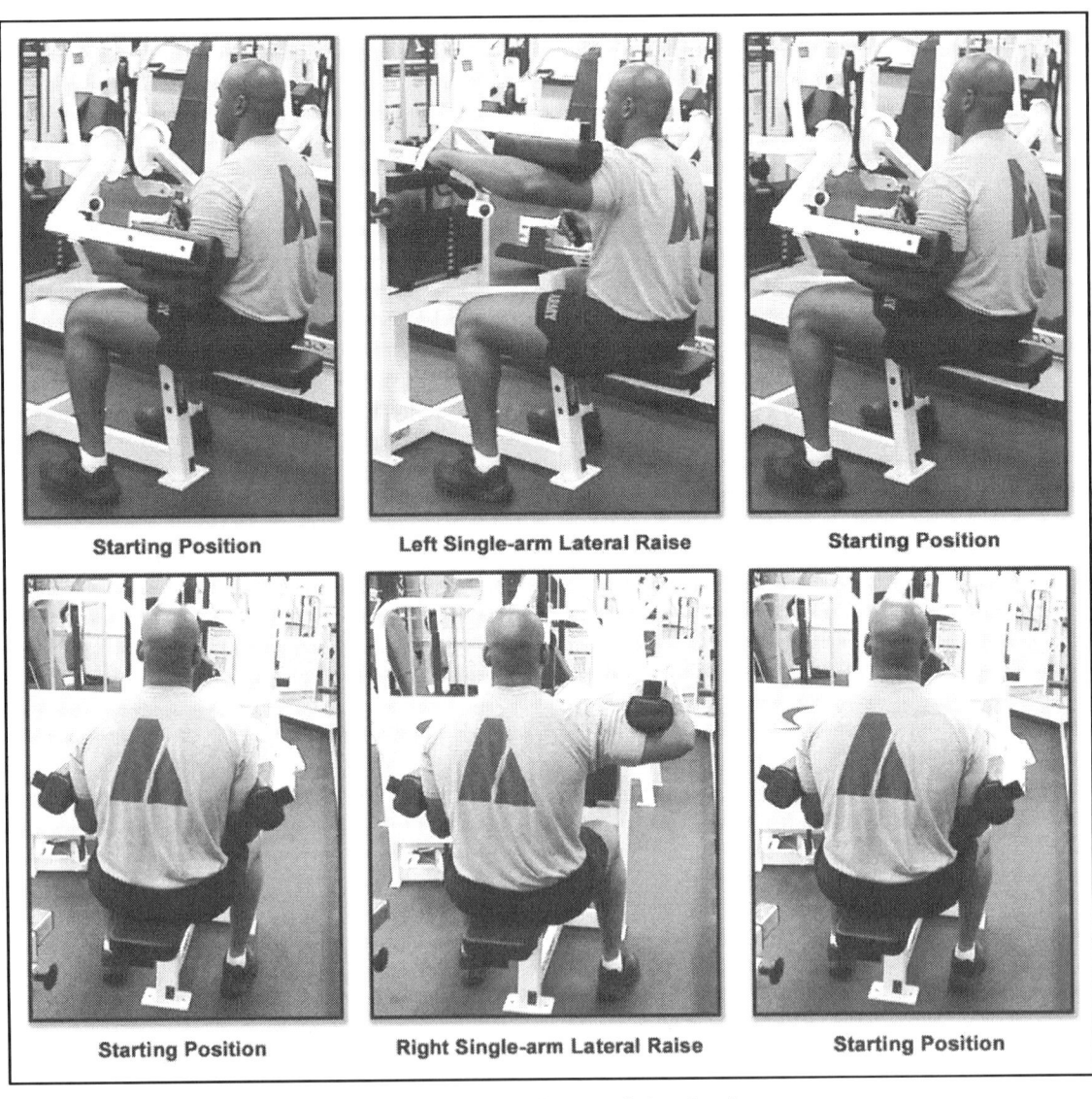

Figure 6-43. Single-arm lateral raise

STRENGTH TRAINING MACHINE DRILL

EXERCISE 9: TRICEPS EXTENSION

Purpose: This exercise develops strength in the triceps muscles (Figure 6-44).

Starting Position (Standing): Straddle stance with a 90-degree angle formed at the upper and lower arms. Select the appropriate weight and ensure the pin is secure in the weight stack. Maintain an erect position, eyes looking straight ahead, grasping the bar with a closed, pronated grip.

Starting Position (Seated): Seated position with the feet firmly on the ground. The seat is adjusted so a 90-degree angle is formed between the upper and lower arms, with elbows shoulder-width apart on the supporting pad, and hands in a closed-grip. The hips and low back are firmly against the seat back with the eyes looking straight ahead. A natural arch is maintained in the lower back. Select the appropriate weight and ensure the pin is secure in the weight stack.

Cadence: SLOW

Count:

1. Push downward until both arms are fully extended, but not locked.
2. Return to the starting position.

Special Conditioning Programs

Figure 6-44. Triceps extension

Check Points:
- Feet remain on the ground, with hips and back firmly on the bench during seated triceps extension.
- Keep the head and neck in a neutral position, looking straight ahead.
- Exhale on count 1 and inhale on count 2.

Precautions: Do not lean forward while performing standing triceps extension. Do not arch the back or allow the hips to rise off the bench during seated exercise.

Chapter 6

STRENGTH TRAINING MACHINE DRILL

MODIFIED EXERCISE 9A: MODIFIED TRICEPS EXTENSION

6-69. This exercise (Figures 6-45 and 6-46) is performed the same as the triceps extension, but the range of motion is much less. The elbows will not fully flex as the resistance is lowered, nor will they fully straighten when the resistance is raised. As the Soldier's condition improves, the range of motion may gradually increase until the exercise is performed to standard. The resistance should not be increased until the Soldier can move through the full range of motion and perform the exercise to standard.

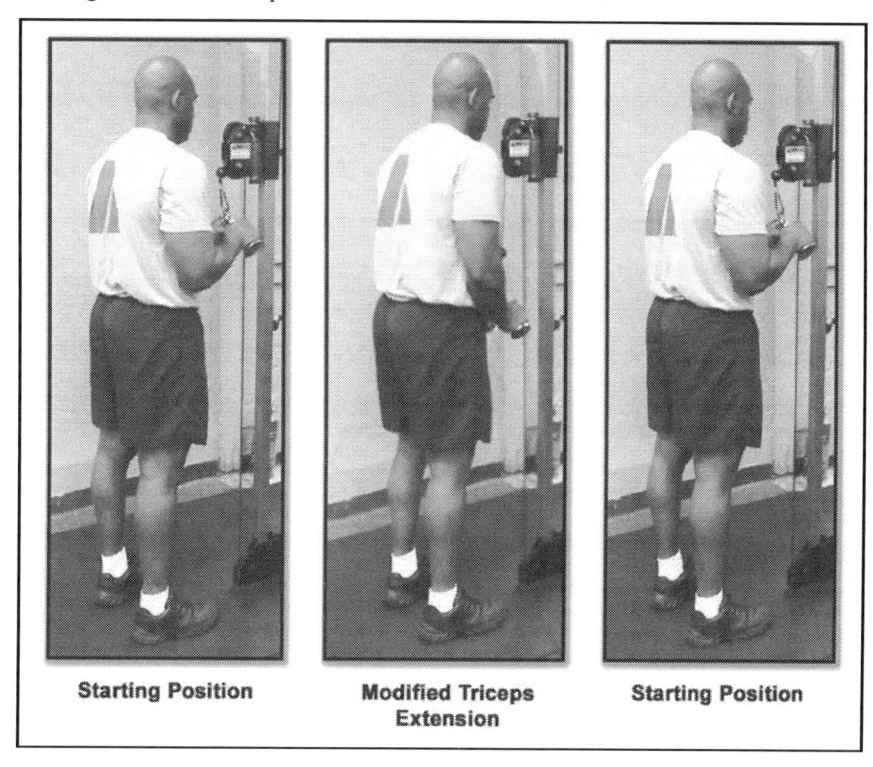

Figure 6-45. Modified triceps extension using a high pulley

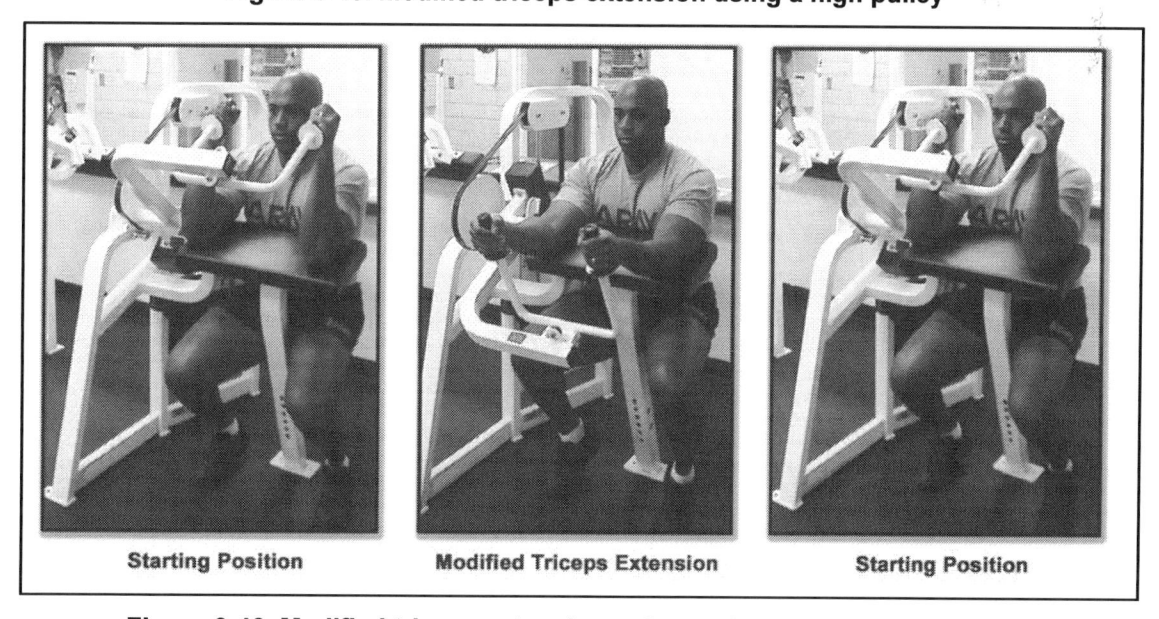

Figure 6-46. Modified triceps extension using a triceps extension machine

STRENGTH TRAINING MACHINE DRILL

MODIFIED EXERCISE 9B: SINGLE-ARM TRICEPS EXTENSION

6-70. This exercise (Figures 6-47 and 6-48) is performed much like the triceps extension, using only one arm at a time. The range of motion and resistance is decreased for the injured side. As the Soldier's condition improves, the range of motion may gradually increase until the exercise is performed to standard. The resistance should not be increased until the Soldier can move through the full range of motion. The single-arm triceps extension is used to maintain a heavy resistance on the good side and/or to reduce the resistance on the injured side.

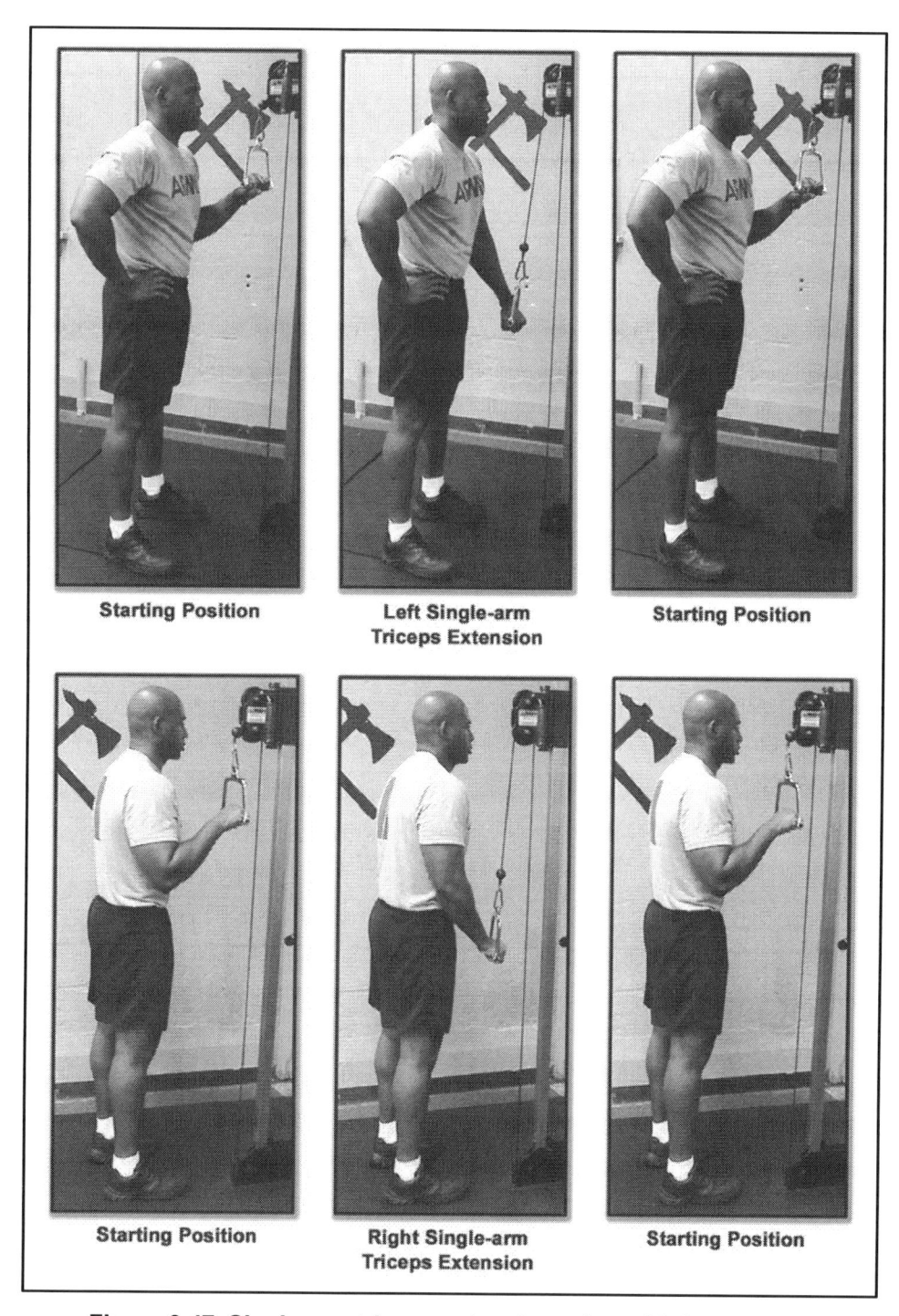

Figure 6-47. Single-arm triceps extension using a high pulley

STRENGTH TRAINING MACHINE DRILL

MODIFIED EXERCISE 9B: SINGLE-ARM TRICEPS EXTENSION (CONTINUED)

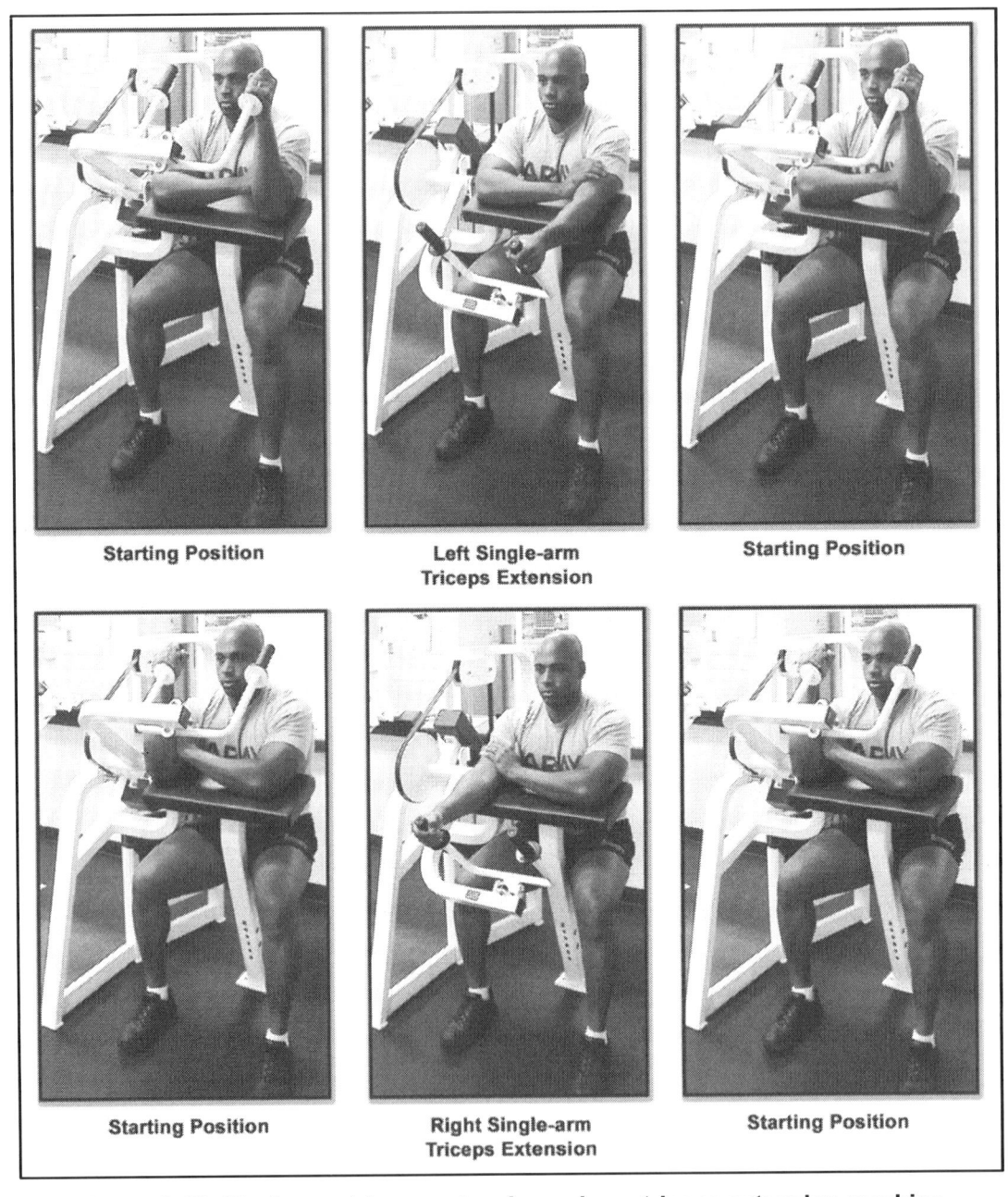

Figure 6-48. Single-arm triceps extension using a triceps extension machine

STRENGTH TRAINING MACHINE DRILL

EXERCISE 10: BICEPS CURL

Purpose: This exercise develops strength in the upper biceps muscles (Figure 6-49).

Starting Position: Seated position with the feet firmly on the ground. The seat is adjusted so the arms are straight, with elbows shoulder-width apart. The back of the upper arms are on the supporting pad with hands in a closed-grip. The hips and low back are firmly against the seat back with the eyes looking straight ahead. A natural arch is maintained in the lower back. Select the appropriate weight and ensure the pin is secure in the weight stack.

Cadence: SLOW

Count:

1. Pull upward until both arms are fully flexed.
2. Return to the starting position.

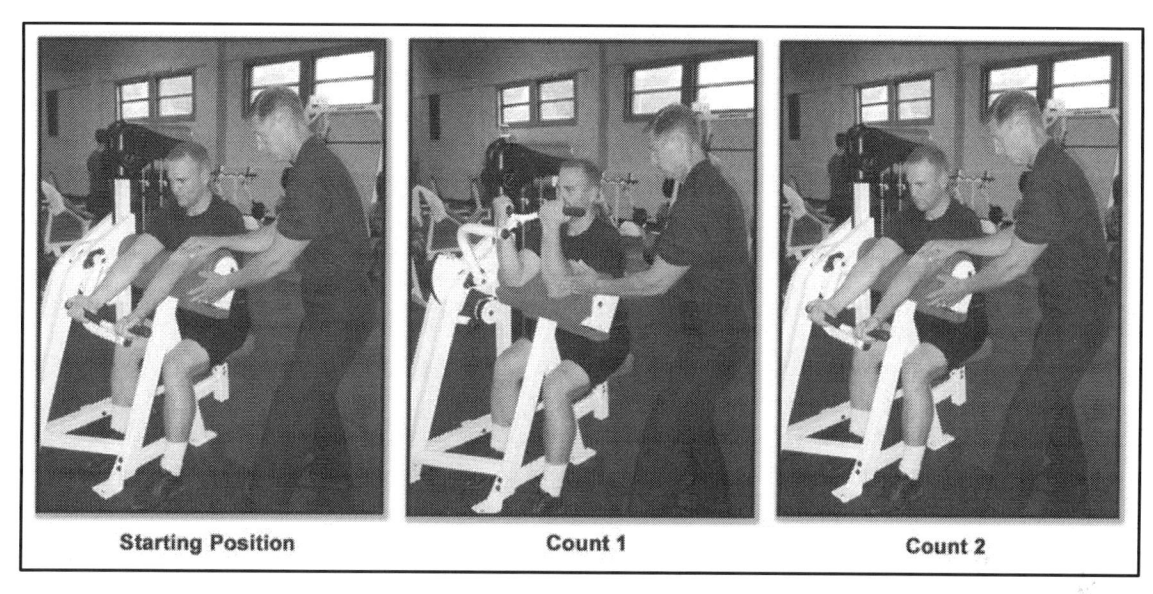

Figure 6-49. Biceps curl

Check Points:

- Feet remain on the ground, with hips and back firmly on the bench during seated triceps extension.
- Keep the head and neck in a neutral position, looking straight ahead.
- Exhale on count 1 and inhale on count 2.

Precautions: Do not arch the back or allow the hips to rise off the bench. Do not arch backward while performing the biceps curl.

STRENGTH TRAINING MACHINE DRILL

MODIFIED EXERCISE 10A: MODIFIED BICEPS CURL

6-71. This exercise (Figure 6-50) is performed the same as the biceps curl, but the range of motion is much less. The elbows will not fully flex as the resistance is raised, nor will they fully straighten when the resistance is lowered. As the Soldier's condition improves, the range of motion may gradually increase until the exercise is performed to standard. The resistance should not be increased until the Soldier can move through the full range of motion and perform the exercise to standard.

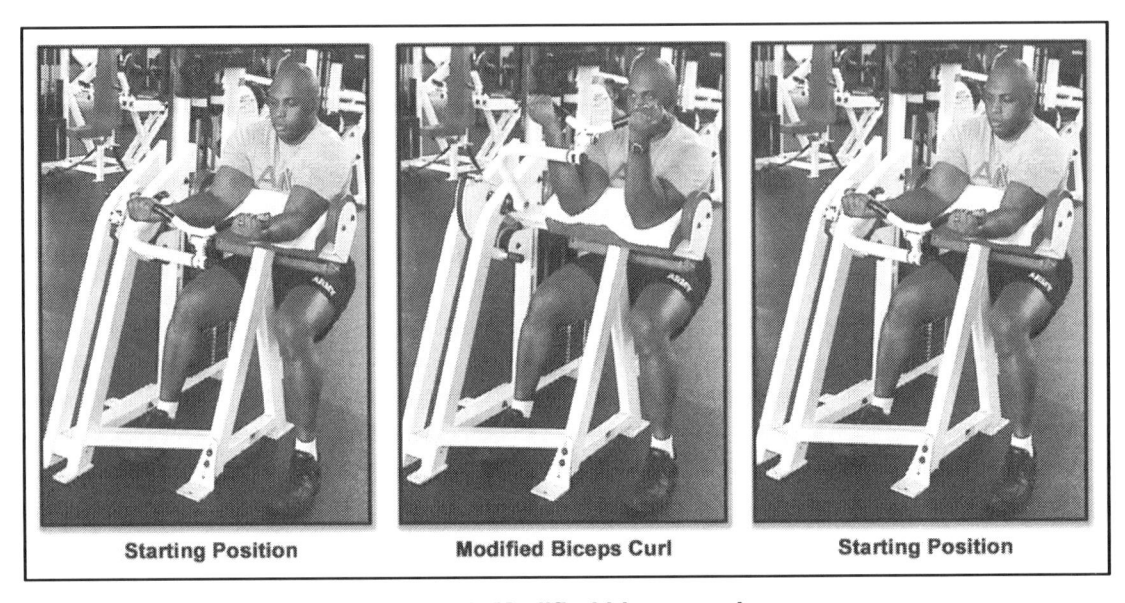

Figure 6-50. Modified biceps curl

STRENGTH TRAINING MACHINE DRILL

MODIFIED EXERCISE 10B: SINGLE-ARM BICEPS CURL

6-72. This exercise (Figure 6-51) is performed much like the biceps curl, using only one arm at a time. The range of motion and resistance is decreased for the injured side. As the Soldier's condition improves, the range of motion may gradually increase until the exercise is performed to standard. The resistance should not be increased until the Soldier can move through the full range of motion. The single-arm biceps curl is used to maintain a heavy resistance on the good side and to reduce resistance on the injured side.

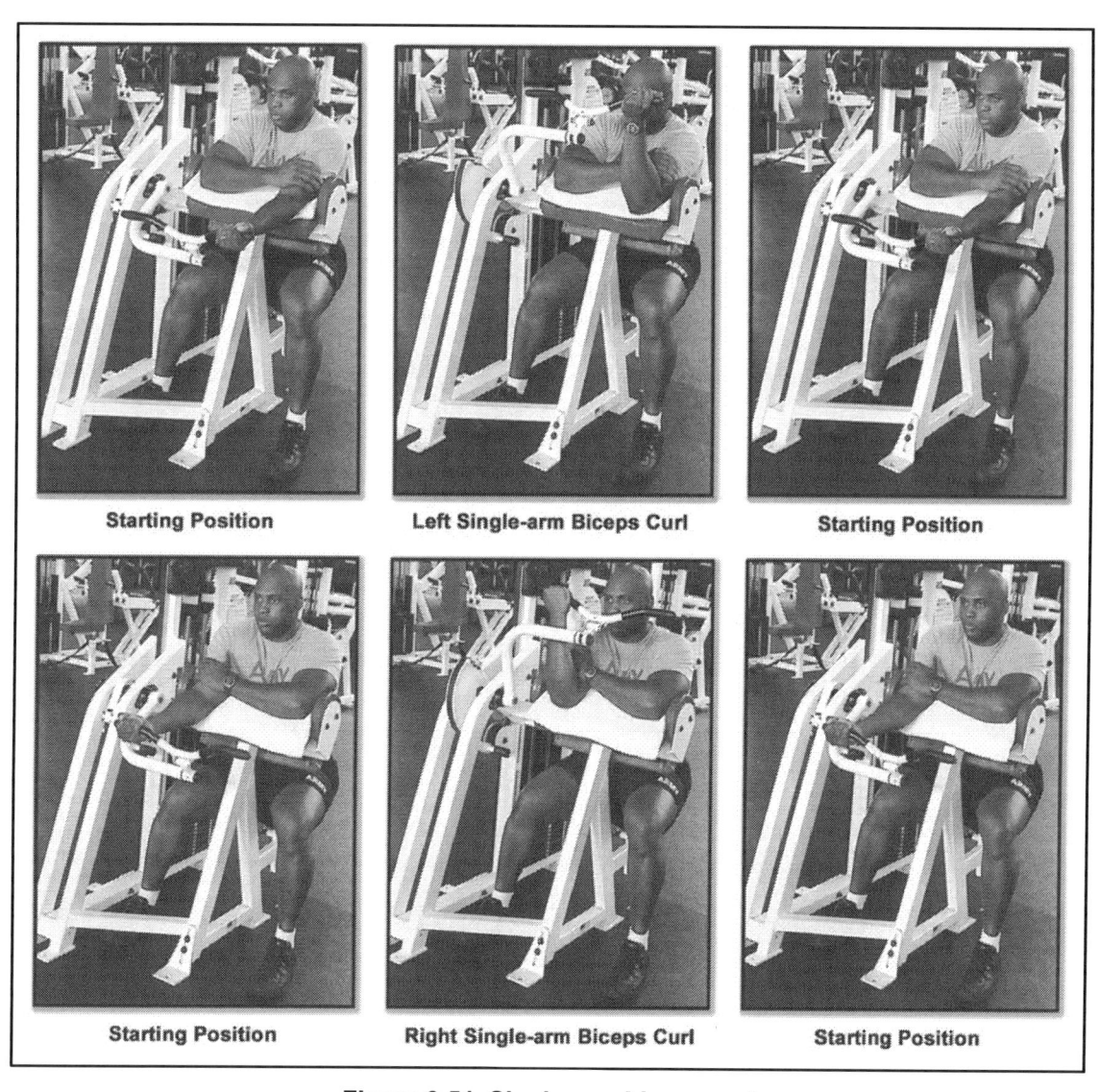

Figure 6-51. Single-arm biceps curl

STRENGTH TRAINING MACHINE DRILL

EXERCISE 11: TRUNK FLEXION

Purpose: This exercise develops strength in the abdominal muscles (Figure 6-52).

Starting Position: Seated position with the feet firmly on the ground. Select the appropriate weight and ensure the pin is secure in the weight stack. The seat is adjusted so the chest pad is located on the upper chest, below the collarbone. The elbows are shoulder-width apart and bent at 90 degrees, with hands in a closed-grip. The hips and low back are firmly against the seat back with the eyes looking straight ahead.

Cadence: SLOW

Count:
1. Bend forward, flexing the trunk, and bringing the chest pad to the thighs.
2. Return to the starting position.

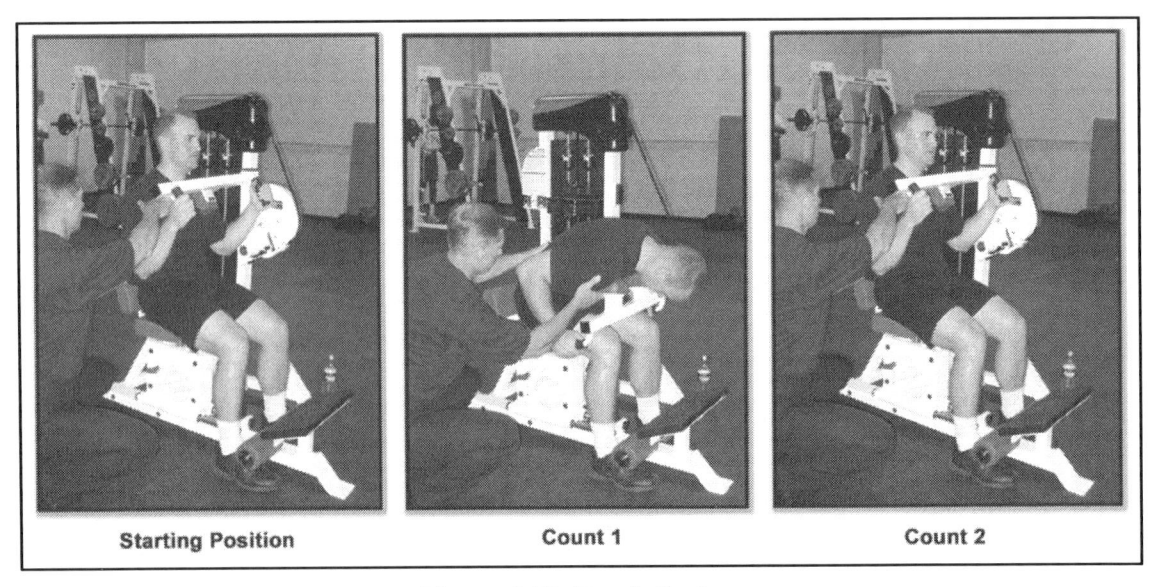

Figure 6-52. Trunk flexion

Check Points:
- Feet remain on the ground, with hips and back firmly on the bench.
- Keep the head and neck in a neutral position.
- Exhale on count 1 and inhale on count 2.

Precautions: Do not jerk into position or allow the hips to rise off the seat.

STRENGTH TRAINING MACHINE DRILL

MODIFIED EXERCISE 11: MODIFIED TRUNK FLEXION

6-73. Physical profiles may limit the range of motion at which Soldiers are able to safely perform trunk flexion exercises. The weight load should be low and the range of motion of the movements should be within the comfort zone of the Soldier (Figure 6-53). Gradually increase the weight load and range of motion as tolerated until the exercise can be performed to standard.

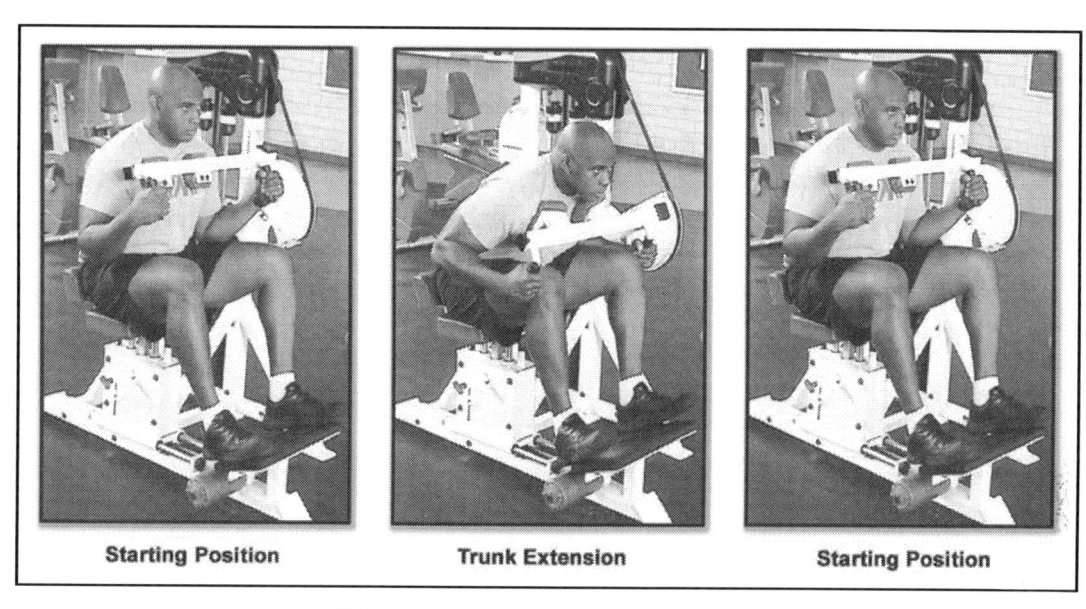

Figure 6-53. Modified trunk flexion

STRENGTH TRAINING MACHINE DRILL

EXERCISE 12: TRUNK EXTENSION

Purpose: This exercise develops strength in the low back muscles (Figure 6-54).

Starting Position: Sit in the machine, leaning slightly forward, with the back firmly against the padded lever arm. Select the appropriate weight and ensure the pin is secure in the weight stack. The hands grip the support bars using a neutral, closed-grip. The head is in a neutral position with the eyes looking straight ahead.

Cadence: SLOW

Count:
1. Raise the upper body and continue extending the trunk, moving to the supine position.
2. Return to the starting position.

Figure 6-54. Trunk extension

Check Points:
- Keep the head and neck in a neutral position.
- Exhale on count 1 and inhale on count 2.

Precautions: Do not jerk into position. Keep the hips and low back in contact with the pads throughout the exercise.

STRENGTH TRAINING MACHINE DRILL

MODIFIED EXERCISE 12: MODIFIED TRUNK EXTENSION

6-74. Physical profiles may limit the range of motion at which Soldiers are able to safely perform trunk extension exercises. The weight load should be low and the range of motion of the movements should be within the comfort zone of the Soldier (Figure 6-55). Gradually increase the weight load and range of motion as tolerated until the exercise can be performed to standard.

Figure 6-55. Modified trunk extension

Special Conditioning Programs

LEVEL II RECONDITIONING DRILLS AND ACTIVITIES

6-75. Soldiers in level II reconditioning are on profile, just off of profile, or cleared to begin level II reconditioning. These Soldiers will perform PRT drills and activities, in some cases, modified to fit the Soldier's specific physical profile or level of injury. See Table 6-3 for the schedule of level II reconditioning drills and activities.

Table 6-3. Reconditioning Level II training schedule

MON	TUE	WED	THU	FRI
PREPARATION: PD	PREPARATION: PD	PREPARATION: PD	PREPARATION: PD	PREPARATION: PD
ACTIVITIES: HSD (5 reps) MMD1 (1 rep) Walk to Run (30 min)	ACTIVITIES: 4C (60 secs) CD 1 (5 reps) CL 1 (5 reps)	ACTIVITIES: HSD (5 reps) MMD1 (1 rep) Walk to Run (30 min)	ACTIVITIES 4C (60 secs) CD 1 (5 reps) CL 1 (5 reps)	ACTIVITIES: HSD (5 reps) MMD1 (1 rep) Walk to Run (30 min)
RECOVERY: RD	RECOVERY: RD	RECOVERY: RD	RECOVERY: RD	RECOVERY: RD

6-76. Preparation, military movement drill 1, CD 1, and recovery will be the same as for unit PRT or may be modified to follow a safe exercise progression. The CL will be performed with spotters as in unit PRT. Spotters must be especially aware of each Soldier's physical limitation. The walk-to-run program safely progresses Soldiers from bouts of walking to increased bouts of continuous running for 30 consecutive minutes. Each week the walking time decreases as the running time increases to reach the 30-minute continuous running goal. (Table 6-4 shows how to conduct the walk-to-run program.)

Table 6-4. Reconditioning walk-to-run progression

Week of Training	Walk	Jog	Repetitions	Total Time
Week I	4 minutes	2 minutes	5 times	30 minutes
Week II	3 minutes	3 minutes	5 times	30 minutes
Week III	2 minutes	4 minutes	5 times	30 minutes
Week IV	1 minutes	5 minutes	5 times	30 minutes
Week V	Run every other day with a goal of reaching thirty consecutive minutes.			

- Perform the activities for each level every other day.
- Spend at least one week at each level. Begin Week V runs with a duration of 15 minutes.
- Walk 5 minutes before and after each session. Progress to 30 consecutive minutes of running over the next 2 to 4 weeks.

Exercise Guidance

6-77. The following exercise guidance is intended for RPLs/ARPLs in the level II reconditioning program. Common sites of pain/injury are given, followed by a discussion of PRT progression. The information below assumes that all profile restrictions have been removed. General exercise guidance is provided for knee injury/pain, foot and ankle injury/pain; lower leg injury/pain, low back injury/pain, and shoulder injury/pain; as well as modifications to exercises based on limitations of various physical profiles. In the pages to follow each of these injury conditions are listed with specific guidance on the conduct of exercise drills and activities as they apply to the knee, foot and ankle, lower leg, back and shoulder pain, and injuries.

Chapter 6

Knee Pain/Injury

6-78. Knee pain/injury may require restrictions. In the post-profile recovery period, progress as follows:

Preparation (PD)

6-79. Resume lunging and squatting movements (to include the high jumper) with a reduced range of motion and fewer repetitions. The high jumper should not be resumed until the Soldier has demonstrated proficiency at all other exercises. Resume the high jumper by only rising to the toes on counts one and three, then gradually progress starting with minimal height and few repetitions. When performing the squat thrust, Soldiers should assume the front leaning rest position by initially stepping into and out of the squat position while bearing most of their body weight with their arms. Soldiers must gradually increase the range of motion and repetitions to meet the standards. Allow Soldiers to use their hands as needed to move into and out of starting and exercise positions on the ground.

Conditioning Drill 1 (CD 1)

6-80. When assuming the starting position for the single-leg push-up, Soldiers should initially step into and out of the squat position to the front leaning rest position. This should be done while bearing most of the body weight with the arms. Allow Soldiers to assume a six-point position if they are unable to maintain good form or keep up with the cadence. Allow Soldiers to use their hands as needed to move into and out of starting and exercise positions on the ground.

Military Movement Drill 1 (MMD1)

6-81. Resume MMD 1 by reducing the distance from 25 to 15 yards and ensure that the Soldier limits the speed and intensity of movement. For laterals, this means decreasing the crouch and stepping the movements instead of maintaining the normal tempo. For verticals, start with minimal air time and gradually progress to more powerful movements. For the shuttle sprint, ensure that the Soldiers are able to negotiate the turns at walking speed before allowing them to run.

Push-up and Sit-up Drill (PSD)

6-82. When performing the squat thrust, Soldiers assume the front leaning rest position by initially stepping into and out of the squat position while bearing most of their body weight with their arms. Allow Soldiers to assume a six-point position for the push-ups if they are unable to maintain good form or keep up with the cadence. To modify the sit-up, allow Soldiers to initially use their hands to move into and out of the supine position.

Climbing Drill 1 (CL 1)

6-83. Proper spotting is essential in the post-profile period. Soldiers performing CL 1 exercise modifications in level II reconditioning depend greatly on their spotters to assist them through the movements of each exercise. Gradually, they will need less help from the spotters. Eventually, they may complete many, if not all the repetitions, with little or no assistance.

Sustained and Speed Running

6-84. If running is restricted, the Soldier will need to maintain conditioning through the use of ETM, the pool, and walking. When the profile ends or allows a return to running, a systematic progression should be followed. The Soldier must be able to walk for 30 minutes without increasing his symptoms before starting the running progression.

Recovery (RD)

6-85. As with all lunges, the amount of knee bend may be restricted for the rear lunge. The starting position for the extend and flex may be assumed as shown for the front leaning rest position. Allow Soldiers to use their hands as needed to move into and out of starting and exercise positions on the ground. In the post-profile period, range of motion for some exercises may still be limited. Gradually increase the range of motion over time and work toward the standard execution of each exercise.

Foot and Ankle Pain/Injury

6-86. PRT activities that involve jumping and landing, running, and single leg weight bearing should be resumed with the most caution. During the post-profile recovery period, progress as follows:

Preparation (PD)

6-87. Resume this drill at a slow cadence with few repetitions. The Soldier should resume the high jumper only after demonstrating proficiency in all other exercises. The Soldier resumes the high jumper by only rising to the toes on counts one and three, and then gradually progressing, starting with minimal height and few repetitions. The instructor monitors lunges closely, since they require most of the body weight to shift to a single leg. The stress of lunges can be limited by reducing the stride and the depth of the lunge. Initially, Soldiers might need to do push-ups by stepping back into the front-leaning rest rather than by performing a squat thrust. The instructor allows the Soldiers to use their hands as needed to move into and out of starting and exercise positions on the ground.

Military Movement Drill 1 (MMD 1)

6-88. Resume MMD 1 by reducing the distance from 25 to 15 yards and ensure that the Soldier limits the speed and intensity of movement. For laterals, this means decreasing the crouch and stepping the movements instead of maintaining the normal tempo. For verticals, start with minimal air time and gradually progress to more powerful movements. For the shuttle sprint, ensure that Soldiers are able to negotiate the turns at walking speed before allowing them to run.

Conditioning Drill 1 (CD 1)

6-89. When assuming the starting position for the single-leg push-up, Soldiers should initially step into and out of the squat position to the front leaning rest position. This should be done while bearing most of the body weight with the arms. Allow Soldiers to assume a six-point position if they are unable to maintain good form or keep up with the cadence. Allow Soldiers to use their hands as needed to move into and out of starting and exercise positions on the ground.

Climbing Drill 1 (CL 1)

6-90. Proper spotting is essential in the post-profile period. Encourage hands on spotting for all participants.

Sustained and Speed Running

6-91. While profiled for running, the Soldier will need to maintain conditioning through the use of ETMs, the pool, and walking. When the profile ends or allows a return to running, a systematic progression must be followed. The Soldier must be able to walk for 30 minutes without increasing his symptoms before starting the running progression.

Recovery (RD)

6-92. The starting position for the extend and flex may be assumed as shown for the front leaning rest position. Allow Soldiers to use their hands as needed to move into and out of starting and exercise positions on the ground. In the post-profile period, range of motion for some exercises may be limited still. Over time, gradually increase the range of motion and work toward the standard execution of each exercise.

Lower Leg Pain/Injury

6-93. PRT activities that involve jumping, landing, and running should be resumed with the most caution. In the post-profile recovery period, progress as follows:

Preparation (PD)

6-94. Resume this drill at a slow cadence with few repetitions. The high jumper should not be resumed until the Soldier has demonstrated proficiency at all other exercises. Resume the high jumper by rising to the toes only on counts one and three, then gradually progress starting with minimal height and few repetitions. Lunges should be monitored closely since they require most of the body weight to shift to a single leg. The stress of lunges can be limited by reducing the stride and the depth of the lunge. Allow Soldiers to use their hands as needed to move into and out of starting and exercise positions on the ground.

Military Movement Drill 1 (MMD 1)

6-95. Resume MMD 1 by reducing the distance from 25 to 15 yards and ensure that the Soldier limits the speed and intensity of movement. For laterals, this means decreasing the crouch and stepping through the movements instead of maintaining the normal tempo. For verticals, start with minimal air time and gradually progress to more powerful movements.

Conditioning Drill 1 (CD 1)

6-96. When assuming the starting position for the single-leg push-up, Soldiers should initially step into and out of the squat position to the front leaning rest position while bearing most of the body weight with the arms. Allow Soldiers to assume a six-point position if they are unable to maintain good form or keep up with the cadence. Allow Soldiers to use their hands as needed to move into and out of starting and exercise positions on the ground.

Climbing Drill 1 (CL 1)

6-97. Proper spotting is essential in the post-profile period. Encourage hands on spotting for all participants.

Sustained and Speed Running

6-98. While profiled for running, the Soldier will need to maintain conditioning through the use of ETMs, the pool, and walking. When the profile ends or allows a return to running, a systematic progression should be followed. Soldiers must be able to walk for 30 minutes without increasing their symptoms before starting the running progression.

Recovery (RD)

6-99. These exercises are generally not restricted, though Soldiers may need to use their hands to move into and out of starting and exercise positions on the ground. In the post-profile period, range of motion for some exercises may still be limited. Over time, gradually increase the range of motion and work toward the standard execution of each exercise.

Special Conditioning Programs

Back Pain or Back Injury

6-100. PRT activities that bend or twist the trunk must be resumed with caution. In the post-profile recovery period, progress as follows:

Preparation (PD)

6-101. Exercises that bend or twist the trunk may have been restricted while on profile. Post-profile, the Soldier starts with a limited range of movement and gradually progresses to the standard positions. Lunges and the squat bender are generally well tolerated, because the trunk remains straight throughout the movement. Post profile, the Soldier resumes the high jumper by rising only to the toes on counts one and three, then gradually progress starting with minimal height and few repetitions. Allow Soldiers to use their hands as needed to move into and out of starting and exercise positions on the ground.

Military Movement Drill 1 (MMD 1)

6-102. The shuttle sprint will normally be restricted by profile. In the post-profile period, resume the shuttle sprint without touching the hand to the ground on turns, and then gradually work toward bending enough to touch the ground. Resume the other MMD 1 exercises by reducing the distance from 25 to 15 yards and ensure that the Soldier limits the speed and intensity of movement. For laterals, this means decreasing the crouch and stepping through the movements instead of maintaining the normal tempo. For verticals, start with minimal air time and gradually progress to more powerful movements.

Conditioning Drill 1 (CD 1)

6-103. When assuming the starting position for the single-leg push-up, Soldiers should initially step into and out of the squat position to the front leaning rest position while bearing most of their body weight with their arms. Allow Soldiers to assume a six-point position if they are unable to maintain good form or keep up with the cadence. Allow Soldiers to use their hands as needed to move into and out of starting and exercise positions on the ground.

Climbing Drill 1 (CL 1)

6-104. Proper spotting is essential in the post-profile period. Encourage hands on spotting for all participants.

Sustained and Speed Running

6-105. If profiled for running, the Soldier will need to maintain conditioning through the use of ETMs, the pool, and walking. When the profile ends or allows a return to running, a systematic progression should be followed. The Soldier must be able to walk for 30 minutes without increasing their symptoms before starting the running progression.

Recovery (RD)

6-106. The extend and flex may be restricted by profile. Post-profile, Soldiers should go to the starting position by stepping back into the front-leaning rest position rather than performing a squat thrust. The other exercises should be tolerated in the post-profile period by starting with a reduced range of motion and gradually working toward the standard. Allow Soldiers to use their hands as needed to move into and out of starting and exercise positions on the ground.

Chapter 6

Shoulder Pain or Shoulder Injury

6-107. PRT activities that involve overhead motion or otherwise stress the shoulder must be resumed with caution. In the post-profile recovery period, progress as follows:

Preparation (PD)

6-108. Exercises that include raising the arms overhead may be restricted by profile. These exercises, unless otherwise restricted by the profile, can still be performed with hands on hips. The push-up will usually be restricted while on profile. After profiling, the Soldier may need to resume the exercise with a modified hand position. Push-up progression may start from the knees. Gradually work toward the standard exercise positions.

Military Movement Drill 1 (MMD 1)

6-109. If this drill is restricted by profile, resume the exercises in the post-profile period by reducing the distance from 25 to 15 yards and ensure that the Soldier limits the speed and intensity of movement. For laterals, this means decreasing the crouch and stepping the movements instead of maintaining the normal tempo. For verticals, start with minimal air time and gradually progress to more powerful movements.

Conditioning Drill 1 (CD 1)

6-110. When assuming the starting position for the single-leg push-up, Soldiers should initially step into and out of the squat position to the front leaning rest position while bearing most of the body weight with the arms. Allow Soldiers to assume a six-point position if they are unable to maintain good form or keep up with the cadence.

Sit-Up (SU)

6-111. Initially, allow Soldiers to use their hands to move into and out of the supine position. An alternate arm position with arms at sides or across the chest may be used.

Climbing Drill 1 (CL1)

6-112. Proper spotting is essential in the post-profile period. Encourage hands-on spotting for all participants.

Recovery (RD)

6-113. The extend and flex is generally the most stressful on the shoulder. The other exercises should be tolerated in the post-profile period by starting with a reduced range-of-motion and gradually working toward the standard. Allow Soldiers to use their hands as needed to move into and out of starting and exercise positions on the ground.

EXERCISE MODIFICATIONS

6-114. The PD, CD 1, military movement drill 1 (MMD 1), and the RD exercises include a wide range of movements requiring strength, endurance, and mobility using standing, seated, prone, and supine postures. Each exercise may be modified to accommodate various physical limitations. This allows Soldiers to work within their physical profiles, gradually progressing to performing each exercise to standard. The following pages describe each drill with exercise modifications to accommodate various physical profile limitations.

PREPARATION DRILL

EXERCISE 1: BEND AND REACH

Purpose: This exercise develops the ability to squat and reach through the legs. It also serves to prepare the spine and extremities for more vigorous movements by moving the hips and spine through full flexion (Figure 6-56).

Starting Position: Straddle stance with arms overhead.

Cadence: SLOW

Count:
1. Squat with the heels flat as the spine rounds forward to allow the straight arms to reach as far as possible between the legs.
2. Return to the starting position.
3. Repeat count 1.
4. Return to the starting position.

Figure 6-56. Bend and reach

Check Points:
- From the starting position, ensure that Soldiers have their hips set, their abdominals tight, and their arms fully extended overhead.
- The neck flexes to allow the gaze to the rear; this brings the head in line with the bend of the trunk.
- The heels and feet remain flat on the ground.
- On counts 2 and 4, they do not go past the starting position.

Precautions: This exercise is always performed at a slow cadence. To protect the back, move into the count 1 position in a slow, controlled manner. Do not bounce into or out of this position, as this may place an excessive load on the back.

PREPARATION DRILL

MODIFIED EXERCISE 1: MODIFIED BEND AND REACH

6-115. The instructor may modify the bend and reach by decreasing the range of motion and limiting the use of the arms. The Soldier may use the modifications shown in Figure 6-57 to exercise within physical profile limitations. The Soldier gradually increases the range of motion and works toward the standard execution of the exercise, then progresses performance to standard.

Figure 6-57. Modified bend and reach

Special Conditioning Programs

PREPARATION DRILL

EXERCISE 2: REAR LUNGE

Purpose: This exercise promotes balance, opens up the hip and trunk on the side of the lunge, and develops leg strength (Figure 6-58).

Starting Position: Straddle stance with hands on hips.

Cadence: SLOW

Count:

1. Take an exaggerated step backward with the left leg, touching down with the ball of the foot.
2. Return to the starting position.
3. Repeat count 1 with the right leg.
4. Return to the starting position.

Figure 6-58. Rear lunge

Check Points:
- Maintain straightness of the back by keeping the abdominal muscles tight throughout the motion.
- After the foot touches down, allow the body to continue to lower. This promotes flexibility of the hip and trunk.
- On counts 1 and 3, step straight to the rear, keeping the feet directed forward. When viewed from the front, the feet maintain their distance apart both at the starting position and at the end of counts 1 and 3.
- Keep the rear leg as straight as possible but not locked.
- Ensure the heel of the rear foot does not touch the ground.

Precautions: This exercise is always performed at a slow cadence. On counts 1 and 3, move into position in a slow, controlled manner. If the cadence is too fast, it will be difficult to go through a full range of motion.

PREPARATION DRILL

MODIFIED EXERCISE 2: MODIFIED REAR LUNGE

6-116. The rear lunge can be modified (Figure 6-59) by decreasing the range of motion at which it is performed. As with all lunges, the amount of knee bend may be restricted for the rear lunge. The feet may be closer together. Concentrate on trying to gradually lower the body in the lunge position. The Soldier gradually increases the range of motion and works toward the standard execution of the exercise, then progresses performance to standard.

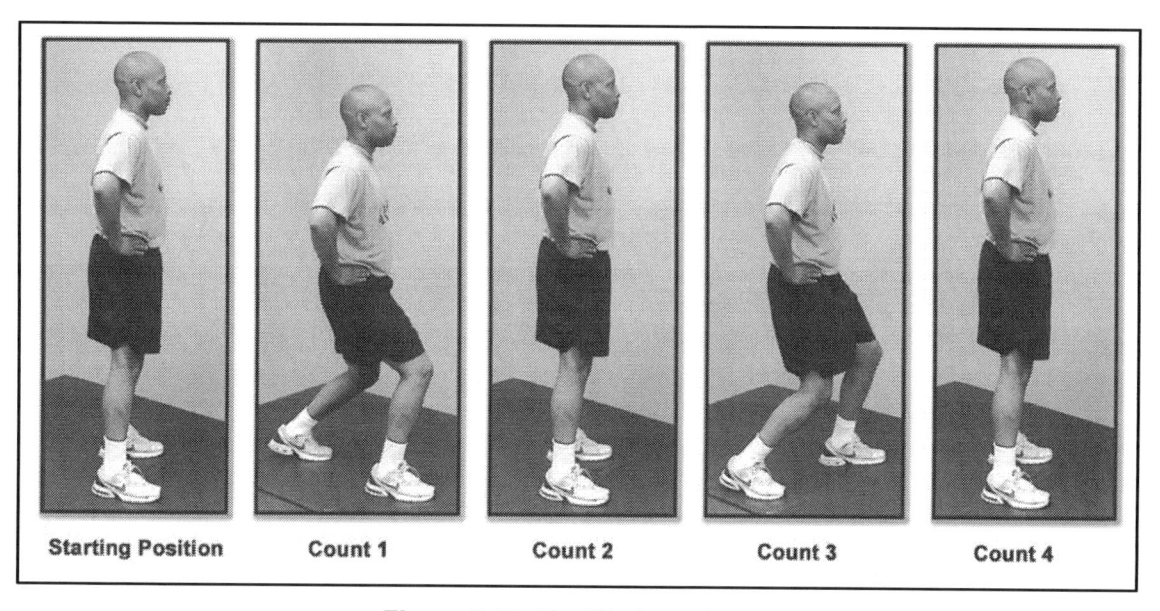

Figure 6-59. Modified rear lunge

PREPARATION DRILL

EXERCISE 3: HIGH JUMPER

Purpose: This exercise reinforces correct jumping and landing, stimulates balance and coordination, and develops explosive strength (Figure 6-60).

Starting Position: Forward leaning stance.

Cadence: MODERATE

Count:

1. Swing arms forward and jump a few inches.
2. Swing arms backward and jump a few inches.
3. Swing arms forward vigorously overhead while jumping forcefully.
4. Repeat count 2. On the last repetition, return to the starting position.

Figure 6-60. High jumper

Check Points:

- At the starting position, the shoulders, the knees, and the balls of the feet should form a straight vertical line.
- On count 1, the arms are parallel to the ground.
- On count 3, the arms should be extended fully overhead. The trunk and legs should also be aligned.
- On each count the Soldier is jumping. On counts 1, 2, and 4 the jumps are 4-6 inches off the ground. On count 3, the Soldier jumps higher (6-10 inches) while maintaining the posture pictured in Figure 6-60.
- On each landing, the feet should be directed forward and maintained at shoulder distance apart. The landing should be "soft" and proceed from the balls of the feet to the heels. The vertical line from the shoulders through the knees to the balls of the feet should be demonstrated on each landing.

Precautions: N/A

Chapter 6

PREPARATION DRILL

MODIFIED EXERCISE 3: MODIFIED HIGH JUMPER

6-117. The instructor may modify the high jumper by decreasing the range of motion and limiting the use of the arms. The Soldier may use the modifications shown in Figure 6-61 to exercise within physical profile limitations. The Soldier gradually increases the range of motion and works toward the standard execution of the exercise, then progresses performance to standard.

Figure 6-61. Modified high jumper (remaining on the ground)

Special Conditioning Programs

PREPARATION DRILL

EXERCISE 4: ROWER

Purpose: This exercise improves the ability to move in and out of the supine position to a seated posture. It coordinates the action of the trunk and extremities while challenging the abdominal muscles (Figure 6-62).

Starting Position: Supine position, arms overhead and feet together, and pointing upward. The chin is tucked and the head is 1-2 inches above the ground. Arms are shoulder-width, palms facing inward, with fingers and thumbs extended and joined.

Cadence: SLOW

Count:

1. Sit up while swinging arms forward and bending at the hip and knees. At the end of the motion, the arms will be parallel to ground, palms facing inward.
2. Return to the starting position.
3. Repeat count 1.
4. Return to the starting position.

Figure 6-62. Rower

Check Points:

- At the starting position, the low back must not be arched excessively off the ground. To prevent this, tighten the abdominal muscles to tilt the pelvis and low back toward the ground.
- At the end of counts 1 and 3, the feet are flat and pulled near the buttocks. The legs stay together throughout the exercise and the arms are parallel to the ground.

Precautions: This exercise is always performed at a slow cadence. Do not arch the back to assume counts 1 and 3.

PREPARATION DRILL

MODIFIED EXERCISE 4: MODIFIED ROWER

6-118. The instructor may modify the rower by decreasing the range of motion or limiting the use of the arms. The Soldier may use the modifications shown in Figure 6-63 and Figure 6-64 to exercise within physical profile limitations. The Soldier gradually increases the range of motion and works toward the standard execution of the exercise, then progresses performance to standard.

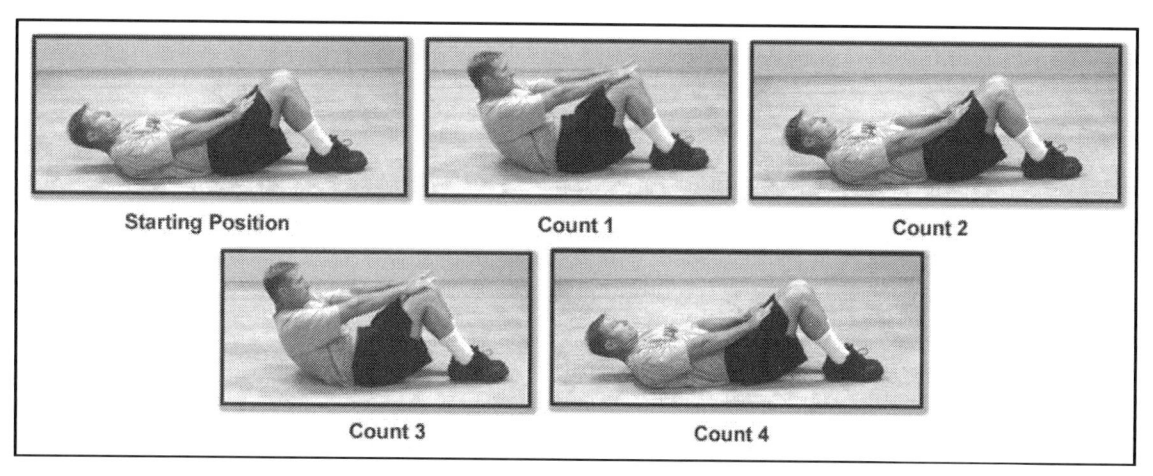

Figure 6-63. Modified rower (limited range of movement)

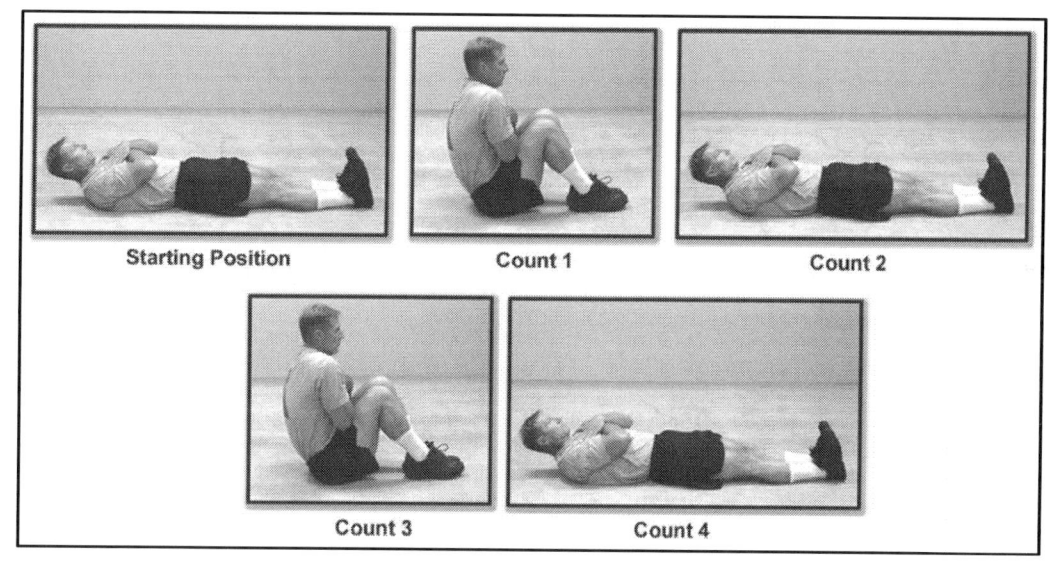

Figure 6-64. Modified rower (without use of arms)

PREPARATION DRILL

EXERCISE 5: SQUAT BENDER

Purpose: This exercise develops strength, endurance, and mobility of the lower back and lower extremities (Figure 6-65).

Starting Position: Straddle stance with hands on hips.

Cadence: SLOW

Count:

1. Squat while leaning slightly forward at the waist with the head up and extend the arms to the front, with arms parallel to the ground and palms facing inward.
2. Return to the starting position.
3. Bend forward and reach toward the ground with both arms extended and palms inward.
4. Return to the starting position.

Figure 6-65. Squat bender

Check Points:

- At the end of count 1, the shoulders, knees and the balls of the feet should be aligned. The heels remain on the ground and the back is straight.
- On count 3, bend forward, keeping the head aligned with the spine and the knees slightly bent. Attempt to keep the back flat and parallel to the ground.

Precautions: This exercise is always performed at a slow cadence. Allowing the knees to go beyond the toes on count 1 increases stress to the knees.

PREPARATION DRILL

MODIFIED EXERCISE 5: MODIFIED SQUAT BENDER

6-119. The instructor may modify the squat bender by decreasing the range of motion and limiting the use of the arms. The Soldier may use the modifications shown in Figure 6-66 to exercise within physical profile limitations. The Soldier gradually increases the range of motion and works toward the standard execution of the exercise, then progresses performance to standard.

Figure 6-66. Modified squat bender

PREPARATION DRILL

EXERCISE 6: WINDMILL

Purpose: This exercise develops the ability to safely bend and rotate the trunk. It conditions the muscles of the trunk, legs, and shoulders (Figure 6-67).

Starting Position: Straddle stance with arms sideward, palms facing down, fingers and thumbs extended and joined.

Cadence: SLOW

Count:

1. Bend the hips and knees while rotating to the left. Reach down and touch the outside of the left foot with the right hand and look toward the rear. The left arm is pulled rearward to maintain a straight line with the right arm.
2. Return to the starting position.
3. Repeat count 1 to the right.
4. Return to the starting position.

Figure 6-67. Windmill

Check Points:
- From the starting position, feet are straight ahead, arms are parallel to the ground, hips set, and abdominals are tight.
- On counts 1 and 3, ensure that the knees bend during the rotation. Head and eyes are directed to the rear on counts 1 and 3.

Precautions: This exercise is always performed at a slow cadence.

Special Conditioning Programs

PREPARATION DRILL

MODIFIED EXERCISE 6: MODIFIED WINDMILL

6-120. The instructor may modify the windmill by decreasing the range of motion and limiting the use of the arms. The modifications to the windmill shown in Figures 6-68, 6-69, and 6-70 may be used to exercise within physical profile limitations and gradually progress performance to standard.

Figure 6-68. Modified windmill (body twist)

Figure 6-69. Modified windmill (hands on hips)

Figure 6-70. Modified windmill (single arm)

PREPARATION DRILL

EXERCISE 7: FORWARD LUNGE

Purpose: This exercise promotes balance and develops leg strength (Figure 6-71).

Starting Position: Straddle stance with hands on hips.

Cadence: SLOW

Count:

1. Take a step forward with the left leg (the left heel should be 3-6 inches forward of the right foot). Lunge forward, lowering the body and allow the left knee to bend until the thigh is parallel to the ground. Lean slightly forward, keeping the back straight.
2. Return to the starting position.
3. Repeat count 1 with the right leg.
4. Return to the starting position.

Figure 6-71. Forward lunge

Check Points:

- Keep the abdominal muscles tight throughout the motion.
- On counts 1 and 3, step straight forward, keeping the feet directed forward. When viewed from the front, the feet maintain their distance apart both at the starting position and at the end of counts 1 and 3.
- On counts 1 and 3, the rear knee may bend naturally, but does not touch the ground. The heel of the rear foot should be off the ground.

Precautions: This exercise is always performed at a slow cadence. On counts 1 and 3, move into position in a controlled manner. Spring off of the forward leg to return to the starting position. This avoids jerking the trunk to create momentum.

PREPARATION DRILL

MODIFIED EXERCISE 7: MODIFIED FORWARD LUNGE

6-121. The instructor may modify the forward lunge by decreasing the range of motion. As with all lunges, this one may restrict knee bend. The Soldier may keep the feet closer together than with the forward lunge. The Soldier concentrates on trying to gradually lower the body in the lunge position (Figure 6-72). Over time, the Soldier gradually increases his range of motion and works toward standard execution of the exercise.

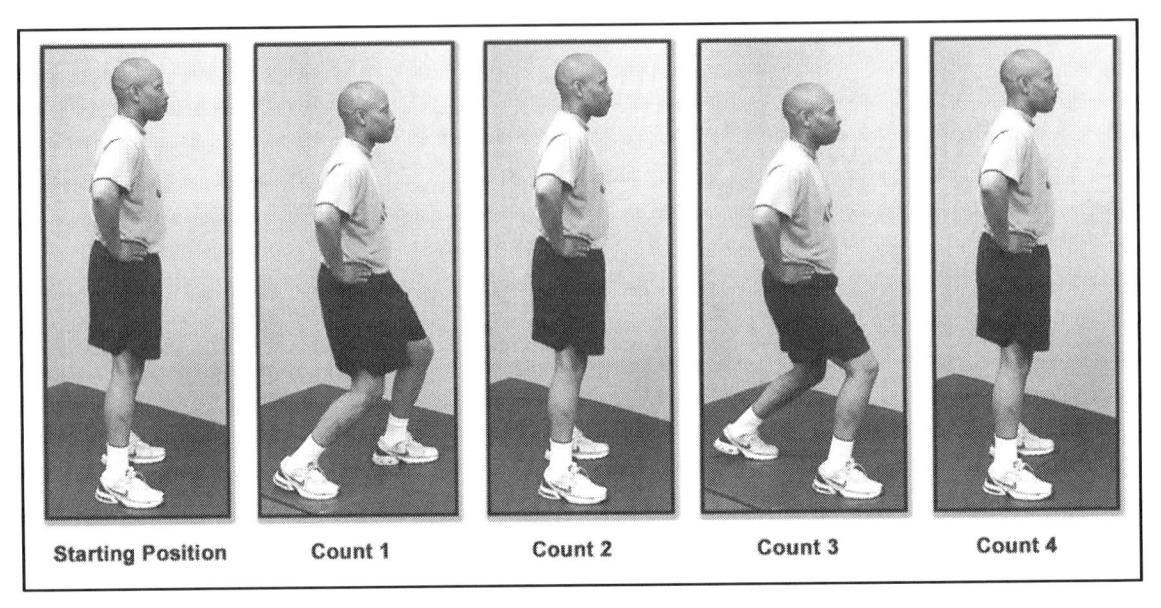

Figure 6-72. Modified forward lunge

PREPARATION DRILL

EXERCISE 8: PRONE ROW

Purpose: This exercise develops strength of the back and shoulders (Figure 6-73).

Starting Position: Prone position with the arms overhead, palms facing downward 1-2 inches off the ground, and toes pointed to the rear.

Cadence: SLOW

Count:

1. Raise the head and chest slightly while lifting the arms and pulling them rearward. Hands make fists as they move toward the shoulders.
2. Return to the starting position.
3. Repeat count 1.
4. Return to the starting position.

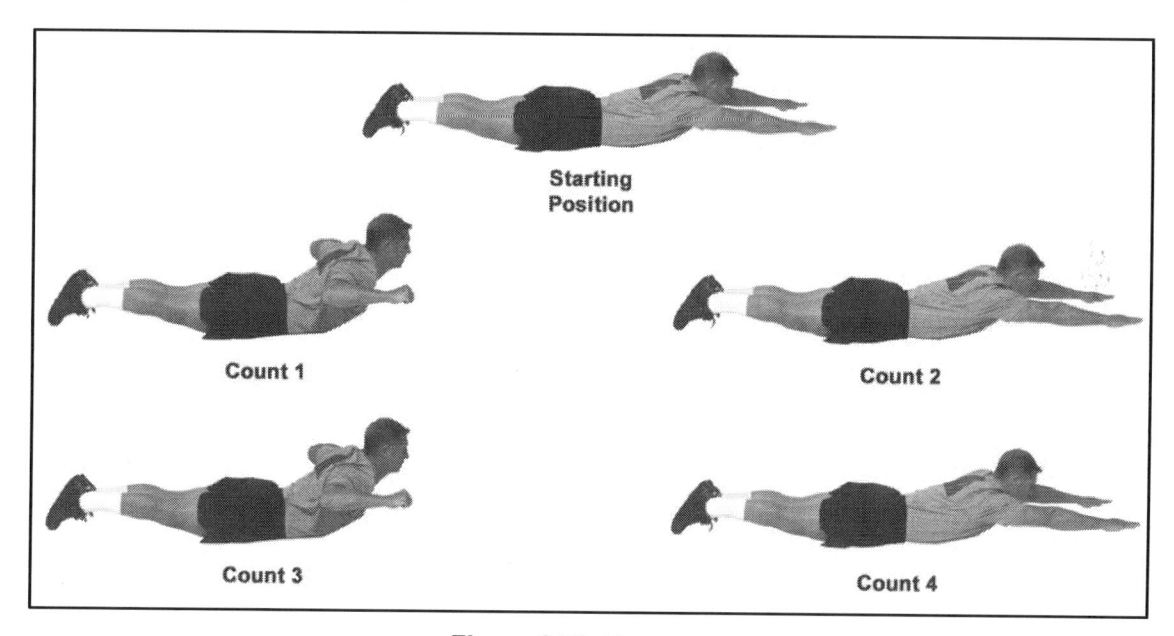

Figure 6-73. Prone row

Check Points:

- At the starting position, the abdominal muscles are tight and the head is aligned with the spine.
- On counts 1 and 3, the forearms are parallel to the ground and slightly higher than the trunk.
- On counts 1 and 3, the head is raised to look forward but not skyward.
- Throughout the exercise, the legs and toes remain in contact with the ground.

Precautions: This exercise is always performed at a slow cadence. Prevent overarching of the back by maintaining contractions of the abdominal and buttocks muscles throughout the exercise.

PREPARATION DRILL

MODIFIED EXERCISE 8: MODIFIED PRONE ROW

6-122. The instructor may modify the prone row by decreasing the range of motion and limiting the use of the arms. The Soldier assumes the starting position using his hands to assist in lowering the body, and then steps back into the six-point stance before lowering the body to the ground. He uses the modifications shown in Figures 6-74 and 6-75 to exercise within physical profile limitations. The Soldier works toward standard execution of the exercise.

Figure 6-74. Modified prone row (assuming starting position)

Figure 6-75. Modified prone row (using the arms)

PREPARATION DRILL

EXERCISE 9: BENT-LEG BODY TWIST

Purpose: This exercise strengthens trunk muscles and promotes control of trunk rotation (Figure 6-76).

Starting Position: Supine position with the hips and knees bent to 90 degrees, arms sideward, palms down with fingers spread. Knees and feet are together, and head is raised two or three inches off the ground with the chin slightly tucked.

Cadence: SLOW

Count:

1. Rotate the legs to the left while keeping the upper back and arms in place.
2. Return to the starting position.
3. Repeat count 1 to the right.
4. Return to the starting position.

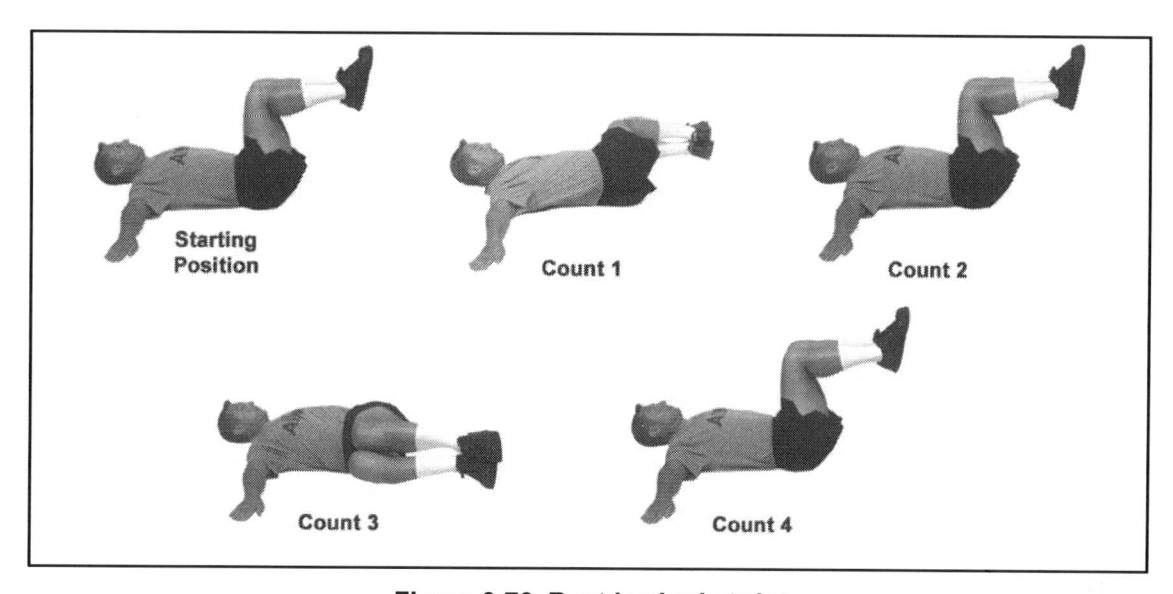

Figure 6-76. Bent-leg body twist

Check Points:

- Tighten the abdominal muscles in the starting position and maintain this contraction throughout the exercise.
- The head should be off the ground with the chin slightly tucked.
- Ensure that the hips and knees maintain 90-degree angles.
- Keep the feet and knees together throughout the exercise.
- Attempt to rotate the legs to about 8-10 inches off the ground. The opposite shoulder must remain in contact with the ground.

Precautions: This exercise is always performed at a slow cadence. Do not rotate the legs to a point beyond which the arms and shoulders can no longer maintain contact with the ground.

Special Conditioning Programs

PREPARATION DRILL

MODIFIED EXERCISE 9: MODIFIED BENT-LEG BODY TWIST

6-123. The starting position for this exercise is the supine position with the arms sideward or at 45 degrees to the body (according to profile limitations). Palms should face downward and knees bent at 90 degrees, with the feet flat on the floor. The head may be on the ground or elevated 1-2 inches depending on profile limitations. The Soldier assumes the starting position as in the bent-leg body twist, leaving the feet flat on the ground. (Figures 6-77 and 6-78).

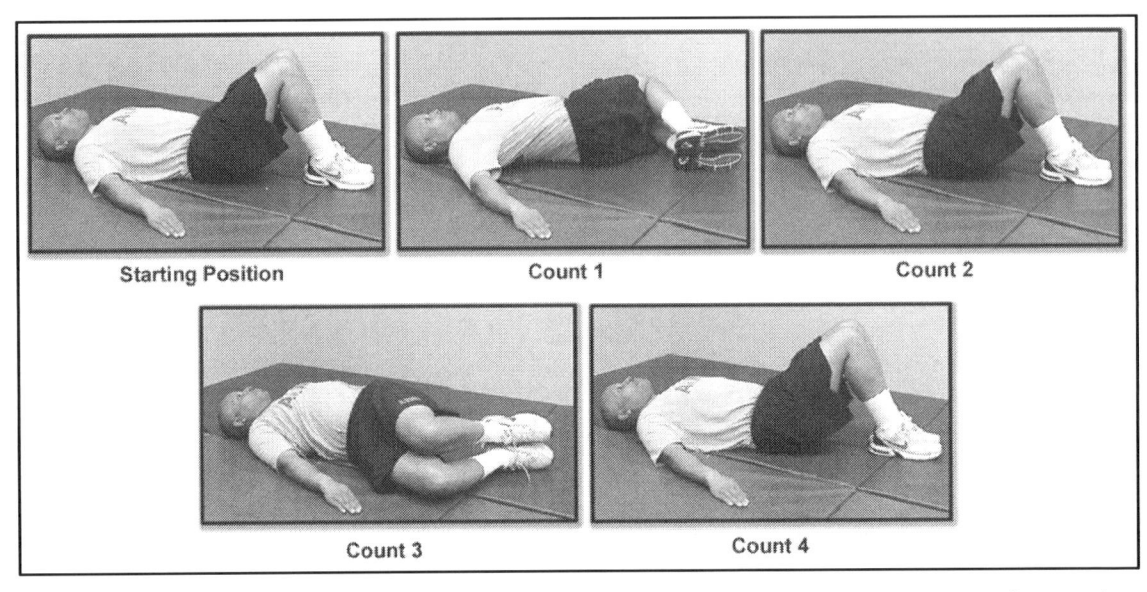

Figure 6-77. Modified bent-leg body twist (head on the ground and arms at 45 degrees)

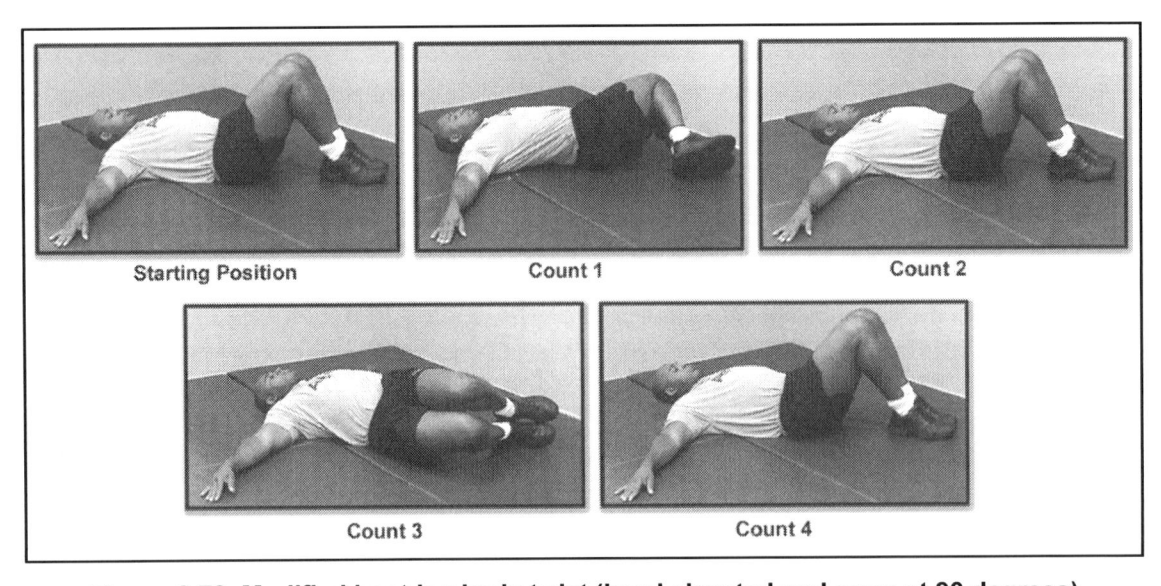

Figure 6-78. Modified bent-leg body twist (head elevated and arms at 90 degrees)

PREPARATION DRILL

EXERCISE 10: PUSH-UP

Purpose: This exercise strengthens the muscles of the chest, shoulders, arms, and trunk (Figure 6-79).

Starting Position: Front leaning rest position.

Cadence: MODERATE

Count:

1. Bend the elbows, lowering the body until the upper arms are parallel with the ground.
2. Return to the starting position.
3. Repeat count 1.
4. Return to the starting position.

Figure 6-79. Push-up

Check Points:

- The hands are directly below the shoulders with fingers spread (middle fingers point straight ahead).
- On counts 1 and 3 the upper arms stay close to the trunk, elbows pointing rearward.
- On counts 2 and 4 the elbows straighten, but do not lock.
- The trunk should not sag. To prevent this, tighten the abdominal muscles while in the starting position and maintain this contraction throughout the exercise.

Precautions: N/A

Variation: Soldiers should assume the six-point stance on their knees when unable to perform repetitions correctly to cadence (Figure 6-80).

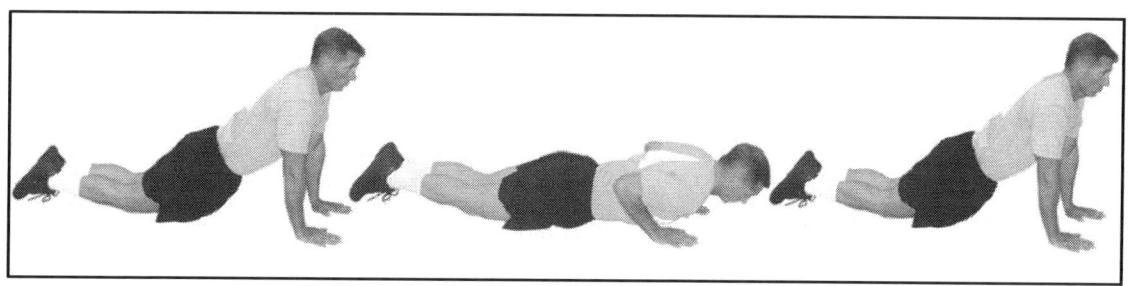

Figure 6-80. Push-up in the 6-point stance

Special Conditioning Programs

PREPARATION DRILL

MODIFIED EXERCISE 10: MODIFIED PUSH-UP

6-124. The Soldier performs the modified push-up in the six-point stance. The Soldier assumes the starting position, using his hands to assist in lowering his body, and then steps back into the six-point stance. Range of movement may be limited throughout the exercise. Over time, the Soldier gradually increases the range of motion and works toward the standard execution of the push-up (Figures 6-81 and 6-82).

Figure 6-81. Modified push-up variation for assuming the 6-point stance

Figure 6-82. Modified push-up

CONDITIONING DRILL 1

EXERCISE 1: POWER JUMP

Purpose: This exercise reinforces correct jumping and landing, stimulates balance and coordination, and develops explosive strength (Figure 6-83).

Starting Position: Straddle stance with hands on hips.

Cadence: MODERATE

Count:

1. Squat with the heels flat as the spine rounds forward to allow the straight arms to reach to the ground, touching with the palms of the hands.
2. Jump forcefully in the air, vigorously raising arms overhead with palms facing inward.
3. Control the landing and repeat count 1.
4. Return to the starting position.

Figure 6-83. Power jump

Check Points:

- At the starting position, tighten the abdominals to stabilize the trunk.
- On counts 1 and 3, keep the back generally straight with the head up and the eyes forward.
- On count 2 the arms should be extended fully overhead. The trunk and legs should also be aligned.
- On each landing, the feet should be directed forward and maintained at shoulder distance apart. The landing should be "soft" and proceed from the balls of the feet to the heels. The vertical line from the shoulders through the knees to the balls of the feet should be demonstrated on each landing.

Precaution: N/A

CONDITIONING DRILL 1

MODIFIED EXERCISE 1: MODIFIED POWER JUMP

6-125. The instructor may modify the power jump by decreasing the range of motion or limiting the use of the arms. The Soldier may use the modifications shown in Figure 6-84 to exercise within physical profile limitations. The Soldier works toward standard execution of the exercise.

Figure 6-84. Modified power jump

CONDITIONING DRILL 1

EXERCISE 2: V-UP

Purpose: This exercise develops the abdominal and hip flexor muscles while enhancing balance (Figure 6-85).

Starting Position: Supine, arms on ground 45-degrees to the side, palms down with fingers spread. The chin is tucked and the head is 1-2 inches off the ground.

Cadence: MODERATE

Count:

1. Raise straight legs and trunk to form a V-position, using arms as needed.
2. Return to the starting position.
3. Repeat count 1.
4. Return to the starting position.

Figure 6-85. V-up

Check Points:

- At the starting position, tighten the abdominal muscles to tilt the pelvis and the lower back toward the ground.
- On counts 1 and 3, the knees and trunk are straight with the head aligned with the trunk.
- On counts 2 and 4, lower the legs to the ground in a controlled manner so as not to injure the feet.

Precautions: To protect the spine, do not jerk the legs and trunk to rise to the V-position.

CONDITIONING DRILL 1

MODIFIED EXERCISE 2: MODIFIED V-UP

6-126. The starting position for this exercise is the supine position with the arms sideward or at 45 degrees to the body (according to profile limitations). Palms are downward and knees are bent at 90 degrees with the feet flat on the floor. The head may be on the ground or elevated 1-2 inches off the ground according to profile limitations. The Soldier assumes the starting position as in the V-up, using the hands as needed to lower the body to the ground. The head is elevated while the back and feet are flat on the ground. On counts 1 and 3, the Soldier lifts the feet off the ground, pulling the knees toward the chest. Then the Soldier lowers the feet to the ground, returning to the starting position on counts 2 and 4 (refer to Figure 6-86). Over time, the Soldier gradually increases the range of motion and works to perform the V-up to standard.

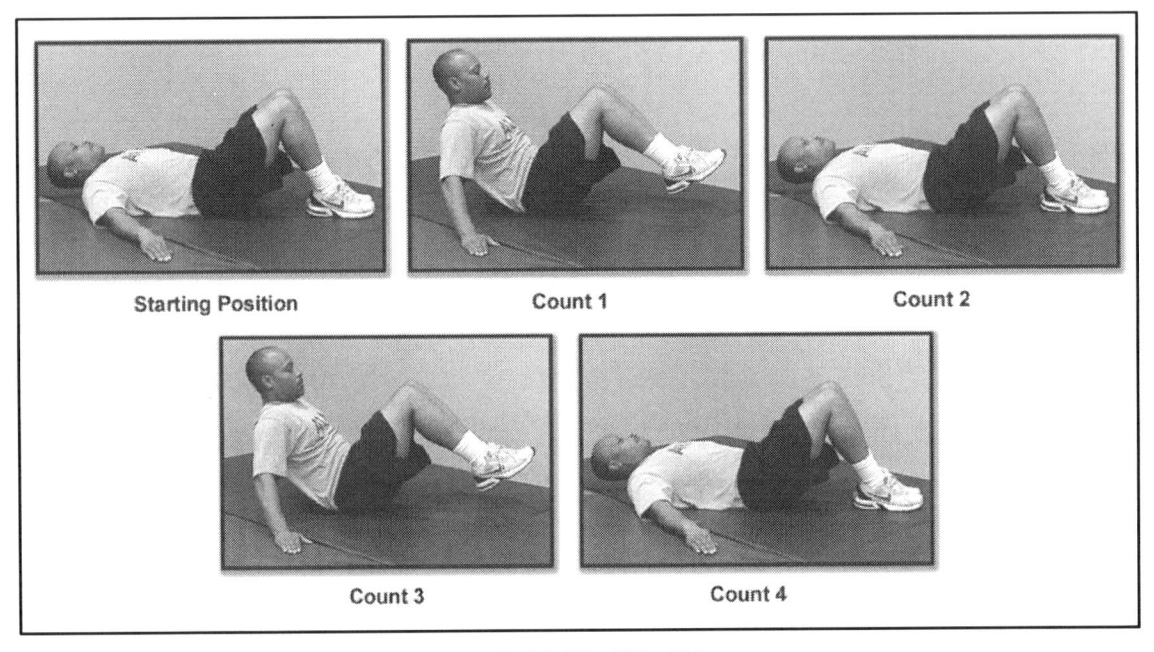

Figure 6-86. Modified V-up

CONDITIONING DRILL 1

EXERCISE 3: MOUNTAIN CLIMBER

Purpose: This exercise develops the ability to quickly move the legs to power out of the front leaning rest position (Figure 6-87).

Starting Position: Front leaning rest position with the left foot below the chest and between the arms.

Cadence: MODERATE

Count:

1. Push upward with the feet and quickly change positions of the legs.
2. Return to the starting position.
3. Repeat the movements in count 1.
4. Return to the starting position.

Figure 6-87. Mountain climber

Check Points:

- Place the hands directly below the shoulders, fingers spread (middle fingers point straight ahead) with the elbows straight, not locked.
- To prevent the trunk from sagging, tighten the abdominal muscles and maintain this contraction throughout the exercise. Do not raise the hips when moving throughout the exercise.
- Align the head with the spine and keep the eyes directed to a point about two feet in front of the body.
- Throughout the exercise, stay on the balls of the feet.
- Move the legs straight forward and backward, not at angles.

Precautions: N/A

CONDITIONING DRILL 1

MODIFIED EXERCISE 3: MODIFIED MOUNTAIN CLIMBER

6-127. The instructor may modify the mountain climber by decreasing the range of motion. The Soldier assumes the starting position, stepping back as in the modified push-up. The Soldier may use the modifications shown in Figure 6-88 to exercise within physical profile limitations. The Soldier gradually increases the range of motion and works toward the standard execution of the exercise, then progresses performance to standard.

Figure 6-88. Modified mountain climber

CONDITIONING DRILL 1

EXERCISE 4: LEG-TUCK AND TWIST

Purpose: This exercise develops trunk strength and mobility while enhancing balance (Figure 6-89).

Starting Position: Seated with trunk straight but leaning backward 45 degrees, arms straight and hands on ground 45 degrees to the rear, palms down. Legs are straight, extended to the front and 8-12 inches off the ground.

Cadence: MODERATE

Count:

1. Raise the legs while rotating on to the left buttock and draw the knees toward the left shoulder.
2. Return to the starting position.
3. Repeat count 1 in the opposite direction.
4. Return to the starting position.

Figure 6-89. Leg-tuck and twist

Check Points:

- At the starting position, tighten the abdominals to stabilize the trunk.
- On all counts, keep the feet and knees together.
- On counts 1 and 3, keep the head and trunk still as the legs move.
- On counts 1 and 3, tuck (bend) the legs and align them diagonal to the trunk.

Precautions: To protect the back on counts 1 and 3, do not jerk the legs and trunk to achieve the end position.

CONDITIONING DRILL 1

MODIFIED EXERCISE 4: MODIFIED LEG-TUCK AND TWIST

Starting Position: The starting position for this exercise is the seated position with the arms sideward or at 45 degrees to the body (according to profile limitations). Place the palms down and bend the knees 90 degrees. Keep the feet flat on the floor. Assume the starting position as in the leg-tuck and twist, but with the feet flat on the ground.

Count: On counts 1 and 3, lift the feet off the ground and rotate to the left or right side, pulling the knees toward the chest. Lower the feet to the ground, returning to the starting position on counts 2 and 4 (Figure 6-90). Over time, gradually increase the range of motion and work toward the standard execution of the leg-tuck and twist.

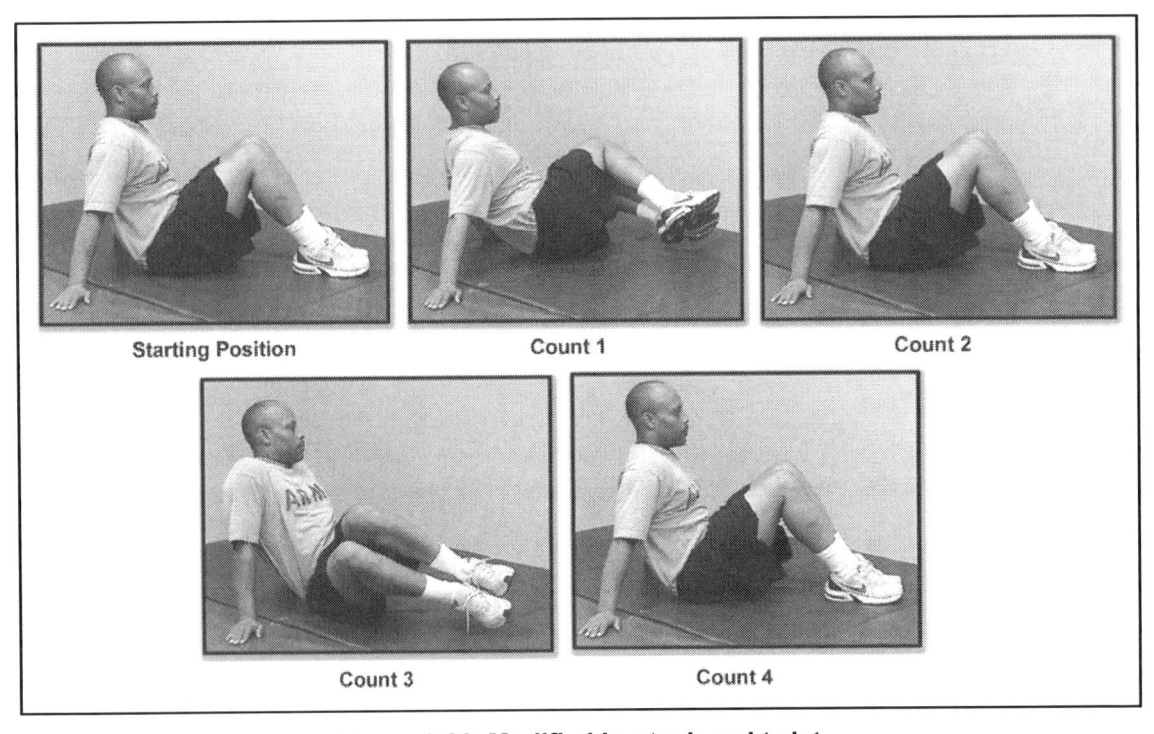

Figure 6-90. Modified leg-tuck and twist

CONDITIONING DRILL 1

EXERCISE 5: SINGLE-LEG PUSH-UP

Purpose: This exercise strengthens muscles of the chest, shoulders, arms, and trunk. Raising one leg while maintaining proper trunk position makes this an excellent trunk stabilizing exercise (Figure 6-91).

Starting Position: Front leaning rest position.

Cadence: MODERATE

Count:

1. Bend the elbows, lowering the body until the upper arms are parallel with the ground while raising the left leg 8 to 10 inches off the ground.
2. Return to the starting position.
3. Repeat count 1, bringing the right leg to 8 to 10 inches off the ground.
4. Return to the starting position.

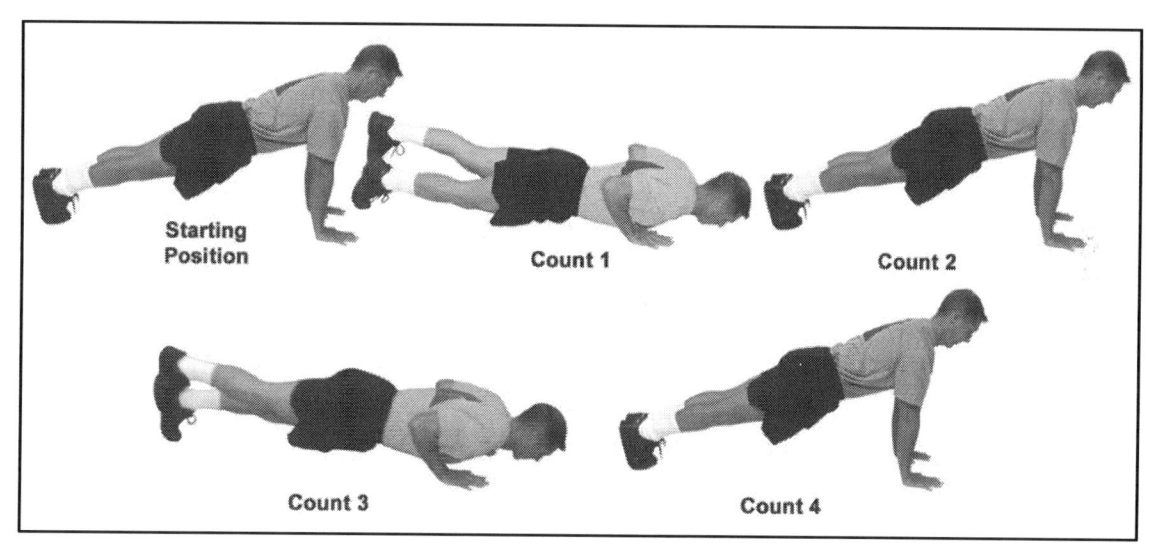

Figure 6-91. Single-leg push-up

Check Points:

- Perform a squat thrust to move into the front leaning rest, and keep the body straight from head to heels. Support the body weight on the hands and the balls of the feet.
- Extend the fingers and spread them so the middle fingers point straight ahead and are directly aligned with the shoulders.
- On counts 1 and 3, the upper arms stay close to the trunk.
- On counts 2 and 4, straighten, but do not lock, the elbows.
- On counts 1 and 3, keep the raised leg straight and aligned with the trunk.
- The trunk must not sag. To prevent this, tighten the abdominal muscles while in the starting position and maintain this contraction throughout the exercise.

Precautions: Do not jerk the leg to be raised past straight alignment with the trunk, as this may place undue stress on the back.

CONDITIONING DRILL 1

MODIFIED EXERCISE 5: MODIFIED SINGLE-LEG PUSH-UP

6-128. The single-leg push-up is modified by performing the modified push-up in the six-point stance. The Soldier assumes the starting position using the hands to assist in lowering the body, and then stepping back into the six-point stance (Figure 6-92). Range of movement may be limited throughout the exercise. The Soldier gradually increases the range of motion (Figure 6-93) and works toward the standard execution of the exercise, then progresses performance to standard.

Figure 6-92. Variation for assuming the 6-point stance

Figure 6-93. Modified single-leg push-up

MILITARY MOVEMENT DRILL 1 EXERCISE MODIFICATIONS

6-129. During level II resume MMD 1 by reducing the distance from 25 to 15 yards and ensure the Soldier limits the speed and intensity of movement.
- For verticals, start with minimal air time and gradually progress to more powerful movements.
- For laterals this means decreasing the crouch and stepping movements instead of maintaining the normal tempo.
- The shuttle sprint is often restricted by profile. When conducting the shuttle sprint, ensure that the Soldier is able to negotiate the turns at walking speed before allowing him to run. In the post-profile period, resume the shuttle sprint without touching the hand to the ground on turns, and then gradually work toward bending enough to touch the ground.

RECOVERY DRILL EXERCISE MODIFICATIONS

6-130. The five exercises that comprise recovery include a wide range of movements that require structural strength, stability, flexibility, and mobility while using standing, seated, prone, and supine postures supported by one or both upper or lower limbs. Allow Soldiers to use their hands as needed to move into and out of starting and exercise positions on the ground. In the post-profile period, range of motion for some exercises may still be limited. Each of the five exercises may be modified to accommodate various physical limitations and gradually progress each exercise to standard.

RECOVERY DRILL

EXERCISE 1: OVERHEAD ARM PULL

Purpose: This exercise develops flexibility of the arms, shoulders, and trunk muscles (Figure 6-94).

Starting Position: Straddle stance with hands on hips.

Position 1: On the command, "Ready, STRETCH," raise the left arm overhead and place the left hand behind the head. Grasp above the left elbow with the right hand and pull to the right, leaning the body to the right. Hold this position for 20-30 seconds.

Starting Position: On the command, "Starting Position, MOVE," assume the starting position.

Position 2: On the command, "Change Position, Ready, STRETCH," raise the right arm overhead and place the right hand behind the head. Grasp above the right elbow with the left hand and pull to the left, leaning the body to the left. Hold this position for 20-30 seconds. On the command, "Starting Position, MOVE," return to the starting position.

Figure 6-94. Overhead arm pull

Check Points:
- Throughout the exercise, keep the hips set and the abdominals tight.
- In positions 1 and 2, lean the body straight to the side, not to the front or back.

Precautions: N/A

RECOVERY DRILL

MODIFIED EXERCISE 1: MODIFIED OVERHEAD ARM PULL

6-131. The instructor may modify this exercise by decreasing the range of motion. The Soldier reaches overhead and then grasps the wrist with the opposite hand instead of the elbow (Figure 6-95). Another modification is to pull the arm across the front of the chest.

Figure 6-95. Modified overhead arm pull and front arm pull

RECOVERY DRILL

EXERCISE 2: REAR LUNGE

Purpose: This exercise develops mobility of the hip flexors and trunk muscles (Figure 6-96).

Starting Position: Straddle stance, hands on hips.

Position 1: On the command, "Ready, STRETCH," take an exaggerated step backward with the left leg, touching down with the ball of the foot. This is the same position as count 1 of the rear lunge in CD 1. Hold this position for 20-30 seconds.

Starting Position: On the command, "Starting Position, MOVE," assume the starting position.

Position 2: On the command, "Change Position, Ready, STRETCH," take an exaggerated step backward with the right leg, touching down with the ball of the foot. This is the same position as count 3 of the rear lunge in CD 1. Hold this position for 20-30 seconds. On the command, "Starting Position, MOVE," return to the starting position.

Figure 6-96. Rear lunge

Check Points:
- Maintain straightness of the back by keeping the abdominal muscles tight throughout the motion.
- After the foot touches down on positions 1 and 2, allow the body to continue to lower.
- Lunge and step in a straight line, keeping the feet directed forward. Viewed from the front, the feet are shoulder width apart, both at the starting position and at the end of positions 1 and 2.
- Keep the forward knee over the ball of the foot on positions 1 and 2.
- Ensure the heal of the rear foot does not touch the ground.

Precautions: When lunging to the left or right, do not let the knee move forward of the toes.

RECOVERY DRILL

MODIFIED EXERCISE 2: MODIFIED REAR LUNGE

6-132. The instructor can modify the rear lunge by decreasing the range of motion (Figure 6-97). As with all lunges, this one might restrict how far the knee can bend. The Soldier may place his feet closer together than with the rear lunge. The Soldier gradually lowers the body into the lunge position. Over time, the Soldier gradually increases the range of motion and works toward the standard execution of each exercise.

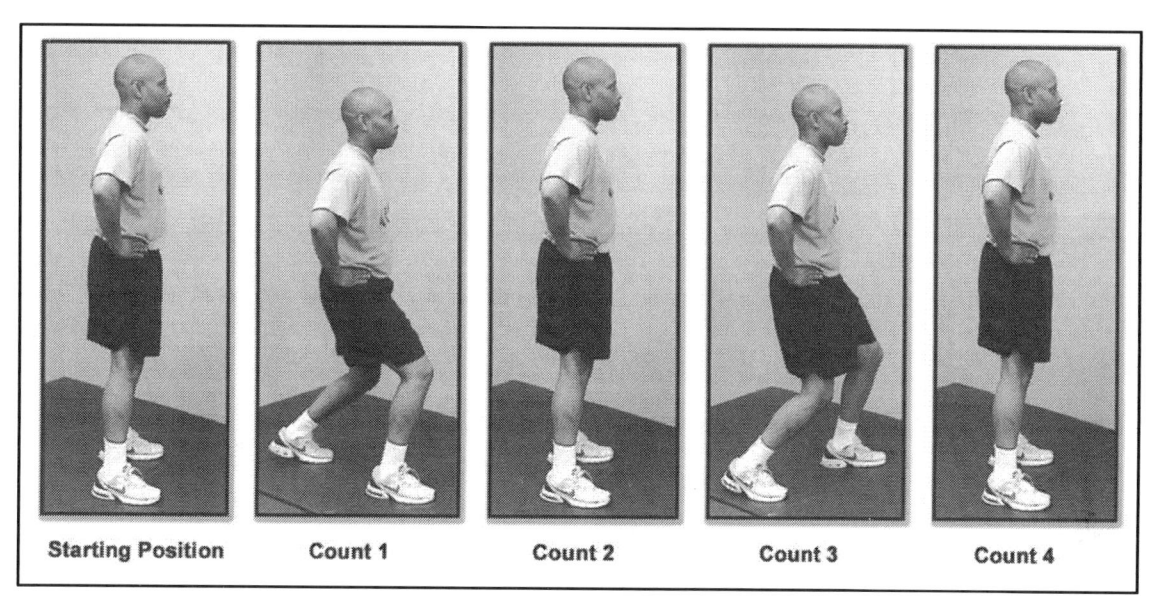

Figure 6-97. Modified rear lunge

RECOVERY DRILL

EXERCISE 3: EXTEND AND FLEX

Purpose: This exercise develops mobility of the hip flexors, abdominals, hip (position 1-extend, Figure 6-98) and the low back, hamstrings, and calves (position 2-flex, Figure 6-98).

Starting Position: The front leaning rest position.

Position 1: On the command, "Ready, STRETCH," lower the body, sagging in the middle, keeping the arms straight and looking upward. Hold this position for 20-30 seconds.

Starting Position: On the command, "Starting Position, MOVE," assume the starting position.

Position 2: On the command, "Change Position, READY, STRETCH," slightly bend the knees and raise the hips upward. Straighten the legs and attempt to touch the ground with the heels. Move the head in line with the arms, forming an A with the body. Keep the feet together and hold this position for 20-30 seconds. On the command, "Starting Position, MOVE," return to the starting position.

Figure 6-98. Extend and flex

Check Points:

- In position 1, the thighs and pelvis rest on the ground. Relax the back muscles while bearing the body weight through the straight arms. Toes point to the rear.
- In position 2, the legs are straight and the arms are shoulder width apart, palms down on the ground. Relax the shoulders and push to the rear with the hands, forming an "A" with the body. Try not to round the shoulders.
- Keep the feet together throughout the exercise.

Precaution: N/A

Variation: Soldiers who cannot extend the trunk in position 1 while keeping the arms straight and hips on the ground may assume the modified position 1 shown above.

RECOVERY DRILL

MODIFIED EXERCISE 3: MODIFIED EXTEND AND FLEX

6-133. The instructor may modify this exercise by using a standing (Figure 6-99) or prone position. The Soldier may assume the starting position for the extend and flex using the prone position. To do so, the Soldier steps back into the front leaning rest position (Figure 6-100) instead of performing a squat thrust. In the post-profile period, range of motion for some exercises may still be limited. Soldiers may modify the extend position by raising up their forearms instead of their hands or by laying prone with the arms alongside the body, palms up (Figure 6-101). Over time, the Soldier gradually increases the range of motion and works toward the standard execution of each exercise.

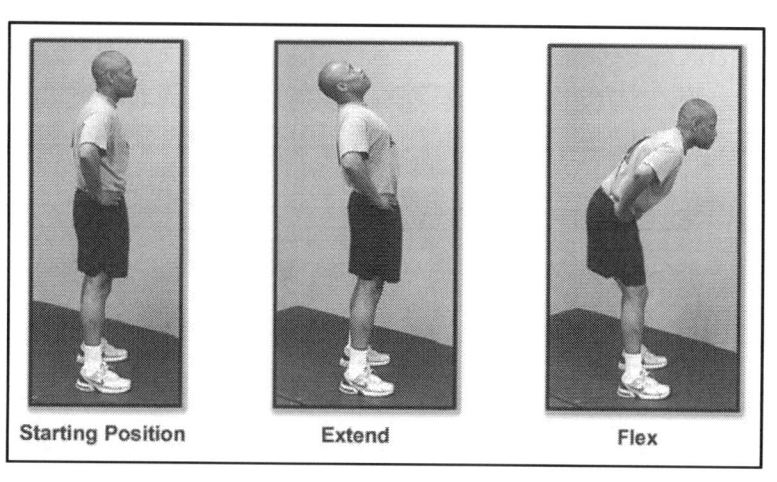

Figure 6-99. Modified extend and flex (standing)

Figure 6-100. Stepping into the modified extend and flex (prone)

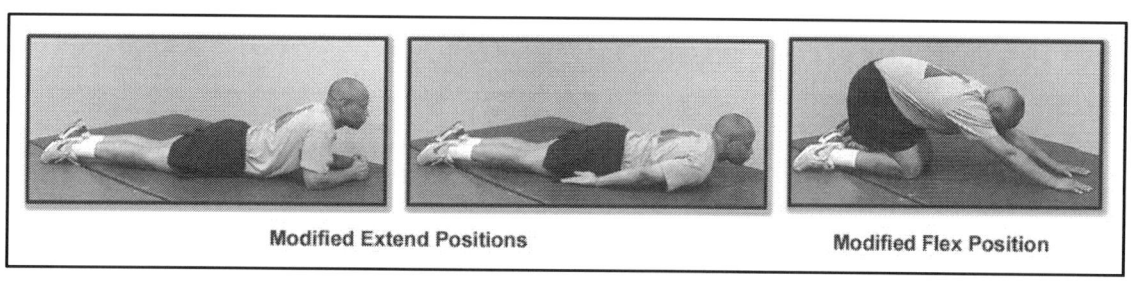

Figure 6-101. Modified extend and flex (prone) starting position

RECOVERY DRILL

EXERCISE 4: THIGH STRETCH

Purpose: This exercise develops flexibility of the front of the thigh and the hip flexor muscles (Figure 6-102).

Starting Position: Seated position, arms at sides and palms on the ground.

Position 1: On the command, "Ready, STRETCH," roll onto the right side and place the right forearm on the ground, perpendicular to the chest. With the right hand, make a fist on the ground with the thumb side up. Grasp the left ankle with the left hand and pull the left heel toward the buttocks and pull the entire leg rearward. Push the left thigh further to the rear with the heel of the right foot. Hold this position for 20-30 seconds.

Starting Position: On the command, "Starting Position, MOVE," assume the starting position.

Position 2: On the command, "Change Position, Ready, STRETCH," lay on the left side and place the left forearm on the ground, perpendicular to the chest. The left hand makes a fist on the ground with the thumb side up. Grasp the right ankle with the right hand and pull the right heel toward the buttocks and pull the entire leg rearward. Push the right thigh further to the rear with the heel of the left foot. Hold this position for 20-30 seconds. On the command, "Starting Position, MOVE," return to the starting position.

Figure 6-102. Thigh stretch

Check Points:
- Keep the abdominal muscles tight throughout this stretch in order to keep the trunk straight.
- Do not pull the heel forcefully to the buttock if there is discomfort in the knee joint.

Precaution: N/A

RECOVERY DRILL

MODIFIED EXERCISE 4: MODIFIED THIGH STRETCH

6-134. The instructor may modify the thigh stretch by decreasing the range of motion. The starting position may be assumed using the hands (Figure 103).

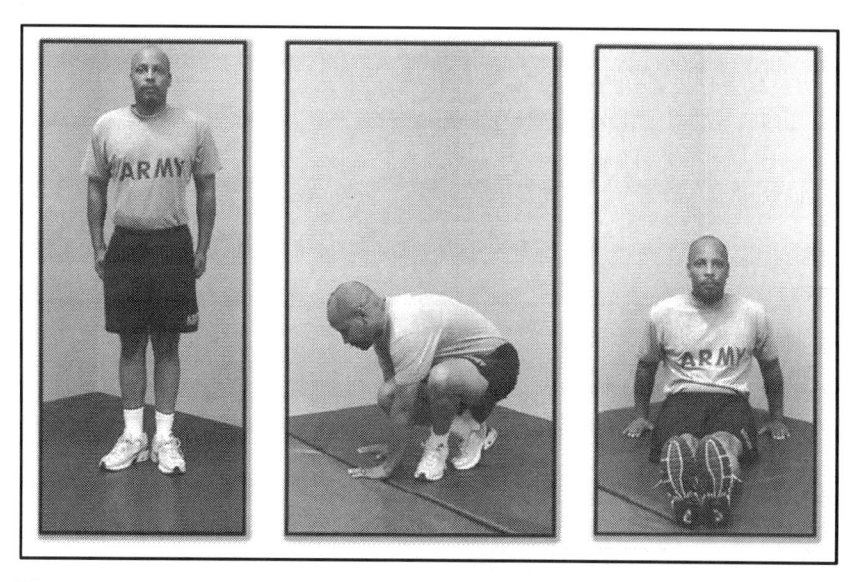

Figure 6-103. Modified thigh stretch (assuming the seated position)

6-135. The knee bend may be restricted so pull the leg slightly toward the front. Over time, the Soldier gradually increases the range of motion and works toward the standard execution of each exercise. The Soldier may also perform this exercise in a kneeling position, assuming the starting position from the modified extend and flex (Figure 6-104).

Figure 6-104. Modified thigh stretch starting positions

RECOVERY DRILL

EXERCISE 5: SINGLE-LEG OVER

Purpose: This exercise develops flexibility of the hips and lower back muscles (Figure 6-105).

Starting Position: Supine position with arms sideward, palms down, feet together, and head on the ground.

Position 1: On the command, "Ready, STRETCH," turn the body to the right, bend the left knee to 90 degrees over the right leg, grasp the outside of the left knee with the right hand and pull toward the right. Hold this position for 20-30 seconds.

Starting Position: On the command, "Starting Position, MOVE," assume the starting position.

Position 2: On the command, "Change Position, Ready, STRETCH," turn the body to the left, bend the right knee to 90 degrees over the left leg, grasp the outside of the right knee with the left hand, and pull toward the left. Hold this position for 20-30 seconds. On the command, "Starting Position, MOVE," return to the starting position.

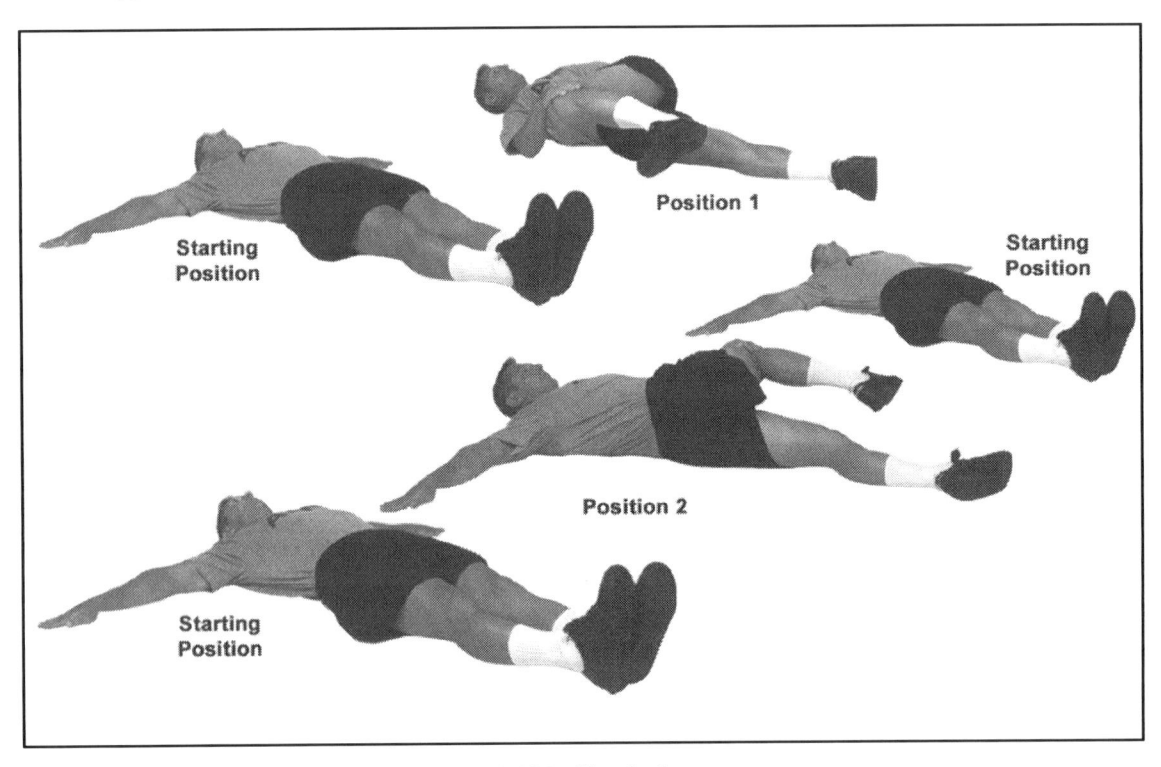

Figure 6-105. Single-leg over

Check Points:
- At the starting position, the arms are directed to the sides at 90 degrees to the trunk; the fingers and thumbs are extended and joined.
- In position 1, keep the left shoulder, arm, and hand on the ground.
- In position 2, keep the right shoulder, arm, and hand on the ground.
- Keep the head on the ground throughout the exercise.

Precaution: N/A

RECOVERY DRILL

MODIFIED EXERCISE 5: MODIFIED SINGLE-LEG OVER

6-136. The starting position for this exercise is supine (Figure 6-106). The Soldier places the arms sideward at 45 degrees to the body, palms downward. The Soldier bends the knees to 90 degrees with the feet flat on the ground. The Soldier rotates the hips and lowers the knees toward the ground.

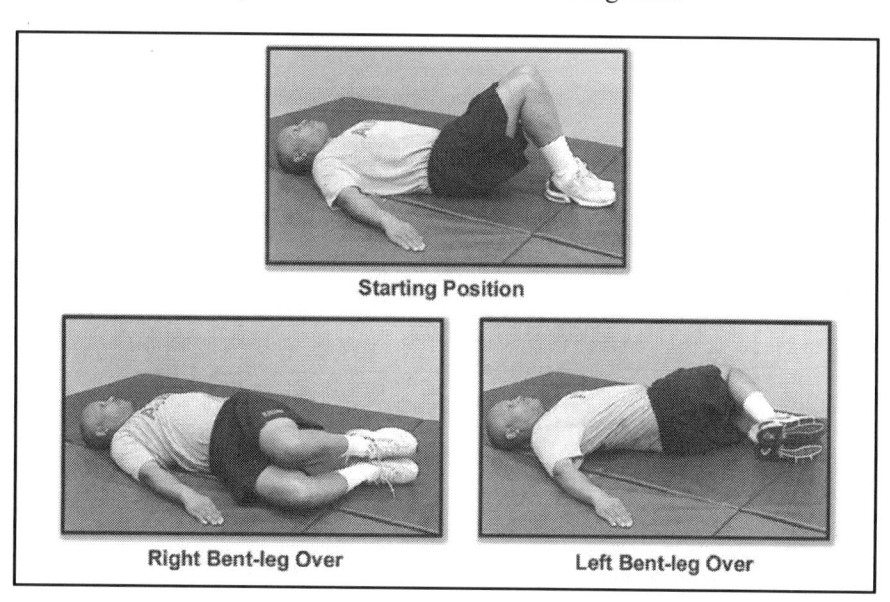

Figure 6-106. Modified single-leg over

6-137. Before being discharged from level II and returning to unit PRT, Soldiers must meet the requirements shown in Table 6-5.

Table 6-5. Reconditioning phase level II exit criteria

PREPARATION	5 REPETITIONS TO STANDARD
MILITARY MOVEMENT DRILL 1	1 REPETITION TO STANDARD
CONDITIONING DRILL 1	5 REPETITIONS TO STANDARD
CLIMBING DRILL 1	5 REPETITIONS TO STANDARD
CONTINUOUS RUNNING	30 MINUTES AT SLOWEST AGR PACE IN THE UNIT
RECOVERY	HOLD EACH STRETCH FOR 20 SECONDS TO STANDARD

Summary

Unit readiness is greatly affected by injuries, illness, and other medical conditions. The Army PRT program is safe and effective. Physical readiness training must challenge Soldiers without breaking them. Some injuries inevitably occur, but units that take measures to control injury risks will have fewer Soldiers on medical profile and more on duty to perform mission requirements. For Soldiers who suffer injuries or are recovering from illness or other medical conditions, effective reconditioning allows them to return to duty at or above their pre-injury level of individual physical readiness. This is what special conditioning programs are all about.

PART THREE

ACTIVITIES

This part discusses the conduct of PRT exercises, drills, and activities.

Chapter 7

Execution of Training

The key to success in PRT execution is skillful leadership with trained AIs who employ command presence, command voice, and organized instruction in the extended rectangular formation. This chapter describes in detail the PRT commands, formations, positions, and counting cadence.

COMMANDS

7-1. This section discusses the importance of proper commands. This cannot be underestimated. Invariably, PRT performance reflects the quality of its commands. Indifferent commands produce indifferent performance. When a command is given distinctly, concisely, with energy, and with proper regard to rhythm, Soldier performance will reflect it. See TC 3-21.5, *Drill and Ceremonies,* for detailed information of command voice, posture, and presence.

TYPES

7-2. The two types of commands used in PRT are preparatory commands and commands of execution. The preparatory command describes and specifies what is required. All preparatory commands are given with rising voice inflection. The command of execution calls into action what has been prescribed. The interval between the two commands should be long enough to permit the Soldier to understand the first one before the second one is given.

COMMAND DELIVERY

7-3. When the PRT leader addresses the formation and is commanding movement or announcing the name of an exercise, he does so from the position of attention. Exceptions are exercises that change position without returning to the position of attention.

7-4. When exercises are performed, Soldiers assume the proper starting position of each exercise on the command "Starting position, MOVE." When conducting exercises, Soldiers are commanded to return to the position of attention from the terminating position of the exercise before they are commanded to assume the starting position for the next exercise. PRT leaders use the command "Position of Attention, MOVE", to bring Soldiers to the position of attention from an exercise terminating position.

7-5. For example, this is how the PRT leader would conduct exercise 4, thigh stretch in the RD.
- From the position of attention, the PRT leader commands, "THE THIGH STRETCH."
- Soldiers respond, "THE THIGH STRETCH."
- From the position of attention, the PRT leader commands, "Starting Position, MOVE."

Chapter 7

- The PRT leader and Soldiers assume the starting position for the thigh stretch.
- From the starting position, the PRT leader commands, "Ready, STRETCH."
- To change position, the PRT leader first commands, "Starting Position, MOVE."
- From the starting position, the PRT leader commands, "Change Position, Ready, STRETCH."
- Upon termination of the exercise, the PRT leader commands, "Starting Position, MOVE."
- The PRT leader assumes the position of attention and commands, "Position of Attention, MOVE."

EXTENDED RECTANGULAR FORMATION

7-6. The Army's traditional formation for PRT activities is the extended rectangular formation. It is best for platoon- to company-size formations because it is simple and easy to assume.

PLATOON ASSEMBLY

7-7. The PRT leader will position a platoon-size unit in a line formation so that the unit is centered and five paces away from the PRT platform after they have assumed the rectangular formation. Refer to Figure 7-1. The PRT leader gives the following commands:

- "Extend to the left, MARCH." Soldiers in the right flank file stand fast with their left arm extended sideward with palm down, fingers and thumbs extended and joined. All other Soldiers turn to the left and double-time forward. After taking the sufficient number of steps, all Soldiers face the front and extend both arms sideward with palms down, fingers and thumbs extended and joined. The distance between fingertips is about 12 inches and dress is to the right.
- "Arms downward, MOVE." The Soldiers lower their arms smartly to their sides. Soldiers in the right flank file lower their left arms to their sides.
- "Left, FACE." Soldiers execute the left face.
- "Extend to the left, MARCH." Soldiers in the right flank file stand fast with their left arms extended sideward with palm down, fingers and thumbs extended and joined. All other Soldiers turn to the left and double-time forward. After taking the sufficient number of steps, all Soldiers face the front and extend both arms sideward with palms down, fingers and thumbs extended and joined. The distance between fingertips is about 12 inches and dress is to the right.
- "Arms downward, MOVE." Soldiers lower their arms smartly to their sides. Soldiers in the right flank file lower their left arms to their sides.
- "Right, FACE." Soldiers execute the right face.
- "From front to rear, COUNT OFF." The front Soldier in each column turns his head to the right rear, and then calls off, "ONE," and faces the front. Successive Soldiers in each column call off in turn "TWO," "THREE," "FOUR," and so on. The last Soldier in each column will not turn his head and eyes to the right while sounding off.
- "Even number to the left, UNCOVER." Even-numbered Soldiers side step to the left squarely in the center of the interval, bringing their feet together. (See Figure 7-2.)

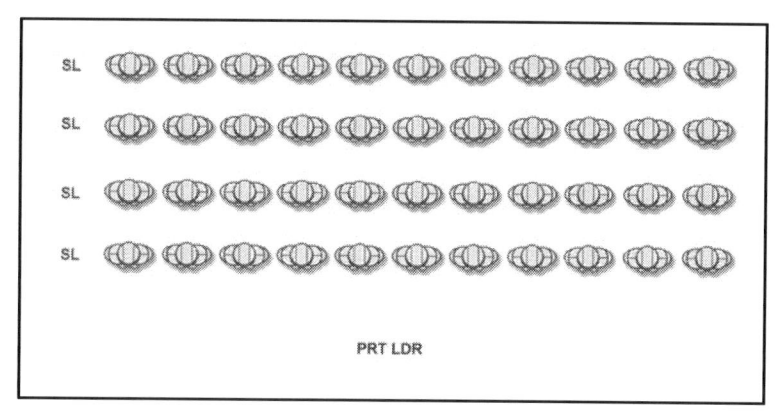

Figure 7-1. Platoon rectangular formation

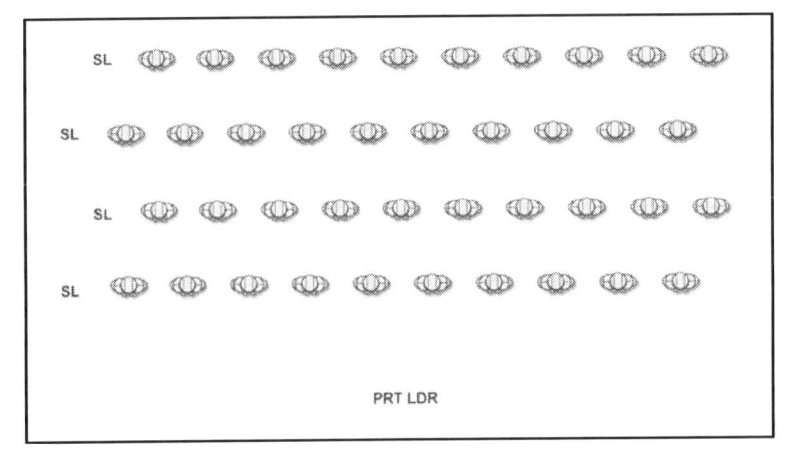

Figure 7-2. Platoon rectangular formation extended and uncovered

PLATOON REASSEMBLY

7-8. To reassemble the formation, the PRT leader commands:

"Assemble to the Right, MARCH." All Soldiers double-time to their original positions in the formation (Figure 7-1).

COMPANY IN LINE WITH PLATOONS IN COLUMN

7-9. The PRT leader will position a company-size unit in the extended rectangular formation from a company in line with platoons in column. He then adjusts the base platoon so that the company will be centered and five paces away from the PRT platform after they have assumed the rectangular formation. Refer to Figure 7-3. The PRT leader gives the commands specified in paragraph 7-7 to extend the formation (Figure 7-4).

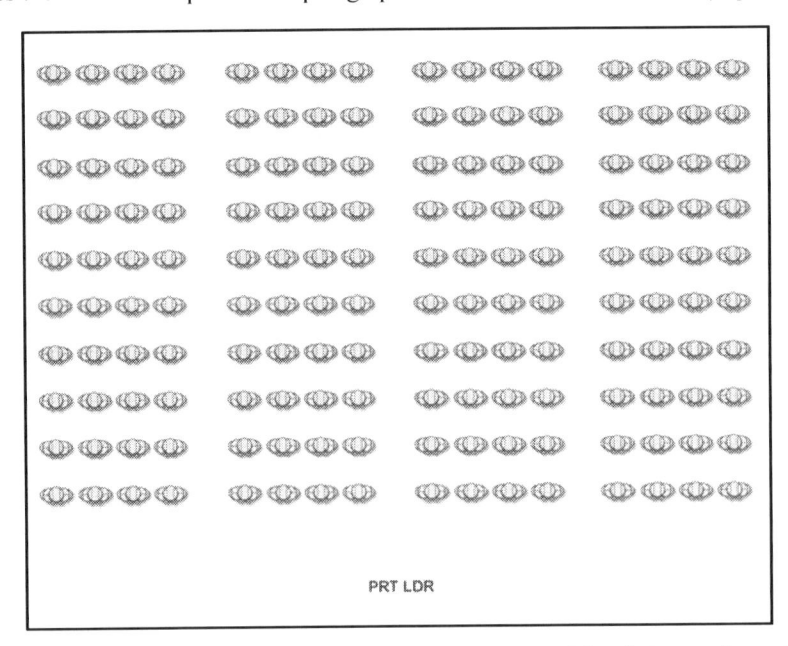

Figure 7-3. Forming a company, company in line with platoons in column

Chapter 7

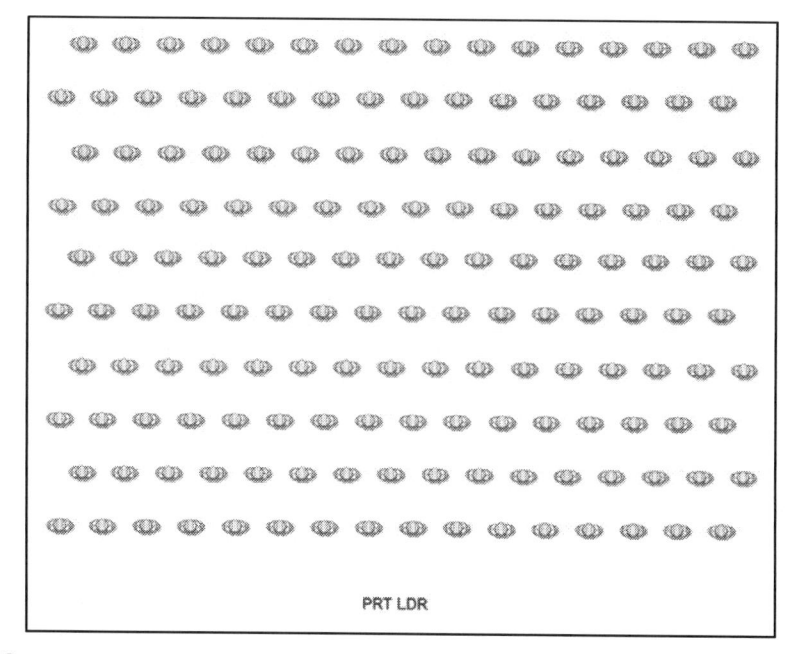

Figure 7-4. Company extended and uncovered, company in line with platoons in column

7-10. To reassemble the formation, the PRT leader commands:

"Assemble to the Right, MARCH." All Soldiers double-time to their original positions in the formation (Figure 7-3).

COMPANY FORMATION EN MASSE

7-11. The PRT leader will position a company-size unit in a rectangular formation. He first adjusts the base platoon so that the company will be centered and five paces away from the PRT platform after they have assumed the rectangular formation. Refer to Figure 7-5. The PRT leader gives the commands specified in paragraph 7-7 to extend the formation (Figure 7-6).

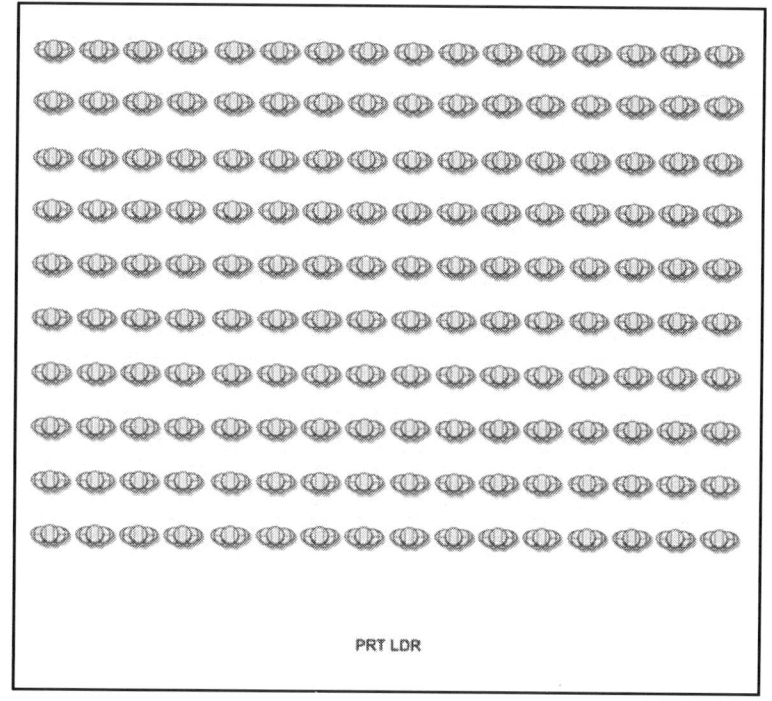

Figure 7-5. Formation of company en masse

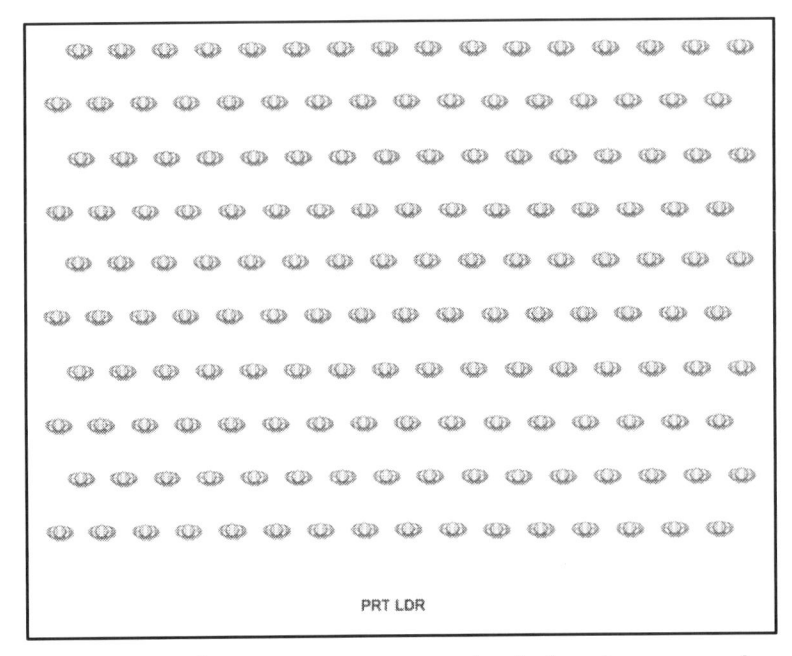

Figure 7-6. Company en masse extended and uncovered

7-12. To reassemble the formation, the PRT leader commands:

"Assemble to the right, MARCH." All Soldiers double-time to their original positions in the formation (Figure 7-5).

PLATOON EXTENDED RECTANGULAR FORMATION, COVERED

7-13. The formation for military movement and GDs is a platoon extended rectangular formation, covered (Figure 7-8). The PRT leader positions the platoon in a line formation so the unit will be centered and five paces away from the PRT platform after it assumes the extended rectangular formation. The PRT leader gives the following commands to extend the platoon formation covered (Figures 7-7 and 7-8).

- "Extend to the left, MARCH." Soldiers in the right flank file stand fast with their left arm extended sideward with palms down, fingers and thumbs extended and joined. All other Soldiers turn to the left and double-time forward. After taking the sufficient number of steps, all Soldiers face the front and extend both arms sideward with palms down, fingers and thumbs extended and joined. The distance between fingertips is about 12 inches and dress is to the right.
- "Arms downward, MOVE." The Soldiers lower their arms smartly to their sides. Soldiers in the right flank file lower their left arms to their sides.
- "Left, FACE." Soldiers execute the left face.
- "Extend to the left, MARCH." Soldiers in the right flank file stand fast with their left arms extended sideward with palms down, fingers and thumbs extended and joined. All other Soldiers turn to the left and double-time forward. After taking the sufficient number of steps, all Soldiers face the front and extend both arms sideward with palms down, fingers and thumbs extended and joined. The distance between fingertips is about 12 inches and dress is to the right.
- "Arms downward, MOVE." Soldiers lower their arms smartly to their sides. Soldiers in the right flank file lower their left arms to their sides.
- "Right, FACE." Soldiers execute the right face.

Chapter 7

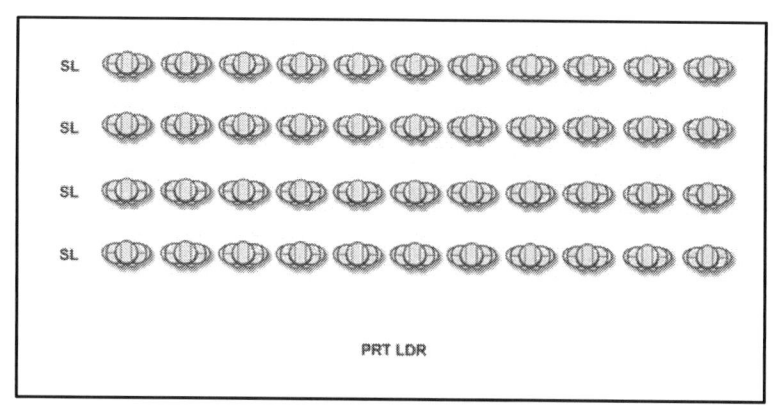

Figure 7-7. Platoon formation en masse

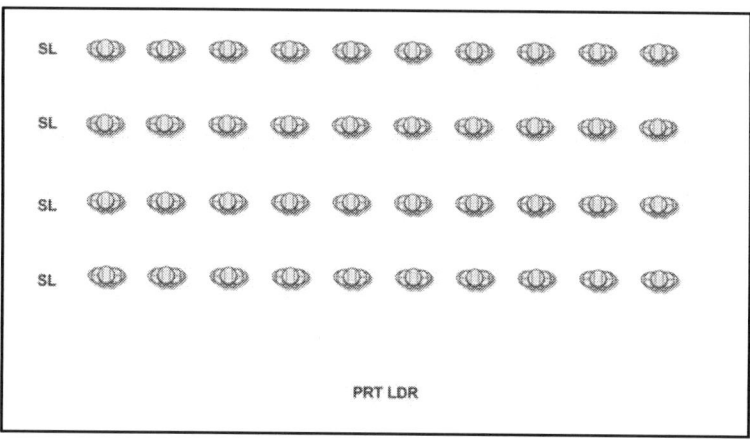

Figure 7-8. Platoon formation extended and covered

7-14. To reassemble the formation, the PRT leader commands:

"Assemble to the Right, MARCH." All Soldiers double-time to their original positions in the formation (Figure 7-7).

POSITIONS

7-15. When a set of conditioning exercises is employed, Soldiers assume the proper starting position of each exercise on the command "Starting Position, MOVE." When conducting exercises, Soldiers are commanded to return to the position of attention from the terminating position of the exercise, before commanded to assume the starting position for the next exercise.

SQUAT POSITION

7-16. To assume the squat position from the position of attention, lower the body by bending the knees and place the hands with palms down and fingers spread, shoulder width in front of the body, and in between the bent legs. Raise the heels, supporting the body weight on the balls of the feet and hands. Direct the head and the eyes to a point about three to four feet in front of the body (Figure 7-9).

Figure 7-9. Squat position

Chapter 7

FRONT LEANING REST POSITION

7-17. The Soldier assumes the front leaning rest position by performing two movements. First, the Soldier moves from the position of attention to the squat position, then thrusts the feet backward to the front leaning rest position. If he has trouble with the squat thrust, he can step back with his left leg—then with his right leg—to get into the front leaning rest position. In the front leaning rest position, maintain straight body alignment from his head to his heels. He supports his body weight on his hands (shoulder width) and on the balls of his feet. He keeps his feet and legs together (Figure 7-10).

Figure 7-10. Front leaning rest position

SIX-POINT STANCE

7-18. Assume the six-point stance by dropping to the knees from the front leaning rest position. Maintain a straight line from the head to the knees (Figure 7-11).

Figure 7-11. Six-point stance

STRADDLE STANCE

7-19. Assume the straddle stance position by standing with the feet straight ahead and aligned with the shoulders (Figure 7-12).

Figure 7-12. Straddle stance

FORWARD LEANING STANCE

7-20. Assume the forward leaning stance by bending the trunk forward 45 degrees, knees bent 45 degrees, with the heels flat on the ground, and the feet aligned with the shoulders. Keep the back straight, maintaining a straight line from the head to the hips (Figure 7-13).

Figure 7-13. Forward leaning stance

Chapter 7

PRONE POSITION

7-21. Assume the prone position by performing three movements: 1) From the position of attention, move to the squat position, 2) thrust the feet backward to the front leaning rest position, and then 3) lower the body slowly to the ground. Keep the elbows close to the body and pointed directly to the rear (Figure 7-14).

Figure 7-14. Prone position

SUPINE POSITION

7-22. To assume the supine position without using the hands, from the standing position, place one foot behind the other and slowly lower the body until the rear knee touches the ground. Sit back onto the buttocks and then lay on the back with feet and legs together (Figure 7-15). When returning to the standing position, sit up and rock forward on one knee. From this position, step up with the other leg and stand without using the hands for assistance (Figure 7-16).

Figure 7-15. Supine position

7-23. If the Soldier has difficulty assuming this position, he can place his hands on the ground as he slowly lower his body to the seated position (Figure 7-16). If he cannot attain the standing position without using his hands, he can place them on the ground to either side of his body and push up while standing from the seated position. To return to a standing position from the supine position, he performs the actions in reverse order (Figure 7-15).

Figure 7-16. Hands down assist to supine position

CADENCE

7-24. The following paragraphs discuss cadence speed and conduct of exercises.

SPEED

7-25. Cadence speed is described as SLOW or MODERATE. The speed of each cadence is listed below:
- SLOW–50 counts per minute.
- MODERATE–80 counts per minute.

EXERCISE NAME

7-26. Once Soldiers have learned the names of the exercises, the PRT leader merely needs to say the exercise name, command the Soldiers to assume the starting position and start them exercising to cadence. For example, this is how the PRT leader begins exercise 1 of preparation, bend, and reach to cadence:
- The PRT leaders states, "the Bend and Reach."
- The Soldiers respond, "the Bend and Reach."
- The PRT leader commands, "Starting Position, MOVE" (Soldiers assume the starting position).
- The PRT leader commands, "In Cadence (Soldiers respond, "In Cadence"), EXERCISE."
- The command, "EXERCISE" initiates movement to the position of count 1.

7-27. The previous command sequence is also used in the conduct of preparation, recovery, CD, and CL exercises.

7-28. Counting cadence ensures that exercises are performed at the appropriate speed. The cadence count indicates termination of movement to each position. The cumulative count is a method of indicating the number of repetitions of an exercise on the fourth count of a 4-count exercise. The use of the cumulative count is required for the following reasons:
- It provides the PRT leader with an excellent method of counting the number of repetitions performed.
- It serves as motivation. Soldiers like to know the number of repetitions they are expected to perform.
- It prescribes an exact amount of exercise for any group.

Chapter 7

COUNTS

7-29. This paragraph describes the conduct of cadence counts:

Two-Count Exercise
- The PRT leader counts, "Up, down."
- The Soldiers respond, "One."
- The PRT leader counts, "Up, down."
- The Soldiers respond, "Two."
- The PRT leader counts, "Up, down."
- The Soldiers respond, "Three," and so forth.

Four-Count Exercise
- The PRT leader counts, "One, two, three."
- The Soldiers respond, "One."
- The PRT leader counts, "One, two, three."
- The Soldiers respond, "Two."
- The PRT leader counts, "One, two, three."
- The Soldiers respond, "Three," and so forth.

Eight-Count Exercise
- The PRT leader counts, "One, two, three, four, five, six, seven."
- The Soldiers respond, "One."
- The PRT leader counts, "One, two, three, four, five, six, seven."
- The Soldiers respond, "Two."
- The PRT leader counts, "One, two, three, four, five, six, seven."
- The Soldiers respond, "Three," and so forth.

Termination

7-30. To terminate an exercise, the PRT leader will raise the inflection of his voice while counting out the cadence of the last repetition. The Soldiers and PRT leader respond with "HALT" upon returning to the starting position.

Two-Count Exercise
- The PRT leader counts, "Up, down."
- The Soldiers respond, "Four."
- The PRT leader counts "Up, down" (with voice inflection).
- The Soldiers and PRT leader respond, "HALT."
- The PRT leader commands "DISMOUNT."
- The Soldiers dismount the climbing bars.
- The PRT leader commands "Position of attention, MOVE."
- The Soldiers assume the position of attention.

Four-Count Exercise
- The PRT leader counts, "One, two, three."
- The Soldiers respond, "Nine."
- The PRT leader counts, "One, two, three" (with voice inflection).
- The Soldiers and PRT leader respond, "HALT."
- The PRT leader commands "Position of attention, MOVE."
- The Soldiers assume the position of attention.

Eight-Count Exercise
- The PRT leader counts, "One, two, three, four, five, six, seven."
- The Soldiers respond, "Four."
- The PRT leader counts, "One, two, three, four, five, six, seven" (with voice inflection on counts five, six, and seven).
- The Soldiers and PRT leader respond, "HALT."
- The PRT leader commands, "Position of attention, MOVE."
- The Soldiers assume the position of attention.

COMMANDS

7-31. The PD consists of ten four-count calisthenic exercises. See paragraph 7-29 for the commands, counting, and cadence instructions used to conduct preparation. Each strength and mobility drill has its own set of commands.

GUERRILLA DRILL

7-32. The GD includes three exercises that are performed from the extended rectangular formation, covered. The commands are followed when performing the shoulder roll, lunge walk, and Soldier carry. The difference for the Soldier carry is that Soldiers change positions at the 25-yard mark and return to the start point. When the PRT leader commands, "the Shoulder Roll," the entire formation repeats, "Shoulder Roll." After this, the leader need not say or repeat the command. The first rank takes one step forward with their left foot, and resume the position of attention:
- On the command, "READY," the first rank moves into the starting position.
- On the command, "GO," the first rank begins the movement.

7-33. In a typical formation with four ranks, the PRT leader directs the front rank remaining in the formation to move forward. He does this immediately after the previous front rank start the movement. The other ranks remain in place, awaiting further instructions. To do this, the PRT leader commands "Next Rank, MOVE FORWARD." Once the rank conducting the movement is about 12 yards into the exercise, the PRT leader commands, "Ready," and the rank moves into the starting position. Immediately after all Soldiers are in the starting position, the leader commands, "GO." He repeats this sequence of commands until all ranks have performed the shoulder roll. Perform the GD in platoon-size or smaller formations (extended and covered). This provides the appropriate work-to-rest ratio of 1:3. If he has less than four ranks for this drill, he enforces a 1:3 work-to-rest ratio.

CONDITIONING DRILLS

7-34. Conditioning drills have four-count and eight-count exercises.

CLIMBING DRILLS

7-35. Climbing drills have two-count exercises.

MILITARY MOVEMENT DRILLS

7-36. Military movement drills 1 and 2 each have three exercises that are performed from the extended rectangular formation, covered. The commands listed below will be followed when performing verticals, laterals, the shuttle sprint, the power skip, crossovers, and the crouch run. During the shuttle sprint, Soldiers will run the first two 25-yard intervals at the pace of the squad leader, then sprint the last 25-yard interval at their own pace. During the crouch run, on the command "Ready," Soldiers move to the starting position of the mountain climber. On the command "GO," Soldiers perform counts 1, 2, and 3 of the mountain climber, then upon finishing count four run forward in the crouch position to the 25-yard mark.

Chapter 7

7-37. For movement drills, the PRT leader states, "verticals" (and the entire formation repeats "verticals"). After this, there is no need to say or repeat "verticals." The first rank takes one step forward with their left foot, and resumes the position of attention.

- On the command, "Ready," the first rank moves into the starting position.
- On the command, "GO," the first rank begins the movement.

7-38. In a typical formation with four ranks, the PRT leader will have the front rank remaining in the formation move forward. He does this immediately after the previous front rank starts the movement. The other ranks should remain in place, awaiting further instructions. To accomplish this, the PRT leader commands "Next Rank, MOVE FORWARD." Once the rank conducting the movement is about 12 yards into the exercise, the PRT leader commands, "Ready," and then the rank moves into the starting position. Immediately after all Soldiers are in the starting position, the leader commands, "GO." The PRT leader repeats this sequence of commands until all ranks have performed verticals. Perform the military movement drill in platoon-size or smaller formations (extended and covered). This provides the appropriate work-to-rest ration of 1:3. If there are less than four ranks, the leader ensures that a 1:3 work-to-rest ratio is followed during military movement drills.

RUNNING ACTIVITIES

7-39. Running activities have different sets of commands. Sustained running begins when the PRT leader states, "double time, MARCH," and terminates the run with the commands, "Quick Time, MARCH." The PRT leader begins each repetition with the commands, "Ready, GO," when conducting HR and the 300-yd SR.

7-40. When conducting 30:60s or 60:120s, the PRT leader begins the activity with slow jogging for ¼-mile on the commands of "Double Time, MARCH," and terminates the ¼-mile run with the commands, "Quick Time, MARCH." 30:60s and 60:120s begin with the PRT leader signaling the start of each work interval (30 or 60 seconds) with one short whistle blast. Two short whistle blasts are used to signal the end of each work interval and the start of the rest intervals (60 or 120 seconds). Upon completion of the scheduled number of repetitions of 30:60s or 60:120s, the PRT leader will command the formation to continue to walk for at least three minutes before performing additional activities or the RD. The PRT leader will use the same commands specified for the shuttle sprint in MMD1 when conducting the 300-yard SR. The only difference is that Soldiers will perform six 50-yard repetitions to complete 300 yards.

RECOVERY DRILL

7-41. The RD exercises require no verbal cadence. Soldiers move in and out of the starting position and each exercise position on the PRT leader's commands. Soldiers hold each exercise position for 20 seconds in the toughening phase and 20-30 seconds in the sustaining phase during recovery. The leader does not count the seconds aloud. This is how he conducts recovery exercise 1, overhead arm pull:

- The PT leader commands, "THE OVERHEAD ARM PULL." Each Soldier responds, "THE OVERHEAD ARM PULL."
- The PRT leader commands, "Starting Position, MOVE." Each Soldier moves into the starting position, straddle stance with hands on hips.
- The command to begin the stretch is, "Ready, STRETCH." Each Soldier raises his left arm overhead and places his left hand behind his head and grasps above his left elbow with his right hand. He then pulls to the right, leaning his body to the right. He holds this position for 20-30 seconds.
- The PRT leader commands, "Starting Position, MOVE." Each Soldier moves into the starting position.
- The PRT leader gives the command to stretch the other side of the body: "Change Position, Ready, STRETCH." Each Soldier raises his right arm overhead and places his right hand behind his head and grasps above his right elbow with his left hand. He then pulls to his left, leaning his body to the left. He holds this position for 20-30 seconds.
- The PRT leader commands, "Starting Position, MOVE." Each Soldier assumes the starting position.
- The PRT leader assumes the position of attention and commands, "Position of Attention, MOVE." Each Soldier assumes the position of attention.

MIRROR EFFECT

7-42. When leading an exercise in front of the formation, the PRT leader begins the movements in count 1 to the right. He continues to mirror the Soldier's movements while facing them throughout the exercise.

> **Summary**
> Successful execution of PRT depends on the leadership of competent instructors and AIs. PRT leaders must have more than the knowledge, skills, and abilities to execute a PRT session. They must also present a positive image of physical fitness.

Chapter 8
Preparation and Recovery

PRT sessions always include the following elements: preparation, activity, and recovery. Exercises performed during preparation ready Soldiers for more intense PRT activities. Recovery exercises are performed at the end of every PRT session to gradually and safely bring the body back to its pre-exercise state.

PREPARATION

8-1. The purpose of preparation is to ready the Soldier for PRT activities. The PD is performed at the beginning of every PRT session. The PD consists of ten exercises performed for 5-10 repetitions at a slow cadence, with the exception of the high jumper and push-up (which are performed at a moderate cadence). When conducted to standard, preparation will last about 15 minutes. Since PRT sessions are generally limited to one hour, preparation must be brief, yet thorough. The objectives of preparation are to:
- Increase body temperature and heart rate.
- Increase pliability of joints and muscles.
- Increase responsiveness of nerves and muscles.

TRAINING AREA

8-2. Any dry, level area of adequate size is satisfactory for conducting the PD.

UNIFORM

8-3. Soldiers should wear ACUs with boots or the IPFU. The uniform should be appropriate for the activity that will follow the PD. For example, when the activity is the CL 2 or the GD, ACUs with boots will be worn.

EQUIPMENT

8-4. N/A.

FORMATION

8-5. The extended rectangular formation is prescribed for the conduct of the PD.

LEADERSHIP

8-6. A PRT leader and AI are required to lead the PD.

INSTRUCTION AND EXECUTION

8-7. The PRT leader must be familiar with the method of teaching these exercises, the commands, the formations, and the use of AIs as described in Chapter 7, Execution of Training. The calisthenic exercises that comprise the PD are always given in cadence. Soldiers begin and terminate each exercise at the starting position, then move to the position of attention. The goal is to complete the entire drill with only enough pauses between exercises for the PRT leader to indicate the next one by name. This continuous method of conducting the PD intensifies the workload and conserves time. Soldiers should memorize the exercises by name and movement.

Chapter 8

PRECISION

8-8. Preparation loses much of its value unless performed exactly as prescribed. During preparation, the focus is always on quality of movement, not quantity of repetitions or speed of movement. A calisthenic cadence that is too fast will not allow Soldiers to achieve a full range of movement and may not adequately prepare them for the activities that follow. Assistant instructors will help to maintain the ranks at the appropriate pace and offer feedback on form.

PROGRESSION

8-9. Preparation is always performed in all phases of PRT. In the toughening phase, the PD consists of 5 repetitions of 10 exercises performed at a slow cadence, with the exception of the high jumper and the push-up, which are performed at a moderate cadence. In the sustaining phase, Soldiers progress from performing 5 repetitions to 10 repetitions of each exercise in the PD.

INTEGRATION

8-10. Preparation not only prepares the body for activities that follow; it also integrates the components of strength, endurance, and mobility.

COMMANDS

8-11. The commands used to conduct preparation are described in Chapter 7, Execution of Training.

PREPARATION DRILL

8-12. Table 8-1 lists the 10 calisthenic exercises that comprise the PD. These 10 exercises are always performed in the order and at the cadence shown.

Table 8-1. Preparation drill

1. Bend and reach	5-10 repetitions, slow
2. Rear lunge	5-10 repetitions, slow
3. High jumper	5-10 repetitions, moderate
4. Rower	5-10 repetitions, slow
5. Squat bender	5-10 repetitions, slow
6. Windmill	5-10 repetitions, slow
7. Forward lunge	5-10 repetitions, slow
8. Prone row	5-10 repetitions, slow
9. Bent-leg body twist	5-10 repetitions, slow
10. Push-up	5-10 repetitions, moderate

8-13. Table 8-2 shows the body segments trained during the PD.

Table 8-2. Body segments trained in the conduct of the preparation drill

PREPARATION DRILL (PD)	HIPS	THIGHS	LOWER LEGS	CHEST	BACK	TRUNK	SHOULDERS	ARMS
1. BEND AND REACH	X	X			X	X	X	
2. REAR LUNGE	X	X	X		X	X		
3. HIGH JUMPER	X	X	X		X	X	X	
4. ROWER	X	X	X		X	X	X	X
5. SQUAT BENDER	X	X	X	X	X	X	X	X
6. WINDMILL	X	X	X	X	X	X	X	X
7. FORWARD LUNGE	X	X	X		X	X		
8. PRONE ROW	X			X	X	X	X	X
9. BENT-LEG BODY TWIST	X	X			X	X		
10. PUSH-UP	X	X	X	X	X	X	X	X

PREPARATION DRILL

EXERCISE 1: BEND AND REACH

Purpose: This exercise develops the ability to squat and reach through the legs. It also serves to prepare the spine and extremities for more vigorous movements, moving the hips and spine through full flexion (Figure 8-1).

Starting Position: Straddle stance with arms overhead, palms facing inward, fingers and thumbs extended and joined.

Cadence: SLOW

Count:

1. Squat with the heels flat as the spine rounds forward to allow the straight arms to reach as far as possible between the legs.
2. Return to the starting position.
3. Repeat count one.
4. Return to the starting position.

Figure 8-1. Bend and reach

Check Points:

- From the starting position, ensure that Soldiers have their hips set, their abdominals tight, and their arms fully extended overhead.
- The neck flexes to allow the gaze to the rear. This brings the head in line with the bend of the trunk.
- The heels and feet remain flat on the ground.
- On counts 2 and 4, do not go past the starting position.

Precautions: This exercise is always performed at a slow cadence. To protect the back, move into the count one position in a slow, controlled manner. Do not bounce into or out of this position in a ballistic manner, as this may place an excessive load on the back.

PREPARATION DRILL

EXERCISE 2: REAR LUNGE

Purpose: This exercise promotes balance, opens up the hip and trunk on the side of the lunge, and develops leg strength (Figure 8-2).

Starting Position: Straddle stance with hands on hips.

Cadence: SLOW

Count:

1. Take an exaggerated step backward with the left leg, touching down with the ball of the foot.
2. Return to the starting position.
3. Repeat count one with the right leg.
4. Return to the starting position.

Figure 8-2. Rear lunge

Check Points:
- Maintain straightness of the back by keeping the abdominal muscles tight throughout the motion.
- After the foot touches down, allow the body to continue to lower. This promotes flexibility of the hip and trunk.
- On counts 1 and 3, step straight to the rear, keeping the feet directed forward. When viewed from the front, the feet maintain their distance apart both at the starting position and at the end of counts 1 and 3.
- Keep the rear leg as straight as possible but not locked and the rear heel off the ground.

Precautions: This exercise is always performed at a slow cadence. On counts 1 and 3, move into position in a slow, controlled manner. If the cadence is too fast, it will be difficult to go through a full range of motion.

Chapter 8

PREPARATION DRILL

EXERCISE 3: HIGH JUMPER

Purpose: This exercise reinforces correct jumping and landing, stimulates balance and coordination, and develops explosive strength (Figure 8-3).

Starting Position: Forward leaning stance, palms facing inward, fingers and thumbs extended and joined.

Cadence: MODERATE

Count:

1. Swing arms forward and jump a few inches.
2. Swing arms backward and jump a few inches.
3. Swing arms forward and vigorously overhead while jumping forcefully.
4. Repeat count 2. On the last repetition, return to the starting position.

Figure 8-3. High jumper

Check Points:

- At the starting position, the shoulders, the knees, and the balls of the feet should form a straight vertical line.
- On count 1, the arms are parallel to the ground.
- On count 3, the arms should be extended fully overhead. The trunk and legs should also be aligned.
- The Soldier is jumping on each count. On counts 1, 2, and 4, the jumps are only 4-6 inches off the ground. On count 3, the Soldier jumps higher (6-10 inches) while maintaining the posture pictured in Figure 8-3.
- On each landing, the feet should be directed forward and maintained at shoulder distance apart. The landing should be "soft" and proceed from the balls of the feet to the heels. The vertical line from the shoulders through the knees to the balls of the feet should be demonstrated on each landing.

Precaution: N/A

PREPARATION DRILL

EXERCISE 4: ROWER

Purpose: This exercise improves the ability to move in and out of the supine position to a seated posture. It coordinates the action of the trunk and extremities while challenging the abdominal muscles (Figure 8-4).

Starting Position: Supine position, arms overhead, feet together and pointing upward. The chin is tucked and the head is 1-2 inches above the ground. Arms are shoulder-width, palms facing inward with fingers and thumbs extended and joined.

Cadence: SLOW

Count:

1. Sit up while swinging arms forward and bending at the hip and knees. At the end of the motion, the arms will be parallel to the ground with palms facing inward.
2. Return to the starting position.
3. Repeat count 1.
4. Return to the starting position.

Figure 8-4. Rower

Check Points:

- At the starting position, the low back must not be arched excessively off the ground. To prevent this, tighten the abdominal muscles to tilt the pelvis and low back toward the ground.
- At the end of counts 1 and 3, the feet are flat and pulled near the buttocks. The legs stay together throughout the exercise and the arms are parallel to the ground.

Precautions: This exercise is always performed at a slow cadence. Do not arch the back to assume counts 1 and 3.

PREPARATION DRILL

EXERCISE 5: SQUAT BENDER

Purpose: This exercise develops strength, endurance, and flexibility of the lower back and lower extremities (Figure 8-5).

Starting Position: Straddle stance with hands on hips.

Cadence: SLOW

Count:

1. Squat while leaning slightly forward at the waist with the head up and extend the arms to the front, with arms parallel to the ground and palms facing inward.
2. Return to the starting position.
3. Bend forward and reach toward the ground with both arms extended and palms inward.
4. Return to the starting position.

Figure 8-5. Squat bender

Check Points:

- At the end of count 1, the shoulders, knees, and the balls of the feet should be aligned. The heels remain on the ground and the back is straight.
- On count 3, bend forward, keeping the head aligned with the spine and the knees slightly bent. Attempt to keep the back flat and parallel to the ground.

Precaution: This exercise is always performed at a slow cadence. Allowing the knees to go beyond the toes on count 1 increases stress to the knees.

PREPARATION DRILL

EXERCISE 6: WINDMILL

Purpose: This exercise develops the ability to safely bend and rotate the trunk. It conditions the muscles of the trunk, legs, and shoulders (Figure 8-6).

Starting Position: Straddle stance with arms sideward, palms facing down, fingers and thumbs extended and joined.

Cadence: SLOW

Count:

1. Bend the hips and knees while rotating to the left. Reach down and touch the outside of the left foot with the right hand and look toward the rear. The left arm is pulled rearward to maintain a straight line with the right arm.
2. Return to the starting position.
3. Repeat count 1 to the right.
4. Return to the starting position.

Figure 8-6. Windmill

Check Points:
- From the starting position, feet are straight ahead, arms parallel to the ground, hips set, and abdominals tight.
- On counts 1 and 3, ensure that both knees bend during the rotation. Head and eyes are directed to the rear on counts 1 and 3.

Precaution: This exercise is always performed at a slow cadence.

PREPARATION DRILL

EXERCISE 7: FORWARD LUNGE

Purpose: This exercise promotes balance and develops leg strength (Figure 8-7).

Starting Position: Straddle stance with hands on hips.

Cadence: SLOW

Count:

1. Take a step forward with the left leg (the left heel should be 3 to 6 inches forward of the right foot). Lunge forward, lowering the body and allow the left knee to bend until the thigh is parallel to the ground. Lean slightly forward, keeping the back straight.
2. Return to the starting position.
3. Repeat count one with the right leg.
4. Return to the starting position.

Figure 8-7. Forward lunge

Check Points:
- Keep the abdominal muscles tight throughout the motion.
- On counts 1 and 3, step straight forward, keeping the feet directed forward. When viewed from the front, the feet maintain their distance apart both at the starting position and at the end of counts 1 and 3.
- On counts 1 and 3, the rear knee bends, but does not touch the ground. The heel of the rear foot should be off the ground.

Precautions: This exercise is always performed at a slow cadence. On counts 1 and 3, move into position in a controlled manner. Spring off of the forward leg to return to the starting position. This avoids jerking the trunk to create momentum.

PREPARATION DRILL

EXERCISE 8: PRONE ROW

Purpose: This exercise develops strength of the back and shoulders (Figure 8-8).

Starting Position: Prone position with the arms overhead, palms down, fingers and thumbs extended and joined, 1 to 2 inches off the ground and toes pointed to the rear.

Cadence: SLOW

Count:

1. Raise the head and chest slightly while lifting the arms and pulling them rearward. Hands make fists as they move toward the shoulders.
2. Return to the starting position.
3. Repeat count 1.
4. Return to the starting position.

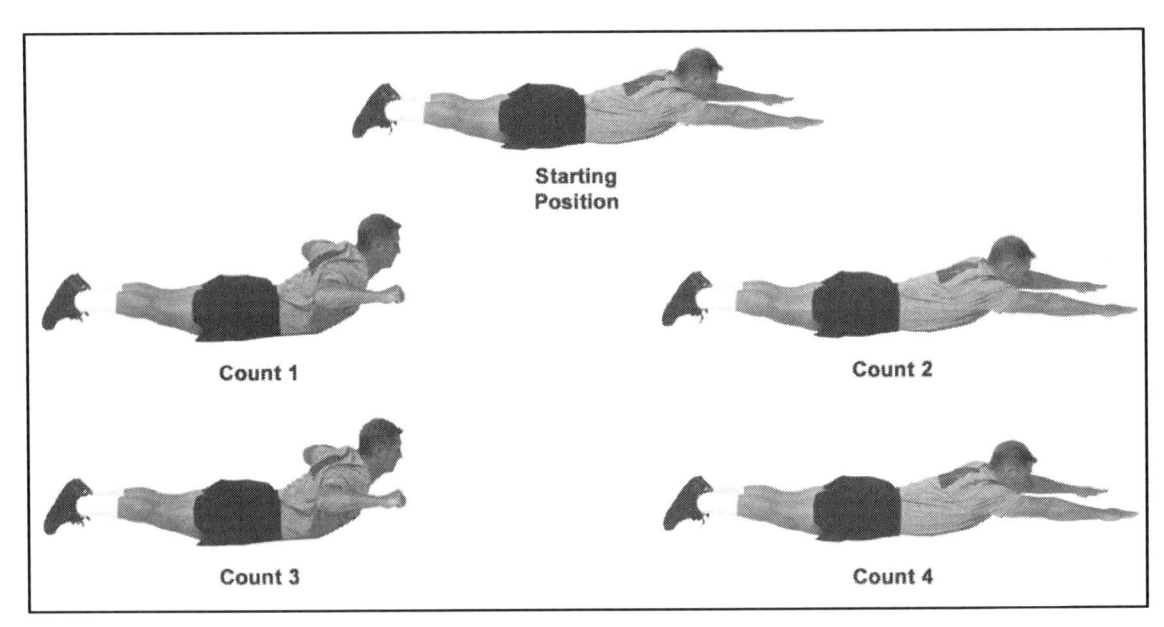

Figure 8-8. Prone row

Check Points:
- At the starting position, the abdominal muscles are tight and the head is aligned with the spine.
- On counts 1 and 3, the forearms are parallel to the ground and slightly higher than the trunk.
- On counts 1 and 3, the head is raised to look forward but not skyward.
- Throughout the exercise, the legs and toes remain in contact with the ground.

Precautions: This exercise is always performed at a slow cadence. Prevent overarching of the back by maintaining contractions of the abdominal and buttocks muscles throughout the exercise.

PREPARATION DRILL

EXERCISE 9: BENT-LEG BODY TWIST

Purpose: This exercise strengthens trunk muscles and promotes control of trunk rotation (Figure 8-9).

Starting Position: Supine position with the hips and knees bent to 90-degrees, arms sideward and palms down. The knees and feet are together.

Cadence: SLOW

Count:

1. Rotate the legs to the left while keeping the upper back and arms in place.
2. Return to the starting position.
3. Repeat count 1 to the right.
4. Return to the starting position.

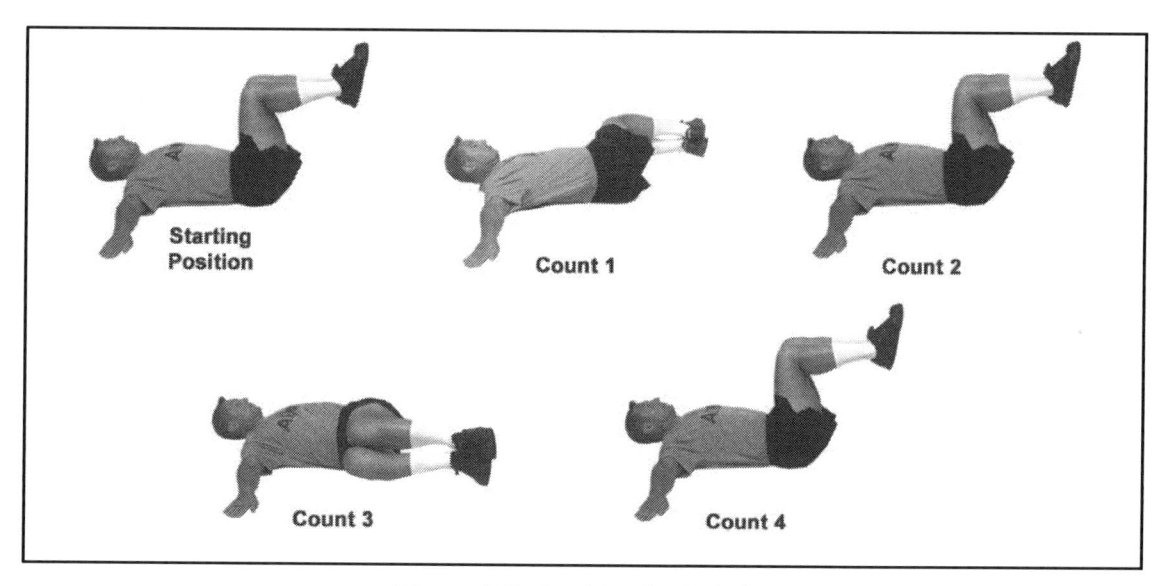

Figure 8-9. Bent-leg body twist

Check Points:

- Tighten the abdominal muscles in the starting position and maintain this contraction throughout the exercise.
- The head should be off the ground with the chin slightly tucked.
- Ensure that the hips and knees maintain 90-degree angles.
- Keep the feet and knees together throughout the exercise.
- Attempt to rotate the legs to about 8 to 10 inches off the ground. The opposite shoulder must remain in contact with the ground.

Precautions: This exercise is always performed at a slow cadence. Do not rotate the legs to a point beyond which the opposite arm and shoulder can no longer maintain contact with the ground.

Chapter 8

PREPARATION DRILL

EXERCISE 10: PUSH-UP

Purpose: This exercise strengthens the muscles of the chest, shoulders, arms, and trunk (Figure 8-10).

Starting Position: Front leaning rest position.

Cadence: MODERATE

Count:

1. Bend the elbows, lowering the body until the upper arms are parallel with the ground.
2. Return to the starting position.
3. Repeat count 1.
4. Return to the starting position.

Figure 8-10. Push-up

Check Points:
- The hands are directly below the shoulders with fingers spread (middle fingers point straight ahead).
- On counts 1 and 3, the upper arms stay close to the trunk, elbows pointing rearward.
- On counts 2 and 4, the elbows straighten but do not lock.
- To prevent the trunk from sagging, tighten the abdominal muscles while in the starting position and maintain this contraction throughout the exercise.

Precaution: N/A

EXERCISE 10A: PUSH-UP USING THE SIX-POINT STANCE

Purpose: Soldiers should assume the six-point stance on their knees, when unable to perform repetitions correctly to cadence (Figure 8-11).

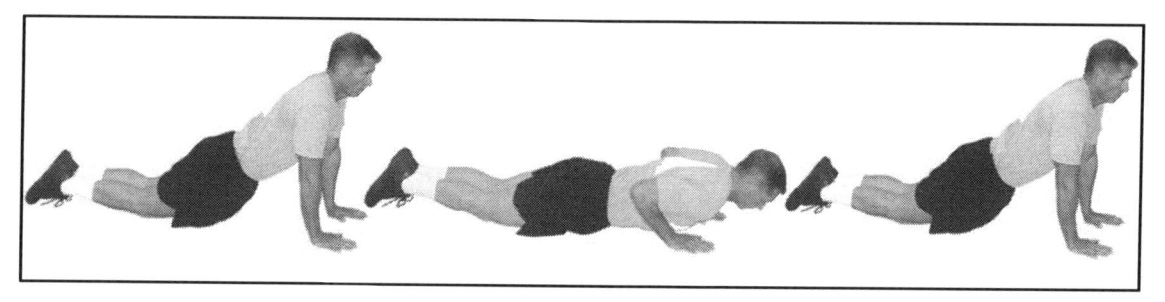

Figure 8-11. Push-up using the six-point stance

RECOVERY

8-14. Recovery serves to gradually slow the heart rate and helps prevent pooling of the blood in the legs and feet. The purpose of the RD is to develop range of motion and stability to enhance performance, control injuries, and gradually bring the body back to its pre-exercise state. To adequately recover from one PRT session to another on consecutive days, Soldiers must restore hydration and energy through proper fluid intake and nutrition. This recovery period also includes receiving adequate rest and sleep to allow the body to physiologically adapt to the physical stresses of PRT.

TRAINING AREA

8-15. Any dry, level area of adequate size is satisfactory for conduct of the RD.

UNIFORM

8-16. Soldiers should wear ACUs with boots or the IPFU. The uniform should be appropriate for the PRT activity that precedes recovery. For example, when the activity is the CL 2 or the GD, ACUs with boots will be worn.

EQUIPMENT

8-17. N/A.

FORMATION

8-18. The extended rectangular formation is prescribed for the conduct of the RD.

LEADERSHIP

8-19. Recovery should last about 15 minutes and occur immediately after the activities of the PRT session. Soldiers should begin recovery after running activities by walking until their heart rates return to less than 100 beats per minute and heavy sweating stops. Walking also may be needed after the end of a strength training circuit activity. Each recovery exercise position will be held for 20-30 seconds. The sequence of exercises listed in Table 8-3 will be performed in its entirety. The RD will be conducted at the end of all PRT sessions, especially after the conduct of the APFT, obstacle course, and foot marching. See Chapter 5, Planning Considerations, for more information.

INSTRUCTION AND EXECUTION

8-20. A PRT leader and AI are required to lead the RD. The PRT leader and AI must be familiar with the method of teaching these exercises, commands, formations, and the use of AIs as described in Chapter 7,

Chapter 8

Execution of Training. Soldiers should memorize the exercises by name and movement. The RD may be conducted by platoon or en masse. Soldiers move in and out of the starting position and exercise positions on the PRT leader's command. Each exercise position is held for 20-30 seconds. Soldiers begin and terminate each exercise at the starting position, then move to the position of attention. The RD is always performed in the order listed. Considerable time and effort must be expended during the early stages to teach precise performance of each exercise. The PRT leader should not execute the RD in cadence and should not count seconds aloud.

PRECISION

8-21. Recovery exercises lose much of their value unless performed exactly as prescribed. PRT leaders and AIs must provide verbal feedback and make spot corrections to ensure that the Soldiers correctly assume the exercise positions.

PROGRESSION

8-22. In the toughening phase Soldiers hold each exercise position for 20 seconds. In the sustaining phase, the Soldier holds each exercise position for 20 seconds and progresses to 30 seconds. For either phase, if time allows, a second set of the RD may be performed.

INTEGRATION

8-23. Recovery integrates the components of strength and mobility by developing stability and flexibility.

COMMANDS

8-24. The commands used to conduct the RD are described Chapter 7, Execution of Training.

RECOVERY DRILL

8-25. Table 8-3 lists the 5, two-position exercises that comprise the RD. These 5 exercises are always performed in the order listed and held for 20 to 30 seconds. The recovery exercises are not given in cadence. Soldiers move in and out of the starting position and exercise positions on the PRT leader's command. The seconds are not counted out loud.

Table 8-3. Recovery drill

1. OVERHEAD ARM PULL	HOLD 20-30 SECONDS
2. REAR LUNGE	HOLD 20-30 SECONDS
3. EXTEND AND FLEX	HOLD 20-30 SECONDS
4. THIGH STRETCH	HOLD 20-30 SECONDS
5. SINGLE-LEG OVER	HOLD 20-30 SECONDS

8-26. Table 8-4 lists the body segments trained in the conduct of RD.

Table 8-4. Body segments trained in the conduct of the recovery drill

RECOVERY DRILL (RD)	MUSCLES							
	HIPS	THIGHS	LOWER LEGS	CHEST	BACK	TRUNK	SHOULDERS	ARMS
1. OVERHEAD ARM PULL					X	X	X	X
2. REAR LUNGE	X	X	X					
3. EXTEND AND FLEX	X	X	X	X	X	X	X	X
4. THIGH STRETCH		X	X	X		X	X	X
5. SINGLE-LEG OVER	X	X			X	X	X	X

RECOVERY DRILL

EXERCISE 1: OVERHEAD ARM PULL

Purpose: This exercise develops flexibility of the arms, shoulders, and trunk muscles (Figure 8-12).

Starting Position: Straddle stance with hands on hips.

Position 1: On the command, "Ready, STRETCH," raise the left arm overhead and place the left hand behind the head. Grasp above the left elbow with the right hand and pull to the right, leaning the body to the right. Hold this position for 20-30 seconds.

Starting Position: On the command "Starting Position, MOVE," assume the starting position.

Position 2: On the command "Change Position, Ready, STRETCH," raise the right arm overhead and place the right hand behind the head. Grasp above the right elbow with the left hand and pull to the left, leaning the body to the left. Hold this position for 20-30 seconds.

Starting Position: On the command "Starting Position, MOVE," return to the starting position.

Figure 8-12. Overhead arm pull

Check Points:
- Throughout the exercise, keep the hips set and the abdominals tight.
- In positions 1 and 2, lean the body straight to the side, not to the front or back.

Precaution: N/A

Preparation and Recovery

RECOVERY DRILL

EXERCISE 2: REAR LUNGE

Purpose: This exercise develops flexibility of the hip flexors and trunk muscles (Figure 8-13).

Starting Position: Straddle stance, hands on hips.

Position 1: On the command "Ready, STRETCH," take an exaggerated step backward with the left leg, touching down with the ball of the foot. This is the same position as count 1 of the rear lunge in the PD. Hold this position for 20-30 seconds.

Starting Position: On the command "Starting Position, MOVE," assume the starting position.

Position 2: On the command "Change Position, Ready, STRETCH," take an exaggerated step backward with the right leg, touching down with the ball of the foot. This is the same position as count 3 of the rear lunge in the PD. Hold this position for 20-30 seconds.

Starting Position: On the command "Starting Position, MOVE," return to the starting position.

Figure 8-13. Rear lunge

Check Points:
- Maintain straightness of the back by keeping the abdominal muscles tight throughout the motion.
- After the foot touches down on positions 1 and 2, allow the body to continue to lower.
- Lunge and step in a straight line, keeping the feet directed forward. Viewed from the front, the feet are shoulder width apart, both at the starting position and at the end of positions 1 and 2.
- Keep the forward knee over the ball of the foot on positions 1 and 2.
- Ensure the heal of the rear foot does not touch the ground.

Precaution: When lunging to the left or right, do not let the knee move forward of the toes.

Chapter 8

RECOVERY DRILL

EXERCISE 3: EXTEND AND FLEX

Purpose: This exercise develops flexibility of the hip flexors, abdominals, hip (position 1–extend, Figure 8-14), and the low back, hamstrings, and calves (position 2–flex, Figure 8-14).

Starting Position: The front leaning rest position.

Position 1: On the command "Ready, STRETCH," lower the body, sagging in the middle, keeping the arms straight and look upward. Hold this position for 20-30 seconds.

Starting Position: On the command "Starting Position, MOVE," assume the starting position.

Position 2: On the command "Change Position, Ready, STRETCH," slightly bend the knees and raise the hips upward. Straighten the legs and try to touch the ground with the heels. Move the head in line with the arms, forming an "A" with the body. Keep the feet together and hold this position for 20-30 seconds.

Starting Position: On the command "Starting Position, MOVE," return to the starting position.

Figure 8-14. Extend and flex

Check Points:
- In position 1, the thighs and pelvis rest on the ground. Relax the back muscles while bearing the bodyweight through the straight arms. Toes point to the rear.
- In position 2, the legs are straight and the arms are shoulder width apart, palms down on the ground. Relax the shoulders and push to the rear with the hands, forming an "A" with the body. Try not to round the shoulders.
- Feet are together throughout the exercise.

Precaution: N/A

Variation: Soldiers, who are unable to extend the trunk in position 1 while keeping the arms straight and hips on the ground, may assume the modified position 1 shown above.

RECOVERY DRILL

EXERCISE 4: THIGH STRETCH

Purpose: This exercise develops flexibility of the front of the thigh and the hip flexor muscles (Figure 8-15).

Starting Position: Seated position, arms at sides and palms on the floor.

Position 1: On the command "Ready, STRETCH," roll onto the right side and place the right forearm on the ground, perpendicular to the chest. The right hand makes a fist on the ground with the thumb side up. Grasp the left ankle with the left hand and pull the left heel toward the buttocks and pull the entire leg rearward. Push the left thigh further to the rear with the heel of the right foot. Hold this position for 20-30 seconds.

Starting Position: On the command, "Starting Position, MOVE," assume the starting position.

Position 2: On the command "Change Position, Ready, STRETCH," lie on the left side and place the left forearm on the ground, perpendicular to the chest. The left hand makes a fist on the ground with the thumb side up. Grasp the right ankle with the right hand and pull the right heel toward the buttocks pulling the entire leg rearward. Push the right thigh further to the rear with the heel of the left foot. Hold this position for 20-30 seconds.

Starting Position: On the command, "Starting Position, MOVE," return to the starting position.

Figure 8-15. Thigh stretch

Check Points:
- Keep the abdominal muscles tight throughout this stretch in order to keep the trunk straight.
- Do not pull the heel forcefully to the buttock if there is discomfort in the knee joint.

Precaution: N/A

Chapter 8

RECOVERY DRILL

EXERCISE 5: SINGLE-LEG OVER

Purpose: This exercise develops flexibility of the hips and lower back muscles (Figure 8-16).

Starting Position: Supine position with arms sideward, palms down, and feet together and head on the ground.

Position 1: On the command, "Ready, STRETCH," turn the body to the right, bend the left knee to 90-degrees over the right leg, and grasp the outside of the left knee with the right hand and pull toward the right. Hold this position for 20-30 seconds.

Starting Position: On the command, "Starting Position, MOVE," assume the starting position.

Position 2: On the command, "Change Position, Ready, STRETCH," turn the body to the left, bend the right knee to 90-degrees over the left leg, and grasp the outside of the right knee with the left hand and pull toward the left. Hold this position for 20-30 seconds.

Starting Position: On the command, "Starting Position, MOVE," return to the starting position.

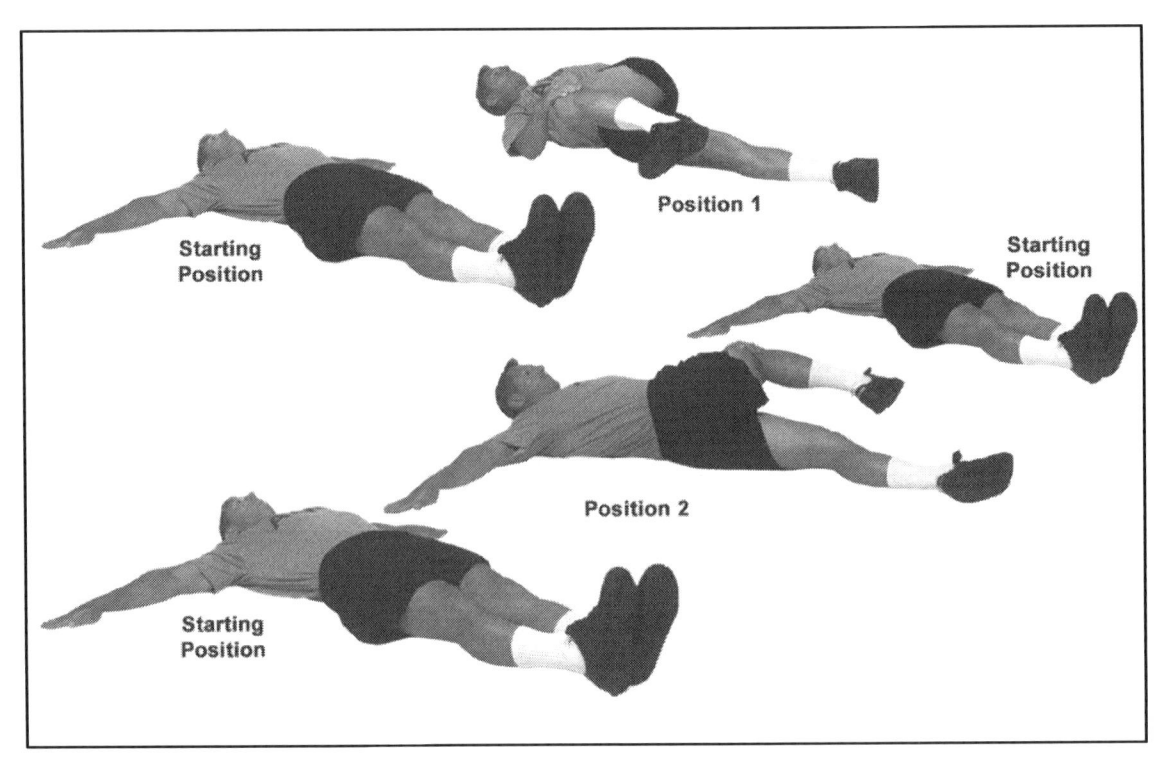

Figure 8-16. Single-leg over

Check Points:
- At the starting position, the arms are directed to the sides at 90-degrees to the trunk; the fingers and thumbs are extended and joined.
- In position 1, keep the left shoulder, arm, and hand on the ground.
- In position 2, keep the right shoulder, arm, and hand on the ground.
- Head remains on the ground throughout the exercise.

Precaution: N/A

Summary

Preparation and recovery are essential elements of every PRT session. Conducting PRT activities without preparation may adversely affect performance and increase the risk of injury. Recovery enhances mobility and gradually brings the body back to its pre-exercise state. Recovery should also carry over until the next PRT session is performed. Restoring adequate hydration and energy through proper nutrition and getting adequate sleep allow the body to refuel, rest, and adapt to the stresses of training.

Chapter 9
Strength and Mobility Activities

"The race is to the swift; the battle to the strong."

John Davidson, 19th Century Poet

This chapter describes strength and mobility exercises, drills, and activities designed for Soldiers in the toughening and sustaining phases of PRT. The purpose of strength and mobility activities is to improve functional strength, postural alignment, and body mechanics as they relate to the performance of WTBDs.

EXERCISE DRILLS

9-1. The regular and precise execution of strategically organized and sequenced exercise drills will develop the body management competencies needed to successfully accomplish WTBDs (Figure 9-1). Table 9-1 describes all strength and mobility drills and activities presented in this chapter. Table 9-2 describes strength and mobility drills and activities and the prescription of intensity, duration, and volume within the toughening and sustaining phases. In addition, Chapter 5, Planning Considerations, provides the template for commanders and PRT leaders to implement strength and mobility activities into their PRT programs.

Figure 9-1. Strength and mobility-related WTBDs

Chapter 9

Table 9-1. Strength and mobility drills and activities

Conditioning Drill 1 (CD 1)	Conditioning Drill 1 consists of basic and intermediate calisthenic exercises that develop foundational fitness and body management by challenging strength, endurance, and mobility through complex functional movement patterns.
Conditioning Drill 2 (CD 2)	Conditioning Drill 2 consists of intermediate and advanced calisthenic exercises that are designed to functionally train the total-body muscular strength and endurance needed to successfully perform WTBDs.
Conditioning Drill 3 (CD 3)	Conditioning Drill 3 consists of advanced calisthenic and plyometric exercises that are designed to functionally train agility, coordination, and the lower-body muscular strength and endurance needed to successfully perform WTBDs.
Push-up and Sit-up Drill (PSD)	The Push-up and Sit-up Drill consists of push-up and sit-up exercises performed in alternating timed sets (30 to 60 seconds each) to enhance upper-body muscular strength and endurance for improved APFT performance.
Climbing Drill 1 (CL 1)	Climbing Drill 1 consists of exercises performed on a high bar or climbing bars. This drill develops upper body and trunk strength and mobility while manipulating body weight off the ground.
Climbing Drill 2 (CL 2)	Climbing Drill 2 consists of exercises performed on a high bar or climbing bars. This drill improves upper body and trunk strength and mobility needed for manipulating body weight while under fighting load.
Strength Training Circuit (STC)	The Strength Training Circuit consists of sequenced exercise stations using strength training equipment and climbing exercises performed for a designated time until all exercises have been performed. Movement and distance between exercise stations may be varied. In the sustaining phase, movement from station to station may include exercises from both military movement drills 1 and/or 2.
Guerrilla Drill (GD)	The Guerrilla Drill consists of dynamic exercises that develop leg power and functional mobility. The emphasis is on improving combative techniques and the ability to carry/evacuate another Soldier.

Strength and Mobility Activities

Table 9-2. Strength and mobility activity prescription

Strength and Mobility Activities					
Activities	Toughening Phase (BCT & OSUT-R/W/B)	Sustaining Phase (AIT & OSUT-B/G)	Sustaining Phase ARFORGEN (Reset)	Sustaining Phase ARFORGEN (Train/Ready)	Sustaining Phase ARFORGEN (Available)
CD1	5 reps	5-10 reps	5-10 reps	5-10 reps	5-10 reps
CD2	5 reps	5-10 reps	5-10 reps	5-10 reps	5-10 reps
CD3	N/A	5-10 reps	5-10 reps	5-10 reps	5-10 reps
PSD	2 sets @ 30-60 sec	2-4 sets @ 30-60 sec	2-4 sets @ 30-60 sec	2-4 sets @ 30-60 sec	2-4 sets @ 30-60 sec
CL1	5 reps	5-10 reps	5-10 reps	5-10 reps	5-10 reps
CL2	N/A	5-10 reps w/load	5-10 reps w/load	5-10 reps w/load	5-10 reps w/load
STC	2-3 rotations @ 60 sec	2-3 rotations @ 60 sec	2-3 rotations @ 60 sec	2-3 rotations @ 60 sec	2-3 rotations @ 60 sec
GD	N/A	1-3 reps	1-3 reps	1-3 reps	1-3 reps
Abbreviations	CD1-Conditioning Drill 1　　　CD2-Conditioning Drill 2　　　CD3-Conditioning Drill 3　　PSD- Push-up/Sit-up Drill　　　CL1-Climbing Drill 1　　　　　CL2-Climbing Drill 2　　STC-Strength Training Circuit　GD- Guerrilla Drill				

CONDITIONING DRILL 1

9-2. Conditioning drill 1 consists of five exercises that develop complex motor skills while challenging strength, endurance, and mobility at a high intensity. All of the exercises in the drill are conducted to cadence, and are always performed in the sequence listed. In the toughening phase, Soldiers should perform no more than five repetitions of each exercise in CD 1. In the sustaining phase, Soldiers progress from 5 to 10 repetitions. If more repetitions are desired, then perform an additional set of the entire drill. Precise execution should never be sacrificed for speed.

TRAINING AREA

9-3. Any level area of adequate size is satisfactory for conduct of CDs.

UNIFORM

9-4. Soldiers will wear the IPFU or ACUs and boots.

EQUIPMENT

9-5. N/A.

FORMATION

9-6. For the most efficient instruction, the ideal unit size is one platoon. Larger units up to a battalion can successfully perform these drills if properly taught and mastered at the small unit level. The extended rectangular formation is prescribed.

LEADERSHIP

9-7. A PRT leader and AI are required to instruct and lead CD 1. The instructor must be familiar with the method of teaching these exercises, commands, counting cadence, cumulative count, formations, starting positions, and use of AIs as described in Chapter 7, Execution of Training. Soldiers should memorize the exercises by name and movement. The exercises are always given in cadence. Soldiers begin and terminate each exercise at the starting position and move to the position of attention before beginning the next exercise. The

Chapter 9

goal is to complete the entire drill with only enough pauses between exercises for the instructor to indicate the next one by name. This continuous method of conducting CD 1 intensifies the workload and conserves time. Considerable time and effort must be expended during the early stages to teach exercises properly to all Soldiers. Teach and practice exercises using a slow cadence (50 counts per minute) until correct form in executing each exercise is achieved.

PRECISION

9-8. Conditioning drill exercises lose much of their value unless performed exactly as prescribed. Precision should never be compromised for quantity of repetitions or speed of movement. A cadence that is too fast will not allow Soldiers to achieve a full range of movement.

PROGRESSION

9-9. Soldiers perform no more than five repetitions of each exercise while learning and practicing CDs. In the toughening phase, CD 1 is performed for five repetitions of each exercise. In the sustaining phase, CD 1 is performed for 5 to 10 repetitions of each exercise. Do not exceed ten repetitions of each exercise. Instead, if more repetitions are desired, perform additional sets of the entire drill.

INTEGRATION

9-10. Conditioning drill 1 integrates the components of strength, endurance, and mobility. This drill builds strength by challenging the control of body weight and promotes endurance without the repetitive motions that often lead to overuse injuries. It also improves mobility by progressively moving the major joints through a full, controlled range of motion.

COMMANDS

9-11. Chapter 7 provides the commands for CD 1.

BODY SEGMENTS TRAINED

9-12. Conditioning drill 1 consists of five 4-count exercises that train the body segments listed in Table 9-3. Instructions for giving commands are listed in Chapter 7, Execution of Training.

Table 9-3. Body segments trained in the conduct of CD 1

CONDITIONING DRILL 1 (CD 1)	MUSCLES							
	HIPS	THIGHS	LOWER LEGS	CHEST	BACK	TRUNK	SHOULDERS	ARMS
1. POWER JUMP	X	X	X		X	X	X	
2. V-UP	X	X	X		X	X	X	X
3. MOUNTAIN CLIMBER	X	X	X	X	X	X	X	X
4. LEG TUCK AND TWIST	X	X	X		X	X	X	X
5. SINGLE LEG PUSH-UP	X	X	X	X	X	X	X	X

CONDITIONING DRILL 1

EXERCISE 1: POWER JUMP

Purpose: This exercise reinforces correct jumping and landing, stimulates balance and coordination, and develops explosive strength (Figure 9-2).

Starting Position: Straddle stance with hands on hips.

Cadence: MODERATE

Count:

1. Squat with the heels flat as the spine rounds forward to allow the straight arms to reach to the ground, attempting to touch with the palms of the hands.
2. Jump forcefully in the air, vigorously raising arms overhead, with palms facing inward.
3. Control the landing and repeat count 1.
4. Return to the starting position.

Figure 9-2. Power jump

Check Points:

- At the starting position, tighten the abdominals to stabilize the trunk.
- On counts 1 and 3, keep the back generally straight with the head up and the eyes forward.
- On count 2, the arms should be extended fully overhead. The trunk and legs should also be in line.
- On each landing, the feet are directed forward and maintained at shoulder distance apart. The landing should be "soft" and proceed from the balls of the feet to the heels. The vertical line from the shoulders through the knees to the balls of the feet should be demonstrated on each landing.

Precaution: N/A

CONDITIONING DRILL 1

EXERCISE 2: V-UP

Purpose: This exercise develops the abdominal and hip flexor muscles while enhancing balance (Figure 9-3).

Starting Position: Supine, arms on ground 45 degrees to the side, palms down. The chin is tucked and the head is 1 to 2 inches off the ground.

Cadence: MODERATE

Count:

1. Raise straight legs and trunk to form a V-position, using arms as needed.
2. Return to the starting position.
3. Repeat count 1.
4. Return to the starting position.

Figure 9-3. V-up

Check Points:
- At the starting position, tighten the abdominal muscles to tilt the pelvis and the lower back toward the ground.
- On counts 1 and 3, straighten the knees and trunk and align the head with the trunk.
- On counts 2 and 4, lower the legs to the ground in a controlled manner so as not to injure the feet.

Precaution: To protect the spine, do not jerk the legs and trunk to rise to the V-position.

CONDITIONING DRILL 1

EXERCISE 3: MOUNTAIN CLIMBER

Purpose: This exercise develops the ability to quickly move the legs to power out of the front leaning rest position (Figure 9-4).

Starting Position: Front leaning rest position with the left foot below the chest and between the arms.

Cadence: MODERATE

Count:

1. Push upward with the feet and quickly change the positions of the legs.
2. Return to the starting position.
3. Repeat the movements in count 1.
4. Return to the starting position.

Figure 9-4. Mountain climber

Check Points:
- The hands are directly below the shoulders with the fingers spread (middle fingers pointing straight ahead) with the elbows straight, not locked.
- To prevent the trunk from sagging, contract and hold the abdominals throughout the exercise. Do not raise the hips and buttocks when moving throughout the exercise.
- Align the head with the spine and direct the eyes to a point about two feet in front of the body.
- Throughout the exercise, remain on the balls of the feet.
- Move the legs straight forward and backward, not at angles.

Precaution: N/A

CONDITIONING DRILL 1

EXERCISE 4: LEG TUCK AND TWIST

Purpose: This exercise develops trunk strength and mobility while enhancing balance (Figure 9-5).

Starting Position: Seated with trunk straight but leaning backward 45 degrees, arms straight and hands on ground 45 degrees to the rear with palms down. Legs are straight, extended to the front, and 8 to 12 inches off the ground.

Cadence: MODERATE

Count:

1. Raise legs while rotating on to the left buttock and draw the knees toward the left shoulder.
2. Return to the starting position.
3. Repeat count 1 in the opposite direction.
4. Return to the starting position.

Figure 9-5. Leg tuck and twist

Check Points:

- At the starting position, tighten the abdominals to stabilize the trunk.
- On all counts, keep the legs and knees together.
- On counts 1 and 3, the head and trunk remain still while the legs move.
- On counts 1 and 3, the legs are tucked (bent) and aligned diagonal to the trunk.

Precaution: To protect the back on counts 1 and 3, avoid jerking the legs and trunk to achieve the end position.

CONDITIONING DRILL 1

EXERCISE 5: SINGLE-LEG PUSH-UP

Purpose: This exercise strengthens muscles of the chest, shoulders, arms, and trunk. Raising one leg while maintaining proper trunk position makes this an excellent trunk stabilizing exercise (Figure 9-6).

Starting Position: Front leaning rest position.

Cadence: MODERATE

Count:

1. Bend the elbows, lowering the body until the upper arms are parallel with the ground while raising the left leg 8-10 inches off the ground.
2. Return to the starting position.
3. Repeat count 1, bringing the right leg to 8-10 inches off the ground.
4. Return to the starting position.

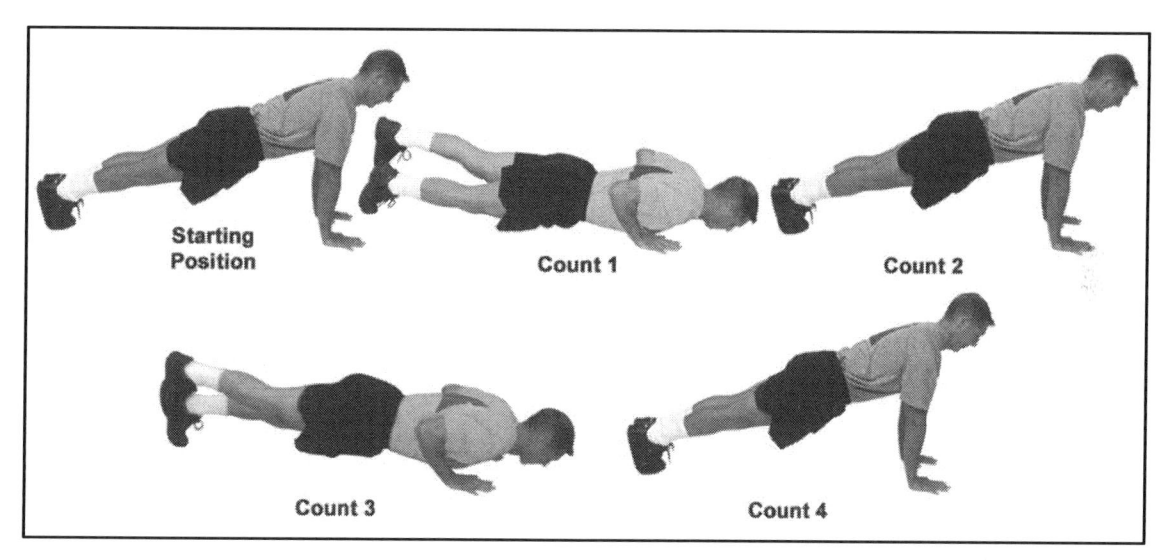

Figure 9-6. Single-leg push-up

Check Points:

- Perform a squat thrust to move into the front leaning rest. Keep the body straight from head to heels. Support the body weight on the hands and balls of the feet.
- The fingers should be extended and spread so the middle fingers point straight ahead and are directly in line with the shoulders.
- On counts 1 and 3, the upper arms stay close to the trunk.
- On counts 2 and 4, straighten but do not lock the elbows.
- On counts 1 and 3, the raised leg is straight and aligned with the trunk.
- To keep the trunk from sagging, tighten the abdominal muscles while in the starting position and maintain this contraction throughout the exercise.

Precautions: Do not jerk the leg being raised on counts 1 and 3. Also do not raise the leg higher than straight alignment with the trunk, as this may place undue stress on the back.

Strength and Mobility Activities

CONDITIONING DRILL 2

9-13. Conditioning drill 2 consists of five advanced exercises that require more complex plyometric and bilateral movement skills, while challenging the components of strength, endurance, and mobility (Table 9-4). Exercises are conducted at a slow (turn and lunge, supine bicycle, and swimmer) or moderate (half jacks and 8-count push-up) cadence. In the toughening phase, Soldiers should perform no more than five repetitions of each exercise in CD 2. In the sustaining phase, Soldiers progress from 5 to 10 repetitions. If more repetitions are desired, then perform an additional set of the entire drill. Precise execution should never be sacrificed for speed.

TRAINING AREA

9-14. Any level area of adequate size is satisfactory for conduct of CDs.

UNIFORM

9-15. Soldiers will wear IPFU or ACUs and boots.

EQUIPMENT

9-16. N/A.

FORMATION

9-17. For the most efficient instruction, the ideal unit size is one platoon. Larger units up to a battalion can successfully perform these drills if properly taught and mastered at the small unit level. The extended rectangular formation is prescribed.

LEADERSHIP

9-18. A PRT leader and an AI are required to instruct and lead CD 2. The instructor must be familiar with the method of teaching the exercises; the commands and counting cadence; cumulative count; formations; starting positions; and the use of AIs as described in Chapter 7. Soldiers should memorize the exercises by name and movement. The exercises are always given in cadence. Soldiers begin and terminate each exercise at the starting position and return to the position of attention before beginning the next exercise. The goal is to complete the entire drill with only enough pauses between exercises for the instructor to indicate the next one by name. This continuous method of conducting CD 2 intensifies the workload and conserves time. Considerable time and effort must be expended during the early stages to teach exercises properly to all Soldiers. Teach and practice exercises using a slow cadence (50 counts per minute) until correct form in executing each exercise is achieved.

PRECISION

9-19. Conditioning drill exercises lose much of their value unless performed exactly as prescribed. Precision should never be compromised for quantity of repetitions or speed of movement. A cadence that is too fast will not allow Soldiers to achieve a full range of movement.

PROGRESSION

9-20. Soldiers perform no more than five repetitions of each exercise while learning and practicing CDs. In the toughening phase, CD 2 is performed for five repetitions of each exercise. In the sustaining phase, CD 2 is performed for five to ten repetitions of each exercise. Do not exceed ten repetitions of each exercise. Instead, perform additional sets of the entire drill if more repetitions are desired.

INTEGRATION

9-21. Conditioning drill 2 integrates the components of strength, endurance, and mobility. This drill builds strength by challenging control of body weight and promotes endurance without the repetitive motions that often lead to overuse injuries. It also improves mobility by progressively moving the major joints through a full, controlled range of motion.

Chapter 9

COMMANDS

9-22. Conditioning drill 2 consists of four 4-count exercises and one 8-count exercise that train the body segments listed in Table 9-4. Chapter 7 provides instructions for giving commands.

Table 9-4. Body segments trained in the conduct of CD 2

CONDITIONING DRILL 2 (CD 2)	MUSCLES							
	HIPS	THIGHS	LOWER LEGS	CHEST	BACK	TRUNK	SHOULDERS	ARMS
1. TURN AND LUNGE	X	X	X	X	X	X	X	X
2. SUPINE BICYCLE	X	X	X	X	X	X	X	X
3. HALF JACKS	X	X	X		X	X	X	
4. SWIMMER	X	X	X		X	X	X	X
5. 8-COUNT PUSH-UP	X	X	X	X	X	X	X	X

Strength and Mobility Activities

CONDITIONING DRILL 2

EXERCISE 1: TURN AND LUNGE

Purpose: This exercise develops the agility needed to rotate, lower, and raise the body for effective changes of direction during military movement drill exercises, the 300-yd SR, and individual movement techniques (Figure 9-7).

Starting Position: Straddle stance with hands on hips.

Cadence: SLOW

Count:
1. Turn 90-degrees to the left, stepping with the left foot, and pivoting on the ball of the right foot. Perform a forward lunge (facing the left) while reaching toward the ground with the right hand. The left arm swings rearward while the left hand reaches rearward at the left side of the body.
2. Stand up, rotate to the right, and return to the starting position, stepping with the right foot and pivoting on the ball of the left foot.
3. Turn 90-degrees to the right, stepping with the right foot and pivoting on the ball of the left foot. Perform a forward lunge (facing the right) while reaching toward the ground with the left hand. The right arm swings rearward while the right arm reaches rearward at the right side of the body.
4. Stand up, rotate to the left, and return to the starting position, stepping with the left foot and pivoting on the ball of the right foot.

Figure 9-7. Turn and lunge

Check Points:
- When changing directions on all counts, the lead footsteps and the rear foot pivots.
- Keep the head in line with the spine throughout the exercise.
- Down positions on counts 1 and 3 are similar to the forward lunge, but with the hand down.

Precaution: N/A

CONDITIONING DRILL 2

EXERCISE 2: SUPINE BICYCLE

Purpose: This exercise strengthens the muscles of the abdomen and controls the rotation of the trunk (Figure 9-8).

Starting Position: Supine position with the fingers interlaced, hands on top of the head. Hips, knees, and ankles are flexed at 90 degrees and lower legs are parallel to the ground. The head is off the ground.

Cadence: SLOW

Count:

1. Bring the left knee toward the chest while flexing and rotating the trunk to the left, attempting to touch the right elbow with the left thigh. As the left knee rises, the right leg extends.
2. Return to the starting position.
3. Bring the right knee toward the chest while flexing and rotating the trunk to the right, attempting to touch the left elbow with the right thigh. As the right knee rises, the left leg extends.
4. Return to the starting position.

Figure 9-8. Supine bicycle

Check Points:

- At the starting position ensure that the hands are on top of the head, not behind the neck.
- Maintain tightness of the abdominals throughout the exercise.
- On counts 1 and 3, attempt to fully extend one leg while bringing the knee of the other to the elbow.

Precaution: On counts 1 and 3, do not jerk the neck or arch the back to assume the up position.

Strength and Mobility Activities

CONDITIONING DRILL 2

EXERCISE 3: HALF JACKS

Purpose: The purpose of this exercise is to jump and land with the legs apart, controlling the landing by laterally braking with the feet, ankles, and legs (Figure 9-9).

Starting Position: Position of attention.

Cadence: MODERATE

Count:
1. Jump and land with the feet shoulder-width apart and pointed straight ahead. The arms are sideward with palms facing down, thumbs and fingers extended and joined.
2. Jump and return to the starting position.
3. Repeat count 1.
4. Repeat count 2, returning to the starting position.

Figure 9-9. Half jacks

Check Points:
- On each landing, the balls of the feet should touch first.
- On counts 1 and 3, do not raise the arms above parallel to the ground.

Precaution: N/A

CONDITIONING DRILL 2

EXERCISE 4: SWIMMER

Purpose: This exercise strengthens the muscles of the low back and the shoulders while promoting quadrilateral coordination of the arms and legs (Figure 9-10).

Starting Position: The prone position with the arms extended, palms facing down, and toes pointed to the rear.

Cadence: SLOW

Count:

1. Raise the left arm and right leg 4 to 6 inches off the ground while arching the back slightly and looking upward.
2. Return to the starting position.
3. Raise the right arm and left leg 4 to 6 inches off the ground, while arching the back slightly and looking upward.
4. Return to the starting position.

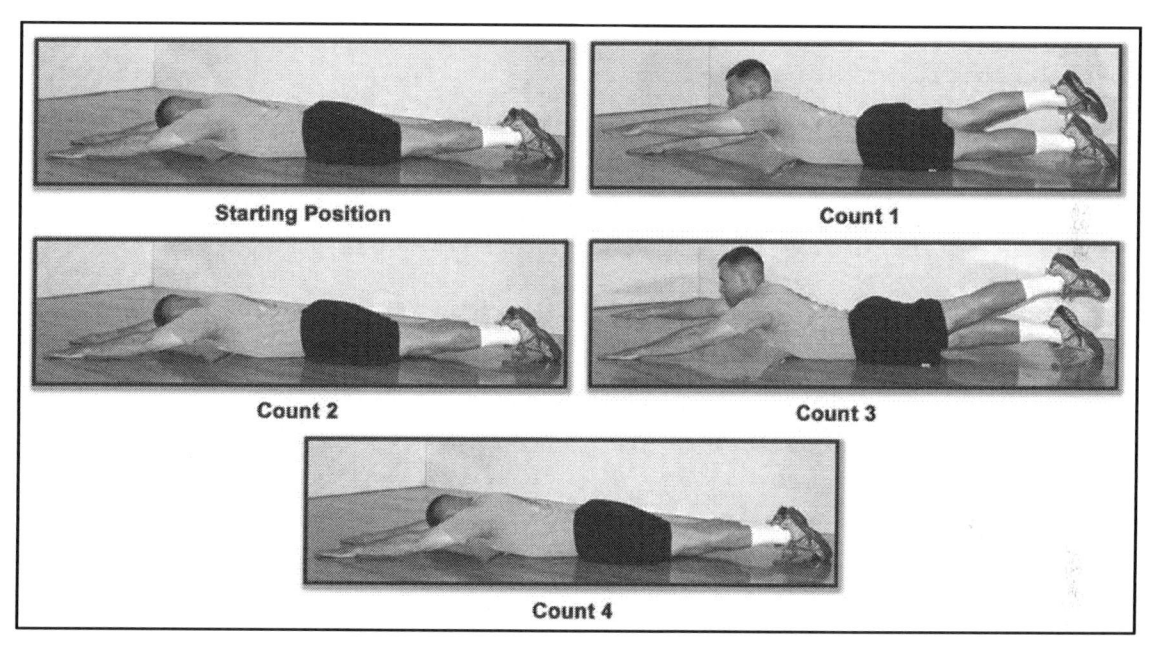

Figure 9-10. Swimmer

Check Points:

- At the starting position and throughout the exercise, maintain tightness in the abdominal and hip muscles.
- On counts 1 and 3, raise the head slightly and look upward.
- Keep the toes pointed throughout the exercise.

Precaution: Do not move into counts 1 and 3 with a jerking motion.

CONDITIONING DRILL 2

EXERCISE 5: 8-COUNT PUSH-UP

Purpose: This exercise combines the functional movements of the squat thrust and push-up to develop total body strength, endurance, and mobility (Figure 9-11).

Starting Position: Position of attention.

Cadence: MODERATE

Count:

1. Assume the squat position.
2. Thrust the legs backward to the front leaning rest position.
3. Bend the elbows, lowering the body until the upper arms are parallel with the ground. Elbows should point to the rear.
4. Return to the front leaning rest position.
5. Repeat count 3.
6. Repeat count 4.
7. Return to the squat position as in count 1.
8. Return to the starting position.

Chapter 9

Figure 9-11. 8-count push-up

Check Points:

- To keep the trunk from sagging, tighten the abdominal muscles while in the starting position and maintain this contraction throughout the exercise.
- On counts 1 through 7, the hands are directly below the shoulders with fingers spread and the middle fingers directed straight forward.
- On counts 1 and 7, keep the heels together and raised.
- On counts 4 and 6, straighten but do not lock the elbows.

Precautions: Allowing the trunk to sag, especially on count 2, strains the back. Avoid this by maintaining a strong abdominal contraction throughout the exercise. If the pushup cannot be performed on counts 2-6 correctly to cadence, quickly assume the 6-point stance before count 3 and return to the front leaning rest position just before performing count 7.

CONDITIONING DRILL 3

9-23. Conditioning drill 3 is conducted in a similar manner to CD 1 and 2; however, the exercises in CD 3 are more difficult and complex. Repeated jumping, landing, and changing of body positions make this a more advanced drill with greater demands placed on the lower extremities. In the toughening phase, Soldiers should not perform CD 3. In the sustaining phase, Soldiers progress from 5 to 10 repetitions. If more repetitions are desired, then perform an additional set of the entire drill. (See Table 9-5.)

Table 9-5. Body segments trained in the conduct of CD 3

CONDITIONING DRILL 3 (CD 3)	HIPS	THIGHS	LOWER LEGS	CHEST	BACK	TRUNK	SHOULDERS	ARMS
1. "Y" SQUAT	X	X	X		X	X	X	X
2. SINGLE-LEG DEAD LIFT	X	X	X		X	X		
3. SIDE-TO-SIDE KNEE LIFTS	X	X	X	X	X	X	X	
4. FRONT KICK ALTERNATE TOE TOUCH	X	X	X	X	X	X	X	X
5. TUCK JUMP	X	X	X	X	X	X	X	X
6. STRADDLE-RUN FORWARD AND BACKWARD	X	X	X			X		
7. HALF-SQUAT LATERALS	X	X	X	X	X	X	X	X
8. FROG JUMPS FORWARD AND BACKWARD	X	X	X	X	X	X	X	X
9. ALTERNATE ¼-TURN JUMP	X	X	X	X	X	X	X	X
10. ALTERNATE-STAGGERED SQUAT JUMP	X	X	X		X	X	X	

CONDITIONING DRILL 3

EXERCISE 1: "Y" SQUAT

Purpose: This exercise develops strength, endurance, and mobility of the lower back and lower extremities (Figure 9-12).

Starting Position: Straddle stance with shoulder blades pulled rearward with arms overhead and palms inward.

Cadence: SLOW

Count:

1. Squat with arms overhead (forming a "Y") without allowing the back to round.
2. Return to the starting position by tightening the buttocks and driving upward.
3. Repeat count 1.
4. Return to the starting position.

Figure 9-12. "Y" squat

Check Points:

- During count 1, lower the body as far as possible without rounding the back, keeping the shoulders drawn rearward, arms forming a "Y" overhead.
- Tighten the buttocks and drive the trunk upward to return to the starting position.
- Heels remain on the ground throughout the exercise.

Precaution: N/A

CONDITIONING DRILL 3

EXERCISE 2: SINGLE-LEG DEAD LIFT

Purpose: This exercise develops strength, endurance, and flexibility of the lower back and lower extremities (Figure 9-13).

Starting Position: Straddle stance with hands on hips.

Cadence: SLOW

Count:

1. Stand maintaining balance on the left foot and bend forward at the waist. Reach straight down toward the ground in front of the body while raising the right leg to the rear.
2. Return to the starting position by tightening the buttocks and driving upward.
3. Stand maintaining balance on the right foot and bend forward at the waist. Reach straight down toward the ground in front of the body while raising the left leg to the rear.
4. Return to the starting position.

Chapter 9

Figure 9-13. Single-leg dead lift

Check Points:

- On counts 1 and 3, the hands are slightly in front of and below the shoulders with fingers spread (middle fingers point straight ahead) with the elbows straight, not locked.
- Maintain a natural arch in the back and move the legs straight forward and backward, not at angles.
- To prevent the trunk from sagging, tighten the abdominal muscles and maintain this contraction throughout the exercise.
- The head is aligned with the spine and the eyes are directed to a point about two feet in front of the body.
- On counts 1 and 3, attempt to keep the heal on the ground.

Precaution: N/A

CONDITIONING DRILL 3

EXERCISE 3: SIDE-TO-SIDE KNEE LIFTS

Purpose: This exercise develops coordination, balance, and explosive strength in the legs (Figure 9-14).

Starting Position: Straddle stance with hands on hips.

Cadence: MODERATE

Count:

1. Hop to the left, landing on the left foot, while simultaneously drawing the right knee toward the chest. The right hand moves comfortably down to the side toward the right ankle and the left hand touches the right knee.
2. Hop to the right, landing on the right foot, while simultaneously drawing the left knee toward the chest, the left hand moves comfortably down to the side toward the left ankle and the right hand touches the left knee.
3. Repeat count 1.
4. Repeat count 2 and return to the starting position on the final repetition.

Chapter 9

Figure 9-14. Side-to-side knee lifts

Check Points:
- At the starting position, tighten the abdominals to stabilize the trunk.
- On all counts, do not allow the back to round; keep the head up and the eyes forward.
- On each landing, the feet should be directed forward and maintained at shoulder distance apart. The landing should be "*soft*" and proceed from the balls of the feet to the heels. The vertical line from the shoulders through the knees to the balls of the feet should be demonstrated on each landing.

Precaution: N/A

CONDITIONING DRILL 3

EXERCISE 4: FRONT KICK ALTERNATE TOE TOUCH

Purpose: This exercise develops balance, coordination, and flexibility of the legs and trunk (Figure 9-15).

Starting Position: Straddle stance with hands on hips.

Cadence: MODERATE

Count:

1. Raise the left leg to the front of the body until it is parallel to the ground while simultaneously bending forward at the waist, extending the right arm forward, and reaching with the right hand toward the left foot, while the left arm reaches rearward.
2. Return to the starting position.
3. Raise the right leg to the front of the body until it is parallel to the ground while simultaneously bending forward at the waist, extending the left arm forward, and reaching with the left hand toward the right foot, while the right arm reaches rearward.
4. Return to the starting position.

Chapter 9

Figure 9-15. Front kick alternate toe touch

Check Points:
- At the starting position, tighten the abdominals to stabilize the trunk.
- On counts 1 and 3, rotate the trunk to reach for the toes keeping the back generally straight.
- Keep the head and the eyes forward throughout the exercise.
- Maintain a slight bend in the knee as it moves parallel to the ground.

Precaution: N/A

CONDITIONING DRILL 3

EXERCISE 5: TUCK JUMP

Purpose: This exercise develops coordination, balance, and explosive strength in the legs (Figure 9-16).

Starting Position: Straddle stance with arms at the sides.

Cadence: SLOW

Count:

1. Perform a half squat, while driving both arms rearward. Jump upward, driving both arms forward, wrapping the hands around the knees, as the knees are drawn toward the chest. Then land in the half-squat position.
2. Return to the starting position.
3. Repeat count 1.
4. Return to the starting position.

Figure 9-16. Tuck jump

Check Points:

- On counts 1 and 3, do not allow the back to round; keep the head up and the eyes forward.
- Cadence is slow to allow for precision and adequate time to properly jump and land; however, each jump on counts 1 and 3 should be performed quickly and explosively.
- On each landing, the feet should be directed forward and maintained at shoulder distance apart. The landing should be "soft" and proceed from the balls of the feet to the heels. The vertical line from the shoulders through the knees to the balls of the feet should be demonstrated on each landing.

Precaution: N/A

CONDITIONING DRILL 3

EXERCISE 6: STRADDLE-RUN FORWARD AND BACKWARD

Purpose: This exercise develops coordination, balance, and explosive strength in the legs (Figure 9-17).

Starting Position: Straddle stance with arms at the sides.

Cadence: MODERATE

Count:

1. Raise the left leg 4 to 6 inches off the ground and bound forward to the left at a 45-degree angle while swinging the right arm forward and left arm rearward.
2. Raise the right leg 4 to 6 inches off the ground and bound forward to the right at a 45-degree angle while swinging the left arm forward and right arm rearward.
3. Repeat count 1.
4. Repeat count 2.
5. Raise the left leg 4 to 6 inches off the ground and bound rearward to the left at a 45-degree angle while swinging the left arm forward and right arm rearward.
6. Raise the right leg 4 to 6 inches off the ground and bound rearward to the right at a 45-degree angle while swinging the right arm forward and left arm rearward.
7. Repeat count 5.
8. Repeat count 6 and assume the starting position.

Figure 9-17. Straddle-run forward and backward

Check Points:

- On all counts, do not allow the back to round; keep the head up and the eyes forward.
- On each landing, the feet should be directed forward and the trail foot moves toward the lead foot, but does not make contact with the ground.

Precaution: N/A

CONDITIONING DRILL 3

EXERCISE 7: HALF-SQUAT LATERALS

Purpose: This exercise develops coordination, balance, and explosive strength in the legs (Figure 9-18).

Starting Position: Straddle stance, slightly crouched, assuming a half-squat, with the back straight, arms at the sides with elbows bent at 90-degrees, and palms facing forward.

Cadence: MODERATE

Count:

1. Maintaining a half-squat step/hop to the left.
2. Maintaining a half-squat step/hop to the right.
3. Maintaining a half-squat step/hop to the right.
4. Maintaining a half-squat step/hop to the left and return to the starting position.

Figure 9-18. Half-squat laterals

Check Points:

- At the starting position, tighten the abdominals to stabilize the trunk.
- On all counts, do not allow the back to round; keep the head up and the eyes forward.
- On each landing, the feet should be directed forward and maintained at shoulder distance apart. The landing should be "soft" and proceed from the balls of the feet to the heels.

Precaution: N/A

CONDITIONING DRILL 3

EXERCISE 8: FROG JUMPS FORWARD AND BACKWARD

Purpose: This exercise develops coordination, balance, and explosive strength in the legs (Figure 9-19).

Starting Position: Straddle stance, slightly crouched, assuming a half-squat, with the back straight, arms at the sides with elbows bent at 90-degrees, and palms facing forward.

Cadence: MODERATE

Count:

1. Maintain a half-squat and hop forward.
2. Maintain a half-squat and hop backward.
3. Repeat count 2.
4. Maintain a half-squat and hop forward, returning to the starting position.

Figure 9-19. Frog jumps forward and backward

Check Points:

- At the starting position, tighten the abdominals to stabilize the trunk.
- On all counts, do not allow the back to round; keep the head up and the eyes forward.
- On each landing, the feet should be directed forward and maintained at shoulder distance apart. The landing should be "soft" and proceed from the balls of the feet to the heels.

Precaution: N/A

CONDITIONING DRILL 3

EXERCISE 9: ALTERNATE ¼-TURN JUMP

Purpose: This exercise develops balance, explosive strength in the legs, and control of trunk rotation (Figure 9-20).

Starting Position: Straddle stance, slightly crouched, assuming a half-squat, with the back straight, arms at the sides with elbows bent at 90-degrees, and palms facing forward.

Cadence: MODERATE

Count:

1. Jump upward and twist the hips, turning the legs 90-degrees to the left.
2. Return to the starting position.
3. Jump upward and twist the hips, turning the legs 90-degrees to the right.
4. Return to the starting position.

Figure 9-20. Alternate ¼-turn jump

Check Points:

- At the starting position, tighten the abdominals to stabilize the trunk.
- On counts 1 and 3, do not allow the back to round; keep the head up and the eyes forward.
- The upper body does not turn; the movement involves only the hips and legs.
- On each landing, the feet should be directed forward and maintained at shoulder distance apart. The landing should be "soft" and proceed from the balls of the feet to the heels. The vertical line from the shoulders through the knees to the balls of the feet should be demonstrated on each landing.

Precaution: N/A

CONDITIONING DRILL 3

EXERCISE 10: ALTERNATE-STAGGERED SQUAT JUMP

Purpose: This exercise develops balance and explosive strength of the legs (Figure 9-21).

Starting Position: Staggered stance with the left leg back and arms at sides; the trunk is generally straight, but tilted slightly forward.

Cadence: SLOW

Count:

1. Squat and touch the ground, between the legs, with the fingertips of the left hand. Jump forcefully into the air, switching legs in mid-air to land with the right leg back and arms at the sides.
2. Squat and touch the ground between the legs with the fingertips of the right hand. Jump forcefully into the air, switching legs in mid-air to land with the left leg back and arms at the sides.
3. Repeat count 1.
4. Repeat count 2 and return to the starting position.

Figure 9-21. Alternate-staggered squat jump

Check Points:
- At the starting position, tighten the abdominals to stabilize the trunk.
- Do not allow the back to round; keep the head up and the eyes forward.
- Cadence is slow to allow for precision and adequate time to properly jump and land; however, each jump should be performed quickly and explosively.
- On each landing, the feet should be oriented to the front. The landing should be "soft" and proceed from the balls of the feet to the heels.

Precaution: N/A

Strength and Mobility Activities

PUSH-UP AND SIT-UP DRILL

9-24. Push-ups and sit-ups develop upper body strength, endurance, and mobility, and specifically prepare Soldiers for APFT performance. Push-ups and sit-ups build upper body and trunk muscular strength and endurance by challenging control of body weight. The PSD promotes muscular endurance without the repetitive motions that often lead to overuse injuries. They improve mobility by progressively moving the major joints through a full, controlled range of motion.

TRAINING AREA

9-25. Any level area of adequate size is satisfactory for conduct of the PSD.

UNIFORM

9-26. Soldiers will wear IPFU or ACUs and boots.

EQUIPMENT

9-27. Stop watch.

FORMATION

9-28. For the most efficient instruction, the ideal unit size is one platoon. Larger units up to a battalion can successfully perform these drills if properly taught and mastered at the small unit level. The extended rectangular formation is prescribed.

LEADERSHIP

9-29. A PRT leader and AI are required to instruct and lead timed sets of push-ups and sit-ups. The leader must know how to teach these exercises. He must know the commands, cadence counts, cumulative count, formations, starting positions, and how to effectively use AIs (Chapter 7).

METHODOLOGY

9-30. The PSD enhances APFT performance in the push-up and sit-up events. The PSD is conducted as follows:

- The first and third ranks conduct the push-up first. The second and fourth ranks count repetitions out loud and monitor technique to ensure the Soldiers perform the push-ups to Army standard (hand placement is determined by the Soldier according to Appendix A) for 30 to 60 seconds. After the first and third ranks complete the push-ups, the ranks swap places: the second and fourth ranks do push-ups and the first and third ranks count and monitor proper technique. After all four ranks complete the first timed set of push-ups; the same process is repeated for sit-ups.
- The sit-up is conducted the same as the push-up: first and third perform, second and fourth count and monitor technique, but also hold the feet of the first and third ranks. Again, when the first and third ranks finish, the ranks swap out again, and the second and fourth ranks perform while the first and third ranks count, monitor technique, and hold the feet.
- Timed sets continue like this, alternating between push-ups and sit-ups and between paired ranks, until all the desired number of timed sets have been completed. The Soldiers should not perform all of their sets of timed push-ups and then perform all of their sets of timed sit-ups. Alternating allows proper work to rest ratio to provide the required recovery. Avoid performing all of one exercise or the other.
- As with any activity, PRT leaders should perform the exercises with the Soldiers in order to determine the appropriate intensity of the PRT session.

Chapter 9

PRECISION

9-31. Push-ups and sit-ups lose much of their value unless performed exactly as prescribed. Precision should never be compromised for quantity of repetitions or speed of movement.

PROGRESSION

9-32. Soldiers perform no more than five repetitions of each exercise while learning and practicing the PSD. They perform timed sets of push-ups and sit-ups during the activity part of the PRT session. They perform as many correct repetitions of push-ups and sit-ups during the 30-second timed sets as they can, progressing to 60-second timed sets. Soldiers that fail with time remaining in the timed set of push-ups will go to their knees and continue to perform the push-up in the six-point stance until time has expired within the timed set.

INTEGRATION

9-33. Performing timed sets of push-ups and sit-ups integrates the components of strength, endurance, and mobility.

COMMANDS

9-34. Follow the procedures in paragraph 9-30.

BODY SEGMENTS TRAINED

9-35. The PSD consists of two exercises that train the body segments listed in Table 9-6. Refer to Appendix A for illustrations and descriptions of the push-up and sit-up according to the APFT.

Table 9-6. Body segments trained in the conduct of PSD

PUSH-UP AND SIT-UP DRILL (PSD)	HIPS	THIGHS	LOWER LEGS	CHEST	BACK	TRUNK	SHOULDERS	ARMS
1. PUSH-UP	X	X	X	X	X	X	X	X
2. SIT-UP	X	X	X		X	X	X	X

CLIMBING DRILLS

9-36. The purpose of the CL is to improve upper body and trunk strength, and the ability to climb and negotiate obstacles. Success in climbing and surmounting obstacles depends on both conditioning and technique. These drills include exercises that condition the muscles of the body that are predominant in climbing. The entire body is involved during climbing by helping to change or stabilize position.

- The hands and feet act as anchor points and initiate movement to the next position.
- The abdominal and back muscles stabilize the body's position.
- The arms push and pull upward with assistance from the much stronger legs.

9-37. Climbing drills, when combined with CDs, the push-up sit-up drill, the GD, and the strength training circuit comprise a well-balanced program of functional strength development. Climbing drills are performed during the activity part of a PRT session.

Strength and Mobility Activities

TRAINING AREA

9-38. The CLs are best performed on climbing bars (Appendix B). To conduct the CLs with multiple Soldiers at one time, allow at least one bar for every three Soldiers. At least one bar is required for every six Soldiers when the CLs are coupled with other strength and mobility drills.

UNIFORM

9-39. Soldiers will wear ACUs and boots or IPFU. Boots and ACUs will be worn when performing the CD in combination with the GD. Additional equipment such as body armor (IOTV), ACH, and weapon will be used when performing CD 2 in the sustaining phase.

EQUIPMENT

9-40. The CDs are best conducted on climbing bars. The thickness of the bars is no more than 1 ½-inch outside diameter. The bars are supported by 6-by-6 inch pressure-treated posts sunk at least 3 feet into the ground and secured with cement. The bar distance from inside post to inside post must be at least 5 feet. The bars should be no more than 8 feet off the ground. A variety of heights or steps should be available to accommodate all Soldiers. Figure 9-22 shows the recommended climbing bar apparatus and Soldier use. (Appendix B provides detailed specifications for constructing climbing bars.)

SPOTTING

9-41. Two spotters are used during CL 1 and 2 to ensure precision, adherence to proper cadence, and safety by assisting Soldiers who are unable to properly perform the desired number of repetitions. All Soldiers performing CL 1 and 2 are required to use spotters, unless they demonstrate the ability to perform 5 repetitions of an exercise unassisted. The Soldier then gives a verbal cue "no spot needed." As Soldiers develop more strength, they will require less assistance from the spotters.

Spotting the Straight-Arm Pull, Pull-Up, and Alternating Grip Pull-Up

9-42. The front spotter assumes a staggered stance with the palms toward the exerciser at approximately chest level. The front spotter's primary role is to support the exerciser if their grip fails. The rear spotter assumes a staggered stance and holds the Soldiers feet on their thighs or abdomen. The hands are placed above the ankles to hold them securely. The role of the rear spotter is to provide a stable base for the exerciser to push against with his legs. When spotting is performed correctly, the rear spotter will neither have to lift nor go up and down with the exerciser. The rear spotter must anticipate the last repetition and release the Soldier when the "down" command is given before the "dismount" command. Soldiers are required to use the foot pegs when mounting and dismounting the bar. This is done to ensure safety and to reduce injuries. Jumping from the mounted position causes compression to the spine and other joints of the body, possibly resulting in injury.

Spotting Heel Hook and Leg Tuck

9-43. One spotter stands on each side of the exerciser in the straddle stance. The rear hand of each spotter is placed in the small of the back and the forward hand is placed beneath the thigh above the back of the knee. Both hands remain in contact with the exerciser throughout the exercise. The rear hand is used primarily to prevent the exerciser from swaying, while the forward hand helps lift the legs into the 'up' position. Soldiers are required to use the foot pegs when mounting and dismounting the bar. This is done to ensure safety and to reduce injuries. Jumping from the mounted position causes compression to the spine and other joints of the body, possibly resulting in injury.

Chapter 9

Figure 9-22. Climbing pod

FORMATION

9-44. Three Soldiers are assigned to each bar. One Soldier exercises and two Soldiers perform as spotters.

LEADERSHIP

9-45. Both a PRT leader and an AI are required in order to instruct and lead CLs. The PRT leader must be familiar with the method of teaching the exercises; the commands and counting cadence; cumulative count; formations; and the use of AIs as described in Chapter 7. The PRT leader must ensure that spotters are properly trained and maintain positive control of the Soldier performing the CL at all times.

9-46. Soldiers should memorize the exercises by name and movement. The exercises are always given in cadence. Soldiers rotate during each exercise until all three have completed the exercise. Only then may the PRT leader move them to the next exercise. Considerable time and effort must be expended during the early stages to teach precise performance of each exercise.

PRECISION

9-47. Climbing drills lose much of their value unless performed exactly as prescribed. Spotters ensure safety and precise execution by helping Soldiers who are tired or unable to properly execute five repetitions on their own. Two spotters help Soldiers though each exercise. As Soldiers become more proficient in each exercise, they will need less assistance and should eventually be able to perform the drill unaided; however, spotters are always present. Spotters help reduce swinging and stabilize body position. Precision should never be compromised for quantity of repetitions or speed of movement. Soldiers should, therefore, perform all movements in a controlled manner without jerking into or out of positions. They should avoid relaxing in the extended hang position, as this can place excessive stress on the shoulder and elbow joints. Soldiers should maintain a contraction in the muscles of the shoulders and upper back to avoid a relaxed, extended hang.

PROGRESSION

9-48. In the sustaining phase, Soldiers progress from 5 to 10 repetitions of each exercise in CL 1 with or without assistance, using only their body weight as resistance. The goal is to perform all five exercises unassisted.

9-49. In the sustaining phase, Soldiers perform CL 2 while under load. Soldiers increase resistance by wearing LBE/LBV, ACH, body armor (IOTV), and individual weapons. Spotters provide assistance until Soldiers can complete all five exercises without help.

INTEGRATION

9-50. The primary emphasis of the CLs is functional strength development. The use of equipment in CL 2 develops the Soldiers' ability to manipulate their body weights under load. The various exercises also involve movements that require mobility. Climbing drills, when combined with CDs, the push-up sit-up drill, the GD, and the strength training circuit comprise a well-balanced program of total body functional strength development.

COMMANDS

9-51. Climbing drills 1 and 2 each consist of five 2-count exercises.

HAND POSITIONS

9-52. Throughout the drills, a variety of hand positions are used to thoroughly train the musculature of the arms, forearms, and hands. Hands are placed shoulder-width apart with thumbs around the bar for the overhand grip. Hands are placed next to each other with thumbs around the bar for the alternating grips (Figure 9-23).

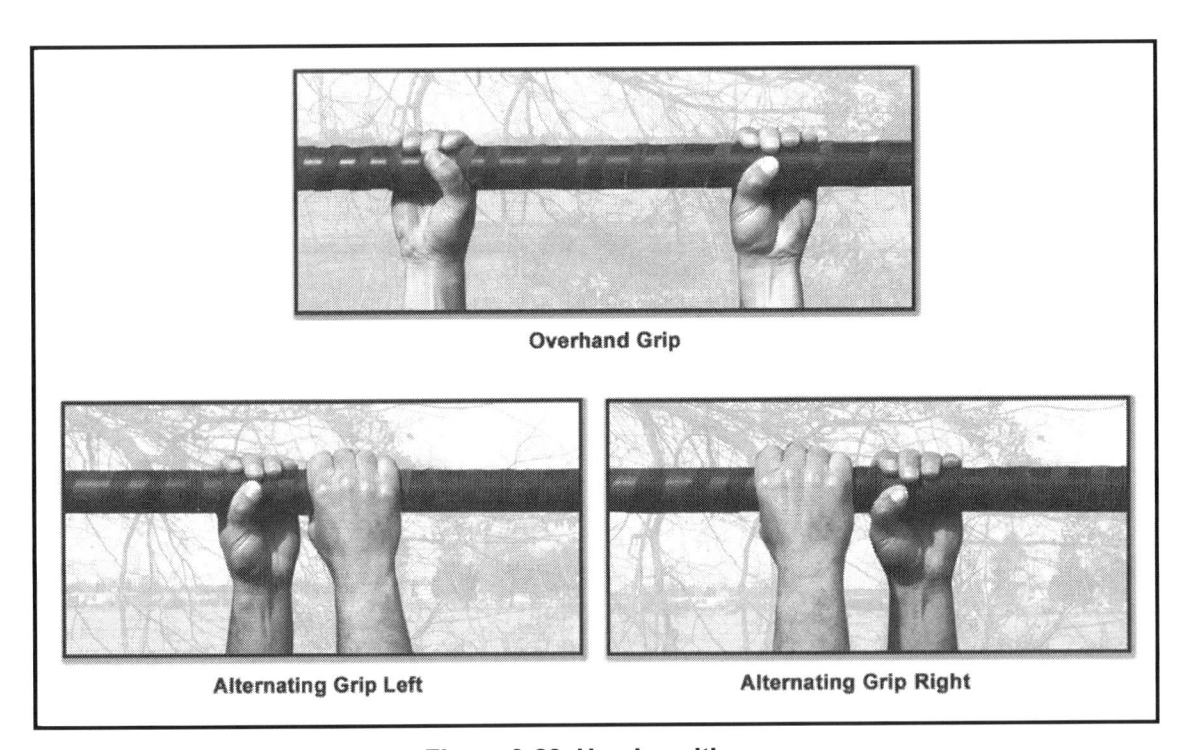

Figure 9-23. Hand positions

CLIMBING DRILL 1

9-53. Climbing drill 1 (CL 1) improves upper body and core strength and the ability to climb and negotiate obstacles. Climbing drill 1 is always performed in its entirety in the order listed. Soldiers perform 5 to 10 repetitions of each exercise in CL 1 with or without assistance, using only their body weight as resistance. The goal is to perform 5 to 10 repetitions of all five exercises unassisted. If a second set is performed, the entire drill is repeated in the order listed. Table 9-7 lists the body segments trained in CL1.

Table 9-7. Body segments trained in the conduct of CL1

CLIMBING DRILL 1 (CL 1)	HIPS	THIGHS	LOWER LEGS	CHEST	BACK	TRUNK	SHOULDERS	ARMS
1. THE STRAIGHT-ARM PULL					X	X	X	X
2. HEEL HOOK	X	X	X	X	X	X	X	X
3. PULL-UP					X	X	X	X
4. LEG TUCK	X	X	X	X	X	X	X	X
5. ALTERNATING GRIP PULL-UP				X	X	X	X	X

CLIMBING DRILL 1

EXERCUSE 1: STRAIGHT-ARM PULL

Purpose: This exercise develops the ability to initiate the pull-up motion and maintain a contraction in the extended hang position (Figure 9-24).

Starting Position: Extended hang using the overhand grip.

Cadence: MODERATE

Count:
1. Keeping the arms straight, pull the body upward using the shoulders and upper back muscles only.
2. Return to the starting position.

Figure 9-24. Straight-arm pull

Check Points:
- Throughout the exercise, keep the arms shoulder width, palms facing away from the body, with the thumbs around the bar.
- Throughout the exercise, keep the elbows straight, but not locked.
- On count 1, pull the body up by engaging the shoulder muscles (squeeze the shoulder blades together).

Precaution: Refer to paragraph 9-42 for spotting.

CLIMBING DRILL 1

EXERCISE 2: HEEL HOOK

Purpose: This exercise develops the ability to raise the legs from a hanging position and hook the feet securely on the bar (Figure 9-25).

Starting Position: Extended hang using the alternating grip, left or right.

Cadence: SLOW

Count:

1. Pull with the arms and curl the lower body toward the bar. Raise the feet above the bar and interlock them securely around the bar.
2. Return to the starting position.

Figure 9-25. Heel hook

Check Points:

- On count 1, initiate movement by first pulling with the arms.
- Secure the feet over the bar by crossing one foot over the other at the ankles.
- On count 2, fully extend the arms to return to the starting position.

Precaution: Refer to paragraph 9-43 for spotting.

CLIMBING DRILL 1

EXERCISE 3: PULL-UP

Purpose: This exercise develops the ability to pull the body upward while hanging (Figure 9-26).

Starting Position: Extended hang using the overhand grip.

Cadence: MODERATE

Count:
1. Keeping the body straight and pull upward with the arms until the chin is above the bar.
2. Return to the starting position.

Figure 9-26. Pull-up

Check Points:
- Throughout the exercise, keep the feet together.
- Throughout the exercise, the arms shoulder-width, palms facing away from the body, with the thumbs around the bar.

Precaution: Refer to paragraph 9-42 for spotting.

CLIMBING DRILL 1

EXERCISE 4: LEG TUCK

Purpose: This exercise develops the abdominal, hip flexor, and grip strength essential to climbing a rope (Figure 9-27).

Starting Position: Extended hang using the alternating grip, left or right.

Cadence: SLOW

Count:

1. Pull up with the arms and raise the knees toward the chest until the elbows touch the thighs just above the knees.
2. Return to the starting position.

Figure 9-27. Leg tuck

Check Points:

- Throughout the exercise, keep the feet together.
- On count 1, the thighs and elbows touch just above the knees.

Precaution: Refer to paragraph 9-43 for spotting.

CLIMBING DRILL 1

EXERCISE 5: ALTERNATING GRIP PULL-UP

Purpose: This exercise develops the muscles used to pull the body upward while using an alternating grip (Figure 9-28).

Starting Position: Extended hang using the alternating grip, left or right.

Cadence: MODERATE

Count:

1. Keep the body straight, pull upward, allowing the head to move to the left or right side of the bar, and touch the left or right shoulder to the bar.
2. Return to the starting position.

Figure 9-28. Alternating grip pull-up

Check Points:

- When using the left alternating grip, Soldiers touch the left shoulder to the bar on count 1. If the right alternating grip is used, Soldiers touch the right shoulder to the bar on count 1.
- On count 2, the arms are fully extended to return to the starting position.
- Keep the feet together, throughout the exercise.

Precaution: Refer to paragraph 9-42 for spotting.

CLIMBING DRILL 2

9-54. Climbing drill 2 (CL 2) is a performance-oriented drill conducted in the sustaining phase that prepares Soldiers for critical tasks under fighting load such as climbing, traversing a rope, and pulling the body up on to a ledge or through a window. Soldiers increase the resistance by performing CL 2 with their LBE/LBV, body armor (IOTV), ACH, and individual weapon. They will hold the UP position of exercise 1, the flexed-arm hang, for five seconds (one repetition, only) and perform five repetitions of each of the remaining four exercises: the heel hook, the pull-up, the leg tuck, and the alternating grip pull-up. Spotters provide assistance until Soldiers can complete all repetitions without assistance. Soldiers may progress from five to ten repetitions and one to two sets of CL 2 once they are able to perform most of the drill unassisted. The goal is to perform 5 to 10 repetitions of all five exercises unassisted. If a second set is performed, the entire drill is repeated in the order listed. Table 9-8 lists the body segments trained in CL 2.

Table 9-8. Body segments trained in CL 2

CLIMBING DRILL 2 (CL 2)	HIPS	THIGHS	LOWER LEGS	CHEST	BACK	TRUNK	SHOULDERS	ARMS
1. FLEXED-ARM HANG					X	X	X	X
2. HEEL HOOK	X	X	X	X	X	X	X	X
3. PULL-UP					X	X	X	X
4. LEG TUCK	X	X	X	X	X	X	X	X
5. ALTERNATING GRIP PULL-UP				X	X	X	X	X

CLIMBING DRILL 2

EXERCISE 1: FLEXED-ARM HANG

Purpose: This exercise develops the ability to hold the body in the flexed-arm hang position (Figure 9-29).

Starting Position: Extended hang using the overhand grip.

Cadence: N/A

Count: This exercise is performed for one repetition of five seconds.

1. On the command UP, keeping the body straight, pull upward with the arms until the chin is above the bar and hold for five seconds.
2. On the command DOWN, return to the starting position.

Figure 9-29. Flexed-arm hang

Check Points:
- Throughout the exercise, the palms are facing away from the body, with the thumbs around the bar.
- Throughout the exercise, keep the feet close together.

Precaution: Refer to paragraph 9-42 for spotting.

CLIMBING DRILL 2

EXERCISE 2: HEEL HOOK

Purpose: This exercise develops the ability to raise the legs from a hanging position and hook the feet securely on the bar (Figure 9-30).

Starting Position: Extended hang using the alternating grip, left or right.

Cadence: SLOW

Count:

1. Pull with the arms and the body toward the bar. Raise the feet above the bar and interlock them securely around the bar.
2. Return to the starting position.

Figure 9-30. Heel hook

Check Points:
- On count 1, initiate movement by first pulling with the arms.
- Secure the feet over the bar by crossing one foot over the other at the ankles.
- On count 2, fully extended the arms to return to the starting position.

Precaution: Refer to paragraph 9-43 for spotting.

CLIMBING DRILL 2

EXERCISE 3: PULL-UP

Purpose: This exercise develops the ability to pull the body upward while hanging (Figure 9-31).

Starting Position: Extended hang using the overhand grip.

Cadence: MODERATE

Count:

1. Keeping the body straight, pull upward with the arms until the chin is above the bar.
2. Return to the starting position.

Figure 9-31. Pull-up

Check Points:

- Throughout the exercise, keep the feet together.
- Throughout the exercise, arms are shoulder-width, palms facing away from the body, with thumbs around the bar.

Precaution: Refer to paragraph 9-42 for spotting.

CLIMBING DRILL 2

EXERCISE 4: LEG TUCK

Purpose: This exercise develops the abdominal, hip flexor, and grip strength essential to climbing a rope (Figure 9-32).

Starting Position: Extended hang using the alternating grip, left or right.

Cadence: SLOW

Count:

1. Pull up with the arms and raise the knees toward the chest until the elbows touch the thighs just above the knees.
2. Return to the starting position.

Figure 9-32. Leg tuck

Check Points:
- Throughout the exercise, keep the feet together.
- On count 1, the thighs and elbows touch just above knees.

Precaution: Refer to paragraph 9-43 for spotting.

CLIMBING DRILL 2

EXERCISE 5: ALTERNATING GRIP PULL-UP

Purpose: This exercise develops the ability to pull the body upward while hanging with an alternating grip (Figure 9-33).

Starting Position: Extended hang using the alternating grip, left or right.

Cadence: MODERATE

Count:

1. Keeping the body straight, pull upward, allowing the head to move to the left or right side of the bar and touch the left or right shoulder to the bar.
2. Return to the starting position.

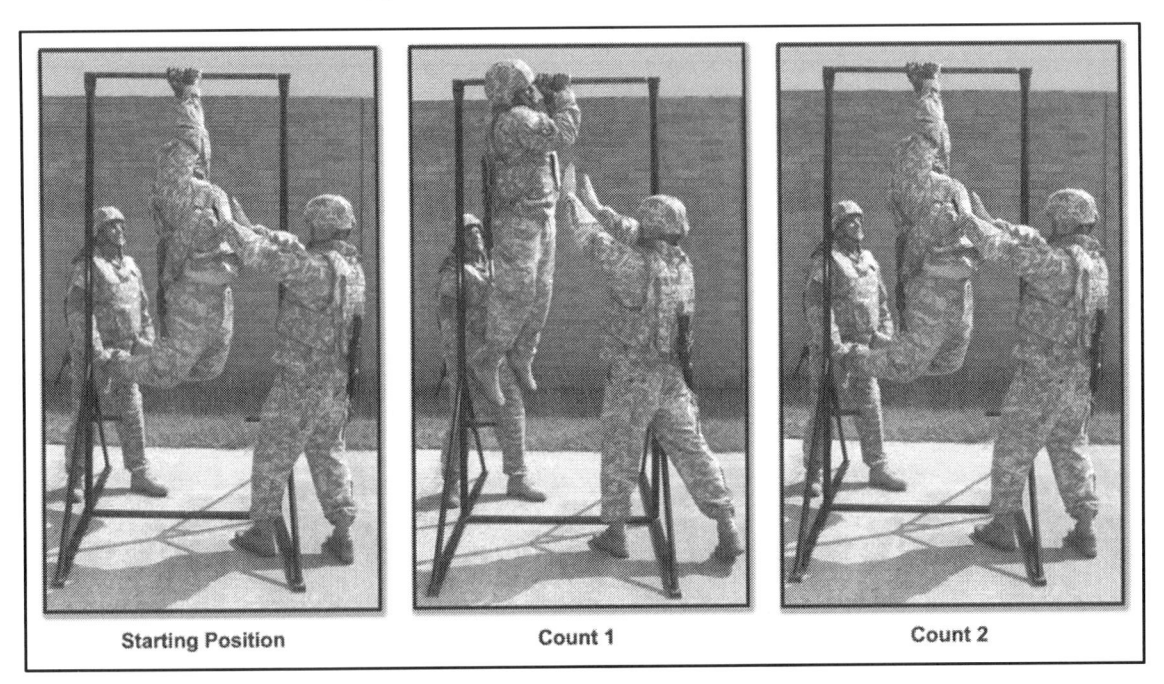

Figure 9-33. Alternating grip pull-up

Check Points:
- If the alternating grip left is used, Soldiers should touch the left shoulder to the bar on count 1.
- If the alternating grip right is used, Soldiers should touch the right shoulder to the bar on count 1.
- On count 2, the arms are fully extended to return to the starting position.
- Throughout the exercise, keep the feet together.

Precaution: Refer to paragraph 9-42 for spotting.

Chapter 9

STRENGTH TRAINING CIRCUIT

9-55. The strength training circuit (STC) produces a total-body training effect for the development of strength and mobility. (See Table 9-9.) A sequence combining a CL, a military movement drill, and kettlebell exercises works every muscle group with active recovery between stations of exercise. The STC is best conducted at platoon level. The STC may be laid out around a running track, field, or any area of adequate size, and with access to climbing bars and kettlebells. This paragraph provides a diagram of the STC, using a running track, climbing bars, and kettlebells. (See Figure 9-34.) Conduct preparation according to Chapter 7 after a walk-through and brief explanation of the STC exercise stations. (See Table 9-10.) The circuit may be completed in three rotations. Soldiers spend 60 seconds at each station. The instructor controls exercise time using a stopwatch and uses a whistle or horn to signal a change of station. At the end of all circuit rotations, recovery is conducted according to instructions in Chapter 7.

Table 9-9. Body segments trained in the conduct of the STC

STRENGTH TRAINING CIRCUIT (STC)	HIPS	THIGHS	LOWER LEGS	CHEST	BACK	TRUNK	SHOULDERS	ARMS
1. SUMO SQUAT	X	X	X	X	X	X	X	X
2. STRAIGHT-LEG DEAD LIFT	X	X	X		X	X	X	X
3. FORWARD LUNGE	X	X	X	X	X	X	X	X
4. 8-COUNT STEP-UP	X	X	X		X	X	X	X
5. PULL-UP OR STRAIGHT-ARM PULL					X	X	X	X
6. SUPINE CHEST PRESS				X	X	X	X	X
7. BENT-OVER ROW					X	X	X	X
8. OVER-HEAD PUSH-PRESS	X	X	X	X	X	X	X	X
9. SUPINE BODY TWIST	X	X	X	X	X	X	X	X
10. LEG TUCK	X	X	X	X	X	X	X	X

Strength and Mobility Activities

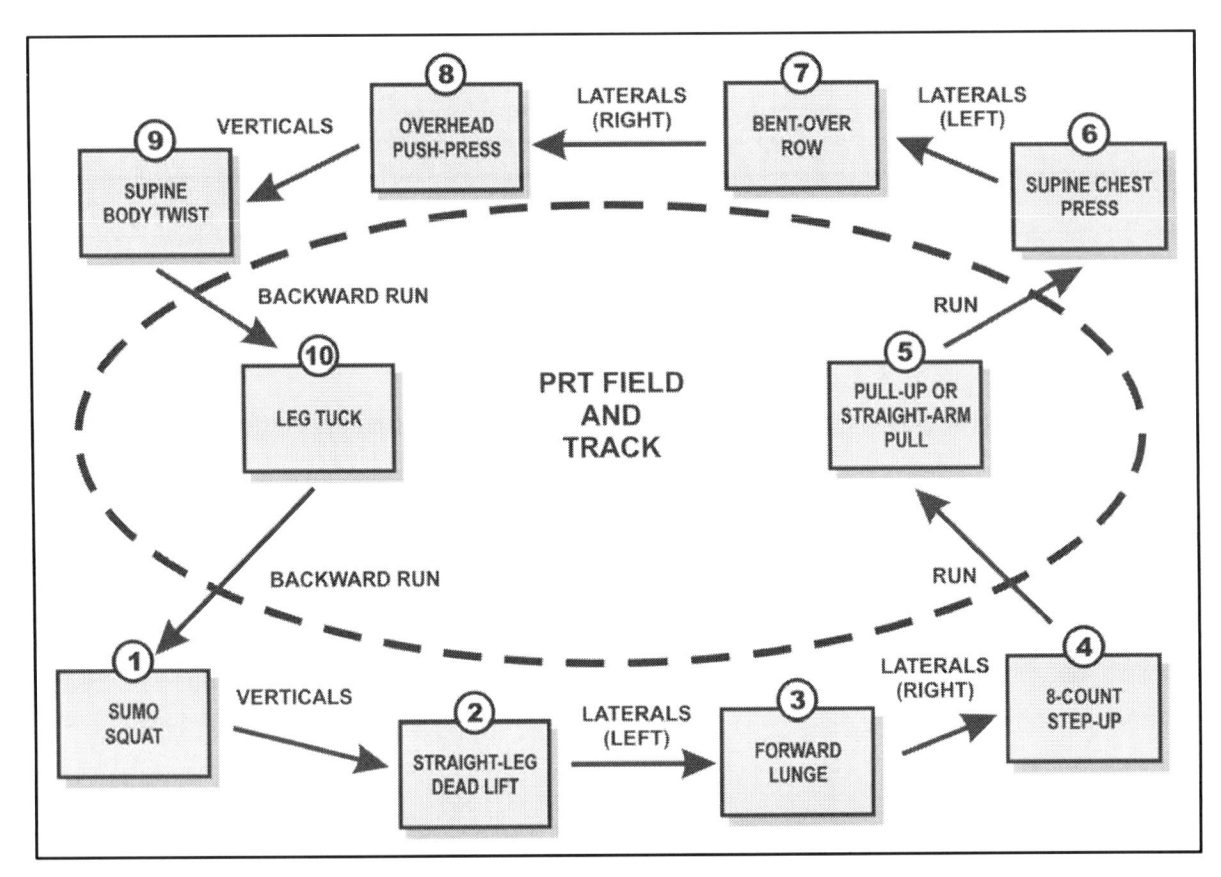

Figure 9-34. Strength training circuit

Chapter 9

Table 9-10. Equipment required to conduct platoon-size STC

Exercise Station	Kettlebells/Weights	Climbing Bars
1. Sumo Squat	6 X 50 lbs 6 X 25 lbs	N/A
2. Straight-leg Deadlift	12 X 40 lbs 12 X 25 lbs	N/A
3. Forward Lunge	12 X 20 lbs 12 X 10 lbs	N/A
4. 8-Count Step-up	12 X 30 lbs 12 X 15 lbs	N/A
5. Pull-up or Straight-arm Pull	N/A	6
6. Supine Chest Press	12 X 40 lbs 12 X 15 lbs	N/A
7. Bent-over Row	12 X 20 lbs 12 X 10 lbs	N/A
8. Overhead Push Press	12 X 30 lbs 12 X 15 lbs	N/A
9. Supine Body Twist	6 X 25 lbs 6 X 10 lbs	N/A
10. Leg Tuck	N/A	6

To conduct the STC at platoon level, 12 climbing bars and 168 kettlebells are required. The following is a list of the number of kettlebells required by weight:
6 @ 50 lbs, 24 @ 40 lbs, 24 @ 30 lbs, 24 @ 25 lbs, 24 @ 20 lbs, 36 @15 lbs and 30 @ 10 lbs.

STRENGTH TRAINING CIRCUIT

STATION 1: SUMO SQUAT

Purpose: This exercise develops strength and mobility of the hips, legs, and lower back muscles (Figure 9-35).

Starting Position: Straddle stance with the feet slightly wider than the shoulders and the toes pointing outward. Hold a single kettlebell with both hands, in front of the body, using a pronated grip (palms facing the body).

Cadence: SLOW

Count:

1. Squat while leaning slightly forward from the waist with the head up. Move downward until the upper legs parallel the ground.
2. Return to the starting position.
3. Repeat count 1.
4. Return to the starting position.

Strength and Mobility Activities

Figure 9-35. Sumo squat

Check Point:
- At the end of count 1, the shoulders, knees, and balls of the feet should be aligned.
- Keep heels on the ground and back straight throughout the exercise.

Precautions: Always perform this exercise at a slow cadence. Do not allow the legs to lower beyond parallel to the ground on count 1. Doing so would increase the stress on the knees.

Movement to Station 2: Verticals (refer to Chapter 10, Endurance and Mobility Activities, Figure 10-4).

Chapter 9

STRENGTH TRAINING CIRCUIT

STATION 2: STRAIGHT-LEG DEAD LIFT

Purpose: This exercise develops strength, endurance, and mobility of the lower back and lower extremities (Figure 9-36).

Starting Position: Straddle stance holding the kettlebells in the front of the legs using a pronated grip (hands facing the thighs). Keep the legs straight, with the knees slightly bent throughout the exercise.

Cadence: SLOW

Count:

1. Lean forward from the waist with the head aligned with the spine. Move downward until the back is flat and parallel to the ground.
2. Return to the starting position.
3. Repeat count 1.
4. Return to the starting position.

Figure 9-36. Straight-leg dead lift

Check Point:

- At the end of count 1, the hips, knees, and the balls of the feet should be aligned.
- Keep heels on the ground and back straight and parallel to the ground.

Precaution: Always perform this exercise at a slow cadence.

Movement to Station 3: Laterals (left). Refer to Chapter 10, Endurance and Mobility Activities, Figure 10-5.

STRENGTH TRAINING CIRCUIT

STATION 3: FORWARD LUNGE

Purpose: This exercise develops functional leg strength, promotes trunk stability and safely trains Soldiers to lift objects off the ground in front of them (Figure 9-37).

Starting Position: Straddle stance holding the kettlebells at the sides using a neutral grip.

Cadence: SLOW

Count:

1. Step forward with the left leg as in the forward lunge as in the PD, allowing the left knee to bend until the left thigh is parallel to the ground. Lean slightly forward from the waist and bring the kettlebells to the left and right sides of the forward leg.
2. Return to the starting position.
3. Repeat count 1 with the right leg.
4. Return to the starting position.

Figure 9-37. Forward lunge

Check Points:

- At the starting position, set the hips and keep the abdominals tight.
- On counts 1 and 3, keep the forward heel flat on the ground and the rear heel up.
- On counts 1 and 3, keep the forward knee directly over the ball of the foot.
- On counts 1 and 3, lean the trunk slightly forward.
- On counts 2 and 4, push off vigorously with the forward leg without jerking the trunk backward.

Precautions: Do not allow the forward knee to go beyond the forward toes or waiver from side to side. Do not jerk the trunk rearward to return to the starting position.

Movement to Station 4: Laterals (right). Refer to Chapter 10, Endurance and Mobility Activities, Figure 10-5.

… Chapter 9

STRENGTH TRAINING CIRCUIT

STATION 4: 8-COUNT STEP-UP

Purpose: This exercise develops strength in the hips and legs and promotes trunk stability (Figure 9-38).

Starting Position: Straddle stance holding the kettlebells at the sides using a neutral grip.

Cadence: SLOW

Count:

1. Step up on a 12 to 18 inch step with the left foot, keeping the kettlebells at the sides of the body.
2. Step up with the right foot.
3. Step down with the left foot.
4. Step down with the right foot.
5. Step up on a 12 to 18 inch step with the right foot, keeping the kettlebells at the sides of the body.
6. Step up with the left foot.
7. Step down with the right foot.
8. Step down with the left foot and return to the starting position.

Strength and Mobility Activities

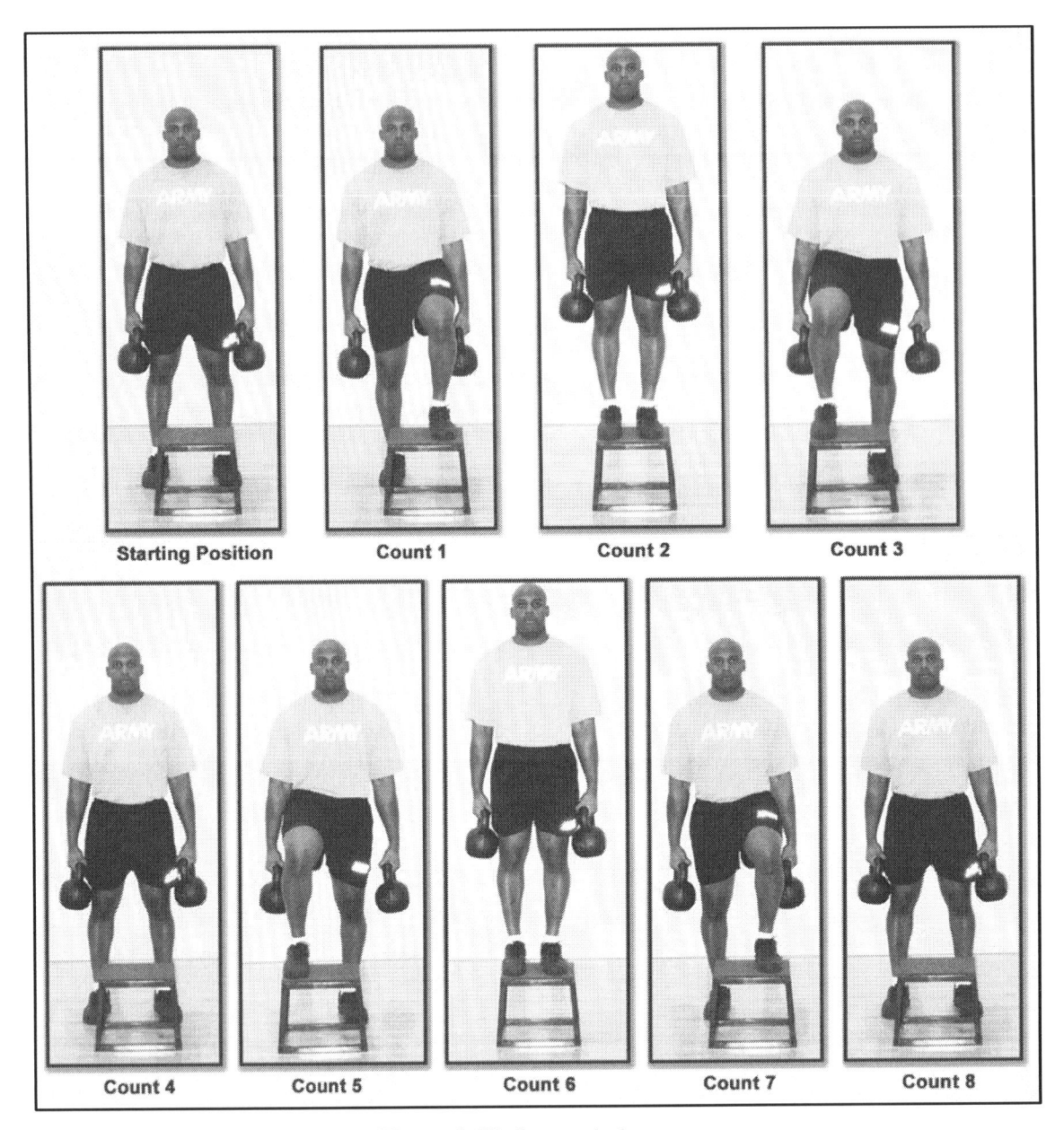

Figure 9-38. 8-count step-up

Check Points:
- On counts 1, 3, 5 and 7, keep the forward knee directly over the ball of the forward foot.
- On counts 1, 3, 5 and 7, lean the trunk slightly forward.
- On counts 1, 3, 5 and 7, push off vigorously with the forward leg without jerking the trunk backward.

Precautions: Do not allow the forward knee to go beyond the toes of the forward foot or waiver from side to side. Do not jerk the trunk rearward to return to the starting position.

Movement to Station 5: Run.

STRENGTH TRAINING CIRCUIT

STATION 5: PULL-UP OR STRAIGHT-ARM PULL

PULL-UP

Purpose: This exercise develops the ability to pull the body upward while hanging (Figure 9-39).

Starting Position: Extended hang using the overhand grip.

Cadence: MODERATE

Count:

1. Keeping the body straight, pull upward with the arms until the chin is above the bar.
2. Return to the starting position.

Figure 9-39. Pull-up

Check Points:

- Throughout the exercise, keep the feet together.
- Throughout the exercise, keep the arms shoulder-width apart, palms facing away from the body, with the thumbs around the bar.

Precaution: N/A

Movement to Station 6: Run.

Strength and Mobility Activities

STRAIGHT-ARM PULL

Purpose: This exercise develops the ability to initiate the pull-up motion and maintain a contraction in the extended hang position (Figure 9-40).

Starting Position: Extended hang using the overhand grip.

Cadence: MODERATE

Count:
1. Keeping the body straight, pull the body upward using the shoulders and back muscles only.
2. Return to the starting position.

Figure 9-40. Straight-arm pull

Check Points:
- Throughout the exercise, keep the arms shoulder width, palms facing away from the body, with the thumbs around the bar.
- Throughout the exercise, keep the elbows straight, but not locked.
- On count 1, pull the body up by engaging the shoulder muscles (squeeze the shoulder blades together).

Precaution: N/A

Movement to Station 6: Run.

STRENGTH TRAINING CIRCUIT

STATION 6: SUPINE CHEST PRESS

Purpose: This exercise strengthens the muscles of the chest, shoulders, and triceps muscles (Figure 9-41).

Starting Position: Supine position with the knees bent at 90 degrees and the feet 8 to 12 inches apart and flat on the ground. Hold the kettlebells in front of the shoulders using a pronated grip (palms facing up). The back of the upper arms should rest on the ground and the forearms should be perpendicular to the ground. The head is on the ground.

Cadence: SLOW

Count:

1. Extend the elbows, raising the kettlebells until the upper arms are straight (not locked) and perpendicular to the ground.
2. Return to the starting position.
3. Repeat count 1.
4. Return to the starting position.

Figure 9-41. Supine chest press

Check Points:
- Keep the head on the ground throughout the exercise.
- On counts 1 and 3, straighten, but do not lock the elbows.

Precaution: N/A

Movement to Station 7: Laterals (left). Refer to Chapter 10, Endurance and Mobility Activities, Figure 10-5.

STRENGTH TRAINING CIRCUIT

STATION 7: BENT-OVER ROW

Purpose: This exercise strengthens the shoulders, upper back, and biceps muscles (Figure 9-42).

Starting Position: Forward leaning stance with the arms hanging straight in front of the legs and the hands holding the kettlebells using a neutral grip (palms facing inward).

Cadence: SLOW

Count:

1. Bend the elbows, raising the kettlebells to the chest.
2. Return to the starting position.
3. Repeat count 1.
4. Return to the starting position.

Figure 9-42. Bent-over row

Check Points:

- On counts 2 and 4 the elbows straighten but do not lock.
- To prevent the trunk from sagging, tighten the abdominal muscles while in the starting position and hold this contraction throughout the exercise.

Precaution: N/A

Movement to Station 8: Laterals (right). Refer to Chapter 10, Endurance and Mobility Activities, Figure 10-5.

STRENGTH TRAINING CIRCUIT

STATION 8: OVERHEAD PUSH PRESS

Purpose: This exercise strengthens the shoulders and triceps muscles (Figure 9-43).

Starting Position: Straddle stance holding the kettlebells at the collar bones using a neutral grip (palms inward).

Cadence: SLOW

Count:

1. From the starting position slightly flex the hips and knees (slight squat) with feet flat on the floor, keeping the upper body and upper arms parallel to the ground. Forcefully extend the hips, knees, and ankles while simultaneously extending the elbows to raise the kettlebells overhead.
2. Return to the starting position by flexing the elbows, controlling the descent to the collar bones and shoulders while simultaneously flexing the hips and knees to reduce the impact of the kettlebells on the shoulders.
3. Repeat count 1.
4. Repeat count 2, returning to the starting position.

Figure 9-43. Overhead push press

Check Points:

- Hold the head and neck in a neutral position, looking straight ahead or slightly upward.
- Perform counts 1 and 3 in a fast, continuous motion.
- Always control the descent of the kettlebells during counts 2 and 4 to avoid injury to the trunk and back.
- On counts 1 and 3, straighten the elbows but do not lock them.
- Tighten the abdominal muscles while in the starting position and hold this contraction throughout the exercise to ensure trunk stability.

Precaution: N/A

Movement to Station 9: Verticals. Refer to Chapter 10, Endurance and Mobility Activities, Figure 10-4.

STRENGTH TRAINING CIRCUIT

STATION 9: SUPINE BODY TWIST

Purpose: This exercise strengthens the trunk muscles and promotes trunk control (Figure 9-44).

Starting Position: Supine position with hips and knees bent at 90 degrees. Hold one kettlebell over the trunk using a neutral grip with the upper arms off the ground and elbows bent 90 degrees. To assume the starting position from the position of attention with the kettlebell on the ground, (1) assume the seated position and grasp a single kettlebell at each side of the handle with palms facing inward, (2) assume the supine position, (3) raise the kettlebell to the starting position over the trunk and raise the feet off the ground.

Cadence: SLOW

Count:

1. Rotate the kettlebell to the left and the legs to the right.
2. Return to the starting position.
3. Rotate the kettlebell to the right and the legs to the left.
4. Return to the starting position.

Figure 9-44. Supine body twist

Check Points:
- On counts 1 and 3, the back is straight and the knees are bent at 90 degrees.
- The head is off the ground and in line with the trunk and the chin is tucked throughout the exercise.
- On counts 1 and 3, the upper arms remain off the ground.

Precaution: Do not swing the arms or arch the back to raise the kettlebell on counts 1 and 3.

Movement to Station 10: Backward run.

STRENGTH TRAINING CIRCUIT

STATION 10: LEG TUCK

Purpose: This exercise develops the abdominal, hip flexor, and grip strength essential to climbing a rope (Figure 9-45).

Starting Position: Extended hang using the alternating grip, left or right.

Cadence: SLOW

Count:

1. Pull up with the arms and raise the knees toward the chest until the elbows touch the thighs just above the knees.
2. Return to the starting position.

Figure 9-45. Leg tuck

Check Points:
- Throughout the exercise, keep the feet together.
- On count 1, the thighs and elbows touch just above the knees.

Precaution: N/A

Movement to Station 1: Backward run.

GUERRILLA DRILL

9-56. The GD, performed in the sustaining phase, consists of three exercises that develop leg power, coordination, and the ability to lift and carry another Soldier. When the Soldiers can precisely execute each exercise and carry, the drill is performed continuously for 1-3 sets. All movement in the carry position is performed at quick time. Each exercise and carry must be taught and demonstrated before Soldiers try to perform the drill. When teaching and demonstrating the GD, use the extended rectangular formation (covered). In the illustrations that follow, Soldier "A" refers to the Soldier performing the carry, and Soldier "B" refers to the Soldier being carried. The drill is always performed in its entirety in the order listed.

TRAINING AREA

9-57. Any dry, level area of adequate size (same as MMD 1) and free from hazards (holes, debris) is satisfactory for conduct of the GD.

UNIFORM

9-58. Soldiers will wear IPFU or ACUs.

EQUIPMENT

9-59. Mark GD area with cones.

FORMATION

9-60. For the most efficient instruction, the unit size should be limited to one platoon. Larger units up to a battalion can successfully perform these drills if properly taught and mastered at the small unit level. The extended rectangular formation (covered) is prescribed.

LEADERSHIP

9-61. A PRT leader and AI are required to instruct and lead the GD. The instructor must be familiar with the method of teaching the exercises; the commands and counting cadence; formations; starting positions; and the use of AIs as described in Chapter 7, Execution of Training. Soldiers should memorize the exercises by name and movement. Considerable time and effort must be expended during the early stages to teach exercises properly to all Soldiers.

PRECISION

9-62. GD exercises lose much of their value unless performed exactly as prescribed. Precision should never be compromised for speed of movement. Moving too fast will not allow Soldiers to perform the exercises with proper technique and may lead to injury. All movement in the carry position is performed at quick time.

PROGRESSION

9-63. Soldiers perform no more than one repetition of each exercise while learning and practicing the GD. The GD is performed in the sustaining phase during the activity part of the PRT session. Soldiers will perform one set of the entire drill, progressing to three sets.

INTEGRATION

9-64. The GD exercises integrate the components of strength, endurance, and mobility through functional movements that relate directly to the performance of WTBDs.

Chapter 9

COMMANDS

9-65. The GD consists of three exercises that are performed in the sequenced prescribed and trains the body segments listed in Table 9-11. The commands for execution of the drill and starting positions for each exercise are described in Chapter 7.

Table 9-11. Body segments trained in the guerrilla drill

GUERRILLA DRILL (GD)	MUSCLES							
	HIPS	THIGHS	LOWER LEGS	CHEST	BACK	TRUNK	SHOULDERS	ARMS
1. SHOULDER ROLL	X	X	X	X	X	X	X	X
2. LUNGE WALK	X	X	X		X	X	X	
3. SOLDIER CARRY	X	X	X	X	X	X	X	X

Strength and Mobility Activities

GUERRILLA DRILL

EXERCISE 1: SHOULDER ROLL

Purpose: This exercise develops the Soldier's ability to safely fall and roll-up to a standing position (Figure 9-46).

Starting Position: Straddle stance.

Movement: Step forward with the left foot, squat down, and make a wheel with the arms by placing the left hand on the ground with the fingers facing to the rear; the right hand is also on the ground with the fingers facing forward. Tuck the chin to avoid injury to the neck. Push off with the right leg and roll over the left shoulder along the left side of the body. To roll to the opposite side, step forward and switch hand and leg positions. Progress to continuously walking and alternating rolling on opposite sides.

Figure 9-46. Shoulder roll

Check Points:

- Lead with the left foot when rolling on the left shoulder and the right foot when rolling on the right shoulder.
- Hands are placed on the ground, facing each other with the middle fingertips of each hand touching at the tips so the arms form a wheel.
- Rotate the upper body so the lead elbow is pointing straight to the front while maintaining a wheel with the arms.
- Tuck the chin so ground contact is made with the arms, shoulder blades, and back, but not with the neck.
- The momentum of the roll brings Soldiers up to their knees. Continue to the feet by pushing off with the rear leg while standing up on the front leg.

Precautions: The PRT leader and an AI(s) must ensure that Soldiers are in the proper squatting position for the roll by ensuring that their hands are on the ground and their chins are tucked before rolling.

GUERRILLA DRILL

EXERCISE 2: LUNGE WALK

Purpose: This exercise develops the leg power needed to move both vertically and horizontally (Figure 9-47).

Starting Position: Straddle stance.

Movement: Walk forward, starting with the left foot, stepping as in the forward lunge. Lightly touch the knee of the rear leg to the ground with each step. Without returning to the starting position, continue to lunge walk to the 25-yard stop point by alternating legs.

Figure 9-47. Lunge walk

Check Points:
- Incorporate arm swing with the arm opposite the forward leg raised parallel to the ground.
- Keep the back straight and the head up.
- Do not allow the knee of the lead leg to move forward of the toes of the forward foot.

Precaution: Do not allow the knee of the rear leg to forcefully contact the ground.

GUERRILLA DRILL

EXERCISE 3: SOLDIER CARRY

Purpose: This exercise develops the Soldier's ability to carry a conscious or unconscious Soldier of comparable size (Figure 9-48).

Starting Position: Soldier "B" assumes the prone position, arms overhead. Soldier "A" straddles Soldier "B" and squats, reaching under Soldier "B's" armpits. Soldier "A" stands lifting Soldier "B" to his knees. Soldier "A" continues to lift Soldier "B" to his feet, leaning him back slightly to lock the legs. Soldier "A" raises one of Soldier "B's" arms overhead and walks under the arm to the front of Soldier "B," oriented sideways to Soldier "B." Soldier "A" bends his knees and leans forward, placing one arm through Soldier "B's" legs. Soldier "A" leans Soldier "B" forward until he lies across Soldier "A's" shoulders. Soldier "A" stands up, lifting Soldier "B" off the ground. Soldier "A," using the hand of his arm that is between Soldier "B's" legs, grasps the wrist of Soldier "B's" arm that is hanging over Soldier "A's" shoulder.

Movement: Soldier "A" moves 25-yards at quick time; then Soldier "B" dismounts; the Soldiers then change positions and return to the starting point.

Figure 9-48. Soldier carry

Check Points:

- Soldier "A" should squat low and grasp Soldier "B" under the arms to lift him from the prone position. Soldier "A" may clasp his hands in front of Soldier "B's" chest to help him lift Soldier "B" to his feet.
- Position Soldier "B" over the shoulder during the carry. Secure the position with one hand, grasping Soldier "B's" forward arm.

Precautions: Keep back straight and use legs to lift Soldier to the carry position. All movement in the carry position is performed at a quick time; no running.

Summary

The purpose of strength and mobility activities is to improve functional strength, postural alignment, and body mechanics as they relate to the performance of WTBDs. The regular and precise execution of the exercise drills listed in this chapter will develop the body management competencies needed to successfully accomplish these tasks.

Chapter 10
Endurance and Mobility Activities

"Each morning in Africa a gazelle wakes up. It knows it must outrun the fastest lion or it will be killed. Every morning in Africa, a lion wakes up. It knows it must run faster than the slowest gazelle or it will starve. It doesn't matter whether you're a lion or a gazelle, when the sun comes up, you'd better be running."

Roger Bannister, first person to run a mile in under four minutes (3:59.4)

Warrior tasks and battle drills require the ability to move quickly on foot. Running short distances at high speed is essential to moving under direct and indirect fire (Figure 10-1).

Figure 10-1. Moving under direct and indirect fire

RUNNING

10-1. The purpose of running is to improve the overall conditioning of the Soldier by developing endurance. Endurance spans a continuum between aerobic and anaerobic systems. Aerobic endurance is developed by performing low to moderate intensity activities for a long duration. Anaerobic endurance is developed by performing high-intensity activities for a short duration, resting, and then repeating the sequence. Aerobic training alone does not fully prepare Soldiers for the functional endurance and strength requirements of WTBDs. The analysis of the physical demands needed to successfully accomplish WTBDs demonstrates a more

Chapter 10

significant requirement for anaerobic endurance. In order to train the complete spectrum of endurance, speed running, sustained running, and foot movement under load must be performed. The running activities described in this chapter may be performed individually or collectively. Table 10-1 describes endurance and mobility activities used in PRT. Table 10-2 describes endurance and mobility activities and the prescription of intensity, duration, and volume within the toughening and sustaining phases. In addition, Chapter 5, Planning Considerations, provides the template for commanders and PRT leaders to implement endurance and mobility activities into their PRT programs.

Table 10-1. Endurance and mobility activities

Endurance and Mobility Activities	
Military Movement Drills 1 and 2 (MMD 1&2)	These drills dynamically prepare the body for more vigorous running activities and develop motor efficiency.
30:60s and 60:120s	30:60s and 60:120s improve the resistance to fatigue of the active muscles by repeatedly exposing them to high intensity effort. As a result of their increased anaerobic and aerobic endurance, Soldiers will be able to sustain performance of physically demanding tasks at a higher intensity for a longer duration.
300-yard Shuttle Run (SR)	The 300-yard Shuttle Run develops the ability to repeatedly sprint after changing direction. It is an indicator of the Soldier's anaerobic endurance, speed, and agility.
Hill Repeats (HR)	Hill repeats are an effective means of developing explosive leg strength, anaerobic power, and speed.
Ability Group Run (AGR)	Ability group runs train Soldiers in groups of near-equal ability to sustain running for improvement in aerobic endurance.
Unit Formation Run (UFR)	Unit formation runs are based on a time and distance that can be achieved with unit integrity and a display of unit cohesion.
Release Run (RR)	Release runs combine the benefits of formation running and individual performance at higher training intensities. Soldiers will run in formation to a specified time (no more than 15 minutes), then are released to run as fast as they can back to the starting point.
Terrain Run (TR)	Terrain running applies the *Train as you will fight* principle to PRT. Running through local training areas, over hills, and around obstacles improves mobility, endurance, and the ability to stop, start, and change direction.
Foot March (FM)	Foot marching as a movement component of maneuver, is a critical Soldier physical requirement. Regular foot marching prepares Soldiers to successfully move under load.
Conditioning Obstacle Course (CDOC)	Running the conditioning obstacle course for time challenges Soldiers' strength, endurance, and mobility, improving individual movement techniques.
Endurance Training Machines (ETM)	Use of endurance training equipment may be based on environmental constraints, safety for Soldiers on physical profile, and isolation of specific muscle groups to be trained during rehabilitation and reconditioning.

Endurance and Mobility Activities

Table 10-2. Endurance and mobility activity prescription

Activities	Toughening Phase (BCT & OSUT-R/W/B)	Sustaining Phase (AIT & OSUT-B/G)	Sustaining Phase ARFORGEN (Reset)	Sustaining Phase ARFORGEN (Train/Ready)	Sustaining Phase ARFORGEN (Available)
MMD 1	1 rep	1 rep	1 rep	1 rep	1 rep
MMD 2	N/A	1 rep	1 rep	1 rep	1 rep
30:60s	6-8 reps	6-10 reps w/wo load	6-10 reps w/wo load	10-15 reps w/wo load	10-15 reps w/wo load
60:120s	6-10 reps	6-10 reps	6-10 reps	6-10 reps	6-10 reps
300-yd SR	1 rep	1-2 reps w/wo load	1-2 reps	1-2 reps w/wo load	1-2 reps w/wo load
HR	N/A	6-8 reps uphill or downhill	6-10 reps uphill or downhill	6-10 reps uphill or downhill	6-10 reps uphill or downhill
AGR	10-30 min	20-30 min	20-30 min	20-30 min	20-30 min
UFR	20-30 min	20-30 min	30 min	30 min	30 min
RR	20-30 min	20-30 min	30 min	30 min	30 min
TR	N/A	20 min	20-30 min	20-30 min	20-30 min
FM	2-15 Km	2-15 Km	10 Km or less	10-30 Km	10-30 Km
CDOC	1 rep	1 rep	1 rep	1 rep	1 rep
ETM	N/A	N/A	20-30 min	20-30 min	20-30 min
Abbreviations	MMD-Military Movement Drill SR-Shuttle Run HR-Hill Repeats AGR-Ability Group Run UFR-Unit Formation Run RR-Release Run TR-Terrain Run FM-Foot March (fl/aml/eaml) CDOC-Conditioning Obstacle Course ETM-Endurance Training Machines				

TRAINING AREA

10-2. Running is conducted over a variety of terrain:
- Hardball (improved and unimproved roads).
- Grassy fields.
- Tracks.
- Wooded areas.
- Hills.
- Tank trails.

UNIFORM

10-3. The commander will specify the appropriate uniform based on the type of running activity to be performed. PRT uniforms appropriate for running include:
- IPFU.
- ACUs and running shoes.
- ACUs and boots.
- ACUs with boots and fighting load.

EQUIPMENT

10-4. Equipment used will be according to installation safety policy requirements (flashlights, reflective vests/bands, traffic cones, AGR route markers placed at ¼ mile intervals). The PRT leader and AI must monitor run time and pace during the conduct of running activities.

Chapter 10

FORMATION

10-5. Formations used in unit running are squad, platoon, company, and battalion in column. Other types of running such as terrain running or speed running will be conducted in one or more columns as determined by the training area and installation safety standing operating procedures.

LEADERSHIP

10-6. The PRT leader and AIs must be able to demonstrate and lead all types of running activities. They must also be familiar with formations, commands, cadence, and placement of Soldiers into ability groups for sustained and speed running.

INSTRUCTION AND EXECUTION

10-7. Running may be performed individually or collectively. When conducting collective training, running is optimized when Soldiers are grouped by near-equal ability. The best way to assign Soldiers to ability groups is by their most recent 1-mile run time assessment. The optimal time and range between each group is 60 seconds. When performing formation sustained running, the PRT leader should be on the left side of the formation and toward the rear to have a full view of all Soldiers and maintain control. Speed running may be conducted individually or collectively by ability group, on a track or designated running area. When conducting speed running, the PRT leader will control running and recovery times from the center of the track or running area using a whistle and stopwatch. Assistant instructors may run with the Soldiers to provide positive motivation and running form corrections.

PRECISION

10-8. Soldiers should be instructed on the running form guidelines in this chapter. Running with optimal body mechanics allows greater efficiency with less chance of injury. Soldiers should strive to demonstrate and maintain proper running form during all running activities.

PROGRESSION

10-9. In the toughening phase, Soldiers perform speed running (30:60s, 60:120s, and the 300-yard SR) and sustained running. Initially, Soldiers perform six repetitions of 30:60s and progress up to 8 repetitions, and then begin performing 60:120s, 6 repetitions progressing to 10 repetitions. The intensity for speed running during the 30- and 60-second work intervals is 75 to 85 percent maximal effort. During the 60- and 120-second recovery intervals, all Soldiers walk until the next work interval begins. At the completion of 30:60s or 60:120s, Soldiers walk two to three minutes before engaging in other PRT activities or recovery. The SR is performed only one time when performed as an activity during a PRT session, in conjunction with 60:120s. The PRT leader designates the number of repetitions and signals the start of each group or individual. Formation running is conducted for no longer than 30 minutes in the toughening phase. All running courses should be marked at ¼-mile intervals so PRT leaders can monitor split times to ensure the maintenance of the appropriate running pace. Sustained running progression is built into the PRT training schedules through the employment of release runs and by moving Soldiers from lower ability groups to higher ability groups.

10-10. In the sustaining phase, Soldiers continue to perform the speed and sustained running activities from the toughening phase. In addition, HR, terrain running, and speed running under load are performed. Hill repeats start with 6 repetitions and add no more than 1 repetition every 2 weeks, not to exceed 10 repetitions. The PRT leader designates the number of repetitions and signals the start of each group or individual. Sustained running should not exceed 30 minutes in the sustaining phase. All running courses should be marked at ¼-mile intervals so PRT leaders can monitor split times to ensure the maintenance of the appropriate running pace. Sustained running progression is accomplished by moving the Soldier from a lower ability group to the next higher ability group. Terrain running is only conducted in the sustaining phase. Distances should generally be 1 mile for densely wooded areas and up to 2 miles on tank trails and open fields. During the sustaining phase, the 300-yard SR may be performed in ACUs and boots, progressing to individual body armor (IOTV) without plates, then with plates, then with fighting load. Caution must be used when determining appropriate progression. Environmental considerations are important in the ramp of progression. Repetitions, pace, load,

Endurance and Mobility Activities

uniform, and total exercise time must be adjusted when exercising at high altitudes and in hot, humid environments. Refer to Appendix D for environmental considerations.

INTEGRATION

10-11. The variety of running activities conducted during the toughening phase (30:60s, 60:120s, the 300-yd shuttle run, release runs, AGR, and unit formation running) and sustaining phase (30:60s, 60:120s, the 300-yd shuttle run, release runs, hill repeats, AGR, and unit formation running) integrate anaerobic and aerobic training. The 300-yard SR, in both the toughening and sustaining phases, and sustaining phase terrain running are also integrated to develop Soldier skills.

COMMANDS

10-12. Calling of cadence and commands is the responsibility of the PRT leader or ability group leader. The command, "Double Time, MARCH" starts the formation running. The command "Quick Time, MARCH" terminates formation running (see TC 3-21.5, *Drill and Ceremonies*). After performance of preparation and any previous PRT activities (military movement drills 1 and/or 2), the Soldiers will jog for about ¼ mile before the first repetition of 30:60s or 60:120s is performed. When conducting 30:60s, 60:120s, or HR en masse, the PRT leader will control work (running phase) and recovery (walking phase) times from the center of the track or running area. The PRT leader will initiate the work (run/hill) interval by signaling with one whistle blast. At the conclusion of the work (run/hill) interval (30 or 60 seconds), the PRT leader will signal with two short whistle blasts. At the conclusion of the recovery (walk) interval (60 or 120 seconds), the PRT leader will again signal with one short whistle blast. This sequence is repeated until the desired number of repetitions is completed. Soldiers of varied abilities run for different numbers of repetitions during the toughening phase. Soldiers who finish early will continue to walk until all Soldiers have completed the activity. At the end of the activity, the entire group will walk for 2 to 3 minutes before performing any subsequent activities or recovery.

FORM

10-13. Running form varies from Soldier to Soldier. Anatomical variations cause a variety of biomechanical manifestations. Many individual variations may be successful. Attempts to force Soldiers to conform to one standard may do more harm than good; however, there are some basic guidelines that may improve running efficiency without overhauling the natural stride. Generally, the form and technique for all types of running is fairly constant. The following information addresses optimal running form for the major body segments (Figure 10-2).

Figure 10-2. Sustained running form

HEAD

10-14. The head should be held high, with the chin pointed forward, neither up nor down. Allowing the head to ride forward puts undue strain on the muscles of the upper back.

Chapter 10

SHOULDERS

10-15. The shoulders should assume a neutral posture, neither rounded forward nor forcefully arched backward. Rounding the shoulders forward is the most common fault in everyday posture while walking and running. The problem is usually associated with tightness of the chest and shoulder muscles. Another problem occurs when the shoulders start to rise with fatigue or increased effort. This position not only wastes energy, but can also adversely affect breathing.

ARMS

10-16. Throughout the arm swing, the elbows should stay at roughly a 90-degree bend. The wrists stay straight and the hands remain loosely cupped. The arm swing should be free of tension, but do not allow the hands to cross the midline of the body.

TRUNK AND PELVIS

10-17. The trunk should remain over its base of support, the pelvis. A common problem with fatigue is allowing the trunk to lean forward of the legs and pelvis. This forces the lower back muscles to expend too much energy resisting further trunk lean to the front.

LEGS

10-18. For distance running, much of the power is generated from below the knee. Energy is wasted as the knees come higher and the large muscles of the hips and thighs are engaged. Practice getting a strong push-off from the ankle of the back leg. This helps to lengthen the stride naturally. Lengthening the stride by reaching forward with the front leg will be counterproductive.

FEET

10-19. The feet should be pointed directly forward while running. With fatigue and certain muscle imbalances, the legs and feet may start to rotate outward. This may hinder performance and create abnormal stresses that contribute to injury.

BREATHING

10-20. Breathing should be rhythmic in nature and coordinated with the running stride.

MILITARY MOVEMENT DRILL 1

10-21. The purpose of MMD1 in the toughening phase (Figure 10-3) is to enhance running form, dynamically prepare the body for more vigorous running activities, and develop motor efficiency. Military movement drill 1 is conducted following preparation and the HSD prior to running activities during the PRT session. Any level area of adequate size is appropriate for conducting MMD1. Beware of hazards such as holes, uneven terrain and rocks. Use caution when conducting MMD1 on wet terrain. This drill is conducted using the extended rectangular formation (covered) and performed by rank. Military movement drill 1 consists of exercises performed at 25-yard intervals: verticals, laterals, and the shuttle sprint. Refer to Table 10-2 for endurance and mobility activities, prescriptions of intensity, duration, and volume within the toughening and sustaining phases. In addition, Chapter 5, Planning Considerations, provides the template for commanders and PRT leaders to implement endurance and mobility activities into their PRT programs.

```
                        Extended Rectangular Formation (covered)

                        RL   S   S   S   S   S   S   S   S   S   S   S
                        RL   S   S   S   S   S   S   S   S   S   S   S
                        RL   S   S   S   S   S   S   S   S   S   S   S
                        RL   S   S   S   S   S   S   S   S   S   S   S
```

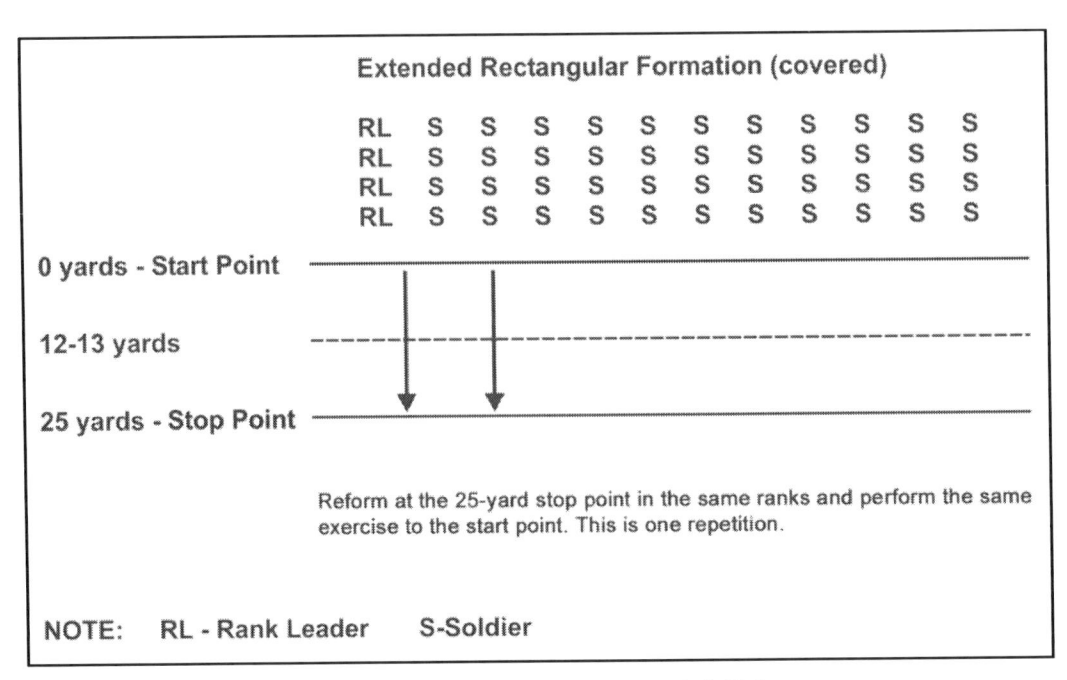

0 yards - Start Point

12-13 yards

25 yards - Stop Point

Reform at the 25-yard stop point in the same ranks and perform the same exercise to the start point. This is one repetition.

NOTE: RL - Rank Leader   S-Soldier

Figure 10-3. Military movement drill 1

MILITARY MOVEMENT DRILL 1

EXERCISE 1: VERTICALS

Purpose: This exercise helps to develop proper running form (Figure 10-4).

Starting Position: Staggered stance with the right foot forward. The right heel is even with the toes of the left foot. The right arm is to the rear with the elbow slightly bent and the left arm is forward. The head is up, looking straight ahead, and the knees are slightly bent.

Movement: Bring the hips quickly to 90-degrees of bend without raising the knees above waist level. Ground contact should be primarily with the balls of the feet. When the left leg is forward, the right arm swings forward and the left arm swings to the rear. When the right leg is forward, the left arm swings forward and the right arm swings to the rear.

Figure 10-4. Verticals

Check Points:

- Arm swing is strong and smooth with the forward arm slightly bent and 90-degrees to the ground and the rearward arm relatively straight.
- Arm swing is from front to rear, not side to side, with the upper part of the forward arm reaching parallel to the ground as it swings to the front.
- Keep a tall stance with a stable, upright trunk. The back remains perpendicular to the ground. There should not be any back swing of the legs.

Precaution: N/A

MILITARY MOVEMENT DRILL 1

EXERCISE 2: LATERALS

Purpose: This exercise develops the ability to move laterally (Figure 10-5).

Starting Position: Straddle stance, slightly crouched, with the back straight, arms at the side with elbows bent at 90-degrees and palms facing forward. Face perpendicular to the direction of movement.

Movement: Step out with the lead leg and then bring the trail leg up and toward the lead leg. The Soldier always faces the same direction so that for the first 25-yards he is moving to the left and for the second 25-yards he is moving to the right.

Figure 10-5. Laterals

Check Points:
- Pick the feet up with each step. Avoid dragging the feet along the ground.
- Crouch slightly while keeping the back straight.
- Avoid hitting the feet and ankles together on each step.
- Rank leaders will face their rank throughout the exercise.

Precaution: N/A

MILITARY MOVEMENT DRILL 1

EXERCISE 3: SHUTTLE SPRINT

Purpose: This exercise develops anaerobic endurance, leg speed, and agility (Figure 10-6).

Starting Position: Staggered stance with the right foot forward. The right heel is even with the toes of the left foot. The right arm is to the rear with the elbow slightly bent and the left arm is forward. The head is up looking straight ahead and the knees are slightly bent.

Movement: Run quickly to the 25-yard mark (as arrow 1 in the following exercise illustration shows). Turn clockwise while planting the left foot and bending and squatting to touch the ground with the left hand. Run quickly back to the starting line (arrow 2) and plant the right foot, then turn counterclockwise and touch the ground with the right hand. Run back to the 25-yard mark (arrow 3) accelerating to near maximum speed.

Check Points:

- Soldiers should slow their movement before planting feet and changing direction.
- Soldiers should squat while bending the trunk when reaching to touch the ground as they change direction.
- Soldiers touch the ground with their left hand on the first turn, then with their right hand on the second turn.
- Accelerate to near maximum speed during the last 25-yard interval.

Precaution: Soldiers should use caution when performing this exercise on wet terrain.

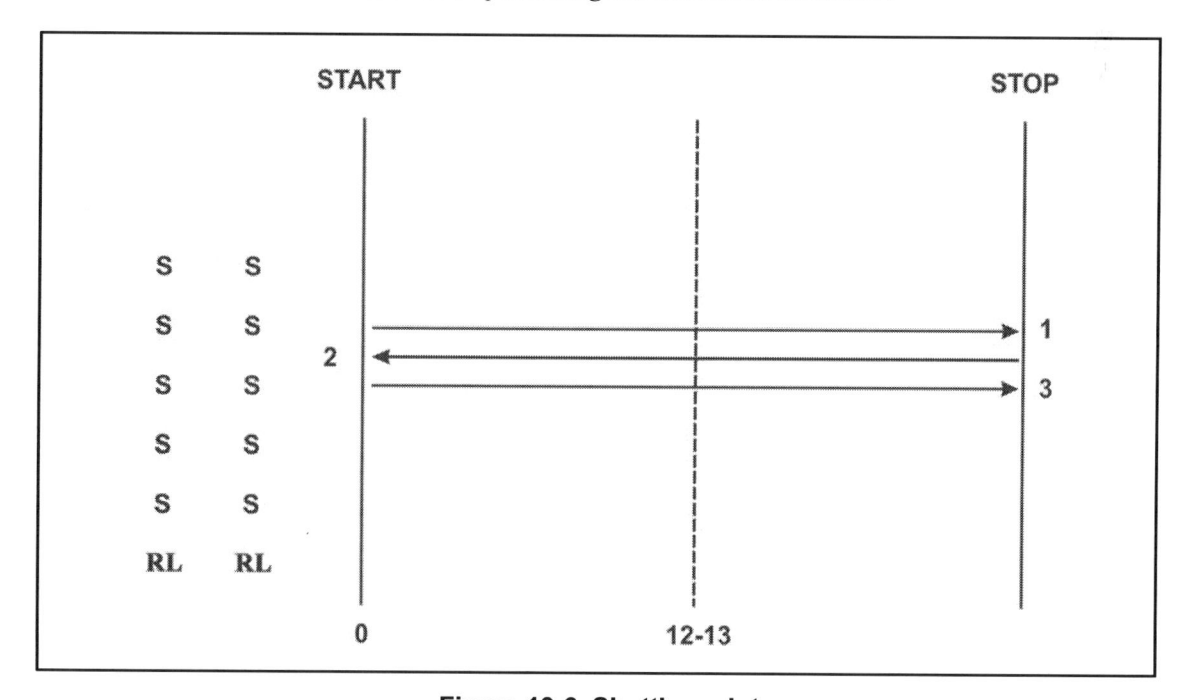

Figure 10-6. Shuttle sprint

MILITARY MOVEMENT DRILL 2

10-22. The purpose of MMD2 in the sustaining phase is to enhance running form, dynamically prepare the body for more vigorous running activities, and develop motor efficiency. Military movement drill 2 is conducted following preparation and the HSD prior to running activities during the PRT session. Military movement 2 contains three dynamic, plyometric exercises that are conducted in the same manner as MMD1. If both drills are conducted, MMD1 should precede MMD2. DO NOT mix exercises between the two drills. Perform the drills as prescribed in this FM. Refer to Table 10-2 for endurance and mobility activities, prescriptions of intensity, and duration and volume within the toughening and sustaining phases. In addition, Chapter 5, Planning Considerations, provides the template for commanders and PRT leaders to implement endurance and mobility activities into their PRT programs.

MILITARY MOVEMENT DRILL 2

EXERCISE 1: POWER SKIP

Purpose: This exercise develops leg power, coordination, and jumping ability from a single leg. It also promotes a powerful extension from the ankle, knee, and hip. (Figure 10-7).

Starting Position: Staggered stance with right foot forward.

Movement: Step with the left foot, then hop and land on the left leg followed by the same action with the opposite leg. When the right leg is forward, the left arm swings forward and the right arm is to the rear. When the left leg is forward, the right arm swings forward and the left arm is to the rear.

Figure 10-7. Power skip

Check Points:

- Start slowly and progress the speed and height of the skip throughout each 25-yard interval.
- Soldiers should gradually incorporate larger arm swings as they jump to get higher elevation. Arm swing is strong and smooth with the forward arm at 90-degrees and the rearward arm relatively straight.
- Arm swing is from front to rear, not side to side, with the upper part of the forward arm reaching parallel to the ground as it swings to the front.

Precaution: N/A

MILITARY MOVEMENT DRILL 2

EXERCISE 2: CROSSOVERS

Purpose: This exercise improves leg coordination and trains Soldiers to move laterally (Figure 10-8).

Starting Position: Straddle stance, slightly crouched, with the back straight, arms at the side with elbows bent at 90-degrees, and palms facing forward or holding weapon. Face perpendicular to direction of movement.

Movement: Cross the trail leg first to the front of the lead leg and step in the direction of travel to return to the starting position. Then cross the trail leg to the rear of the lead leg and step in the direction of travel to return to the starting position. Repeat sequence to the 25-yard stop point. Always face the same direction so that movement of the first 25-yards is to the left and movement of the second 25-yards is to the right.

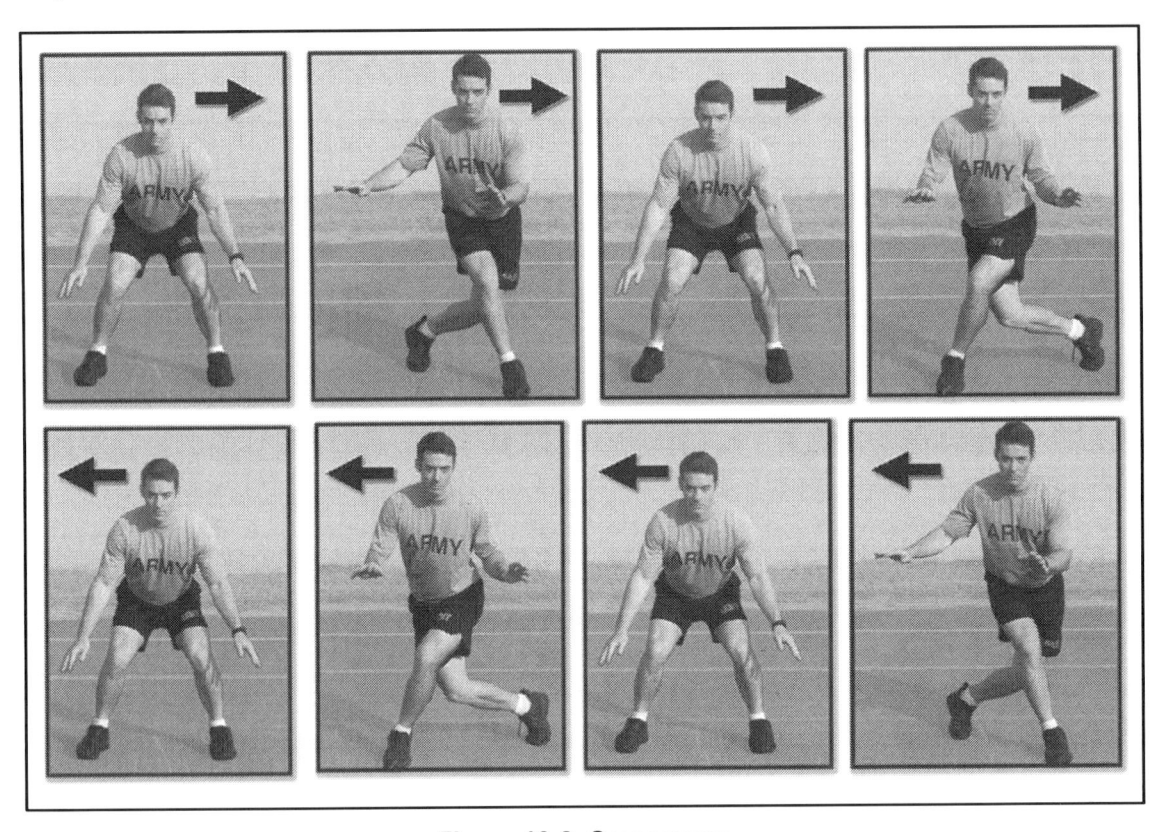

Figure 10-8. Crossovers

Check Points:
- Pick the feet up with each step. Avoid dragging the feet along the ground.
- Crouch slightly while keeping the back straight.
- Maintain the trunk perpendicular to the direction of travel while allowing the hips to move naturally.
- Rank leaders will face their rank throughout the exercise.

Precaution: N/A

MILITARY MOVEMENT DRILL 2

EXERCISE 3: CROUCH RUN

Purpose: This exercise develops the ability to run quickly in a crouched position (Figure 10-9).

Starting Position: Assume the starting position for exercise three of CD 1: mountain climber.

Movement: Power out of the starting position, performing one repetition of mountain climber, then upon finishing count 4, run forward in the crouch position to the 25-yard mark. Turn clockwise while planting the left foot and bending and squatting to touch the ground with the left hand, as in performing the shuttle sprint in MMD1. Crouch run quickly back to the starting line and plant the right foot, turn counter-clockwise and touch the ground with the right hand. Accelerate out of the crouch run to an upright position and sprint back to the 25-yard mark gradually accelerating to near maximum speed.

Figure 10-9. Crouch run

Check Points:

- Move from the crouch run starting position by executing one repetition of mountain climber and firing out of count four with the right leg and swinging the left arm forward to the crouch run.
- On the crouch run, stay low with minimal arm swing.
- Soldiers should slow their movement before planting their feet and changing direction.
- Soldiers should squat while bending the trunk when reaching to touch the ground as they change direction.
- Soldiers touch the ground with their left hand on the first turn, then with their right hand on the second turn.
- Accelerate to near maximum speed during the last 25-yard interval.

Precaution: Soldiers should use caution when performing this exercise on wet terrain.

SPEED RUNNING

10-23. Speed running is based on the training principle that a greater amount of intense work can be performed if the work is interspersed with periods of recovery. Improvements in physical fitness are affected to a greater extent by the intensity of training than by the frequency or duration of the training. During speed running, Soldiers perform a work interval in a specified time for a specific number of repetitions. The work intervals are followed immediately by an active recovery interval. Multiple work intervals cause the onset of fatigue many times during a single training session. Speed running improves the resistance to fatigue of the active muscles by repeatedly exposing them to high intensity effort. As a result of their increased anaerobic and aerobic endurance, Soldiers are able to sustain performance of physically demanding tasks at a higher intensity for a longer duration. The training stimulus associated with speed running occurs from the combination of work and recovery. A very short recovery period may not allow the body to recover sufficiently to perform the next work interval at the desired intensity. A very long recovery period may allow the body to recover too much and some of the training effect would be lost. Generally, duration of the recovery period depends on the intensity and duration of the work interval. An appropriate work to recovery ratio for improving Soldier physical readiness is 1:2. Speed running has three variables: work duration, recovery duration, and the number of repetitions. The speed running activities appropriate for Soldiers to improve physical readiness and APFT 2-mile run performance are 30:60s and 60:120s. Refer to Table 10-2 for appropriate speed running prescriptions for the toughening and sustaining phases. When conducting speed running, the AIs will perform the activity by running with Soldiers in the unit. This allows the AI's to continually monitor and motivate Soldiers throughout the conduct of the exercise. The PRT leader positions himself to supervise the conduct of speed running and uses a stopwatch and a whistle for signaling the "*Start*" and "*Stop*" of each work and rest interval. Refer to Table 10-2 for endurance and mobility activities, prescriptions of intensity, and duration and volume within the toughening and sustaining phases. In addition, Chapter 5, Planning Considerations, provides the template for commanders and PRT leaders to implement endurance and mobility activities into their PRT programs.

30:60s

10-24. Soldiers will perform 30:60s adhering to a work to recovery ratio of 1:2. During the work interval, Soldiers will sprint for 30 seconds. During the recovery interval, Soldiers walk for 60 seconds. This is one repetition of a 30:60. Speed running will cause Soldiers to spread out over the course of the running track during the work interval. If required, the PRT Leader will have Soldiers regroup before the start of the next work interval. Soldiers run at a slow pace (jog) ¼ mile before beginning 30:60s. Table 10-2 shows speed running progression. Soldiers should walk at least 3 minutes before performing additional activities or recovery.

60:120s

10-25. Soldiers perform 60:120s adhering to a work to recovery ratio of 1:2. During the work interval, Soldiers sprint for 60 seconds. During the recovery interval, Soldiers walk for 120 seconds. This is one repetition of a 60:120. Speed running causes Soldiers to spread out over the course of the running track during the work interval. If required, the PRT leader has Soldiers regroup before the start of the next work interval. All ability groups should run at a slow pace (jog) ¼ mile before beginning 60:120s. Table 10-2 shows speed running progression. Soldiers should walk at least 3 minutes before performing additional activities or recovery.

TRAINING AREAS FOR SPEED RUNNING

10-26. Ideally, the training area for the conduct of 30:60s and 60:120s is a ¼-mile or a 400-meter oval running track. The PRT leader should stand in the middle of the training area so he can see all Soldiers. From there, the Soldiers can easily hear his whistled commands to start and stop walking and running intervals. If 30:60s or 60:120s are conducted on a road, the route MUST be wide enough for Soldiers to turn around and not collide. The recommended distances for conducting 30:60s or 60:120s on a straight road course is at least 100 yards and a maximum of 200 yards (Figure 10-10).

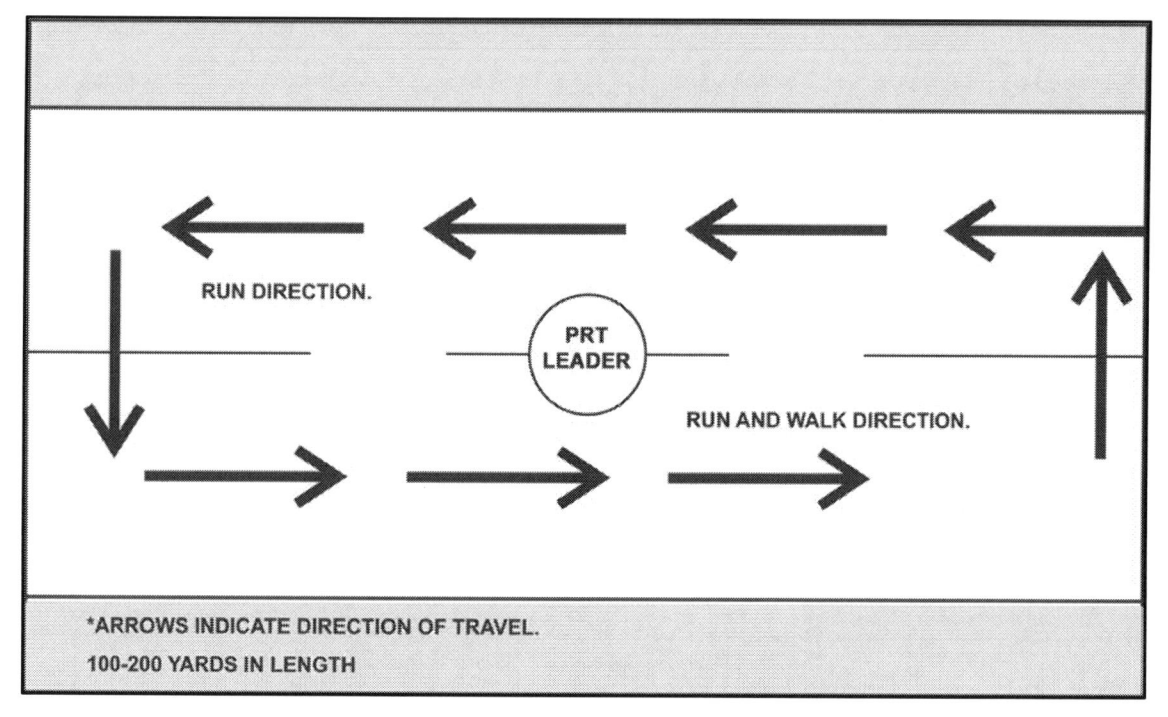

Figure 10-10. Speed running on a straight course

300-YARD SHUTTLE RUN

10-27. The 300-yard SR run develops the ability to repeatedly sprint after changing direction. It is an indicator of the Soldier's anaerobic endurance, speed, and agility. The 300-yard SR is conducted from the extended rectangular formation (covered) as shown in Figure 10-3. On the command, "Ready," one Soldier in each column will move behind the starting line and assume the ready position (staggered stance). On the command, "GO," Soldiers sprint to a line 25-yards from the starting line. They must touch the line or beyond it with their left hand, then return to touch the starting/finish line with their right hand. This is considered one repetition. Soldiers will perform six repetitions alternating touching the lines with opposite hands. On the last (sixth) repetition, Soldiers sprint past the starting/finish line without touching it. The PRT leader and AIs ensure that Soldiers sprint in their own lanes and run with their heads up to watch for other Soldiers who may be moving in the opposite direction. Figure 10-11 shows the running patterns and requirements of the 300-yard SR. Refer to Table 10-2 for endurance and mobility activities, prescriptions of intensity, duration, and volume within the toughening and sustaining phases. In addition, Chapter 5, Planning Considerations, provides the template for commanders and PRT leaders to implement endurance and mobility activities into their PRT programs.

Endurance and Mobility Activities

Checkpoints:
- Soldiers should slow their movement before planting feet and changing direction.
- Soldiers should both bend the trunk and squat when reaching to touch the ground as they change direction.
- Soldiers touch the ground with their left hand on the first turn, then with their right hand on the second turn and continue to alternate hand touches on each turn.
- Soldiers must sprint with their heads up and watch for other soldiers who may be moving in an opposite direction.

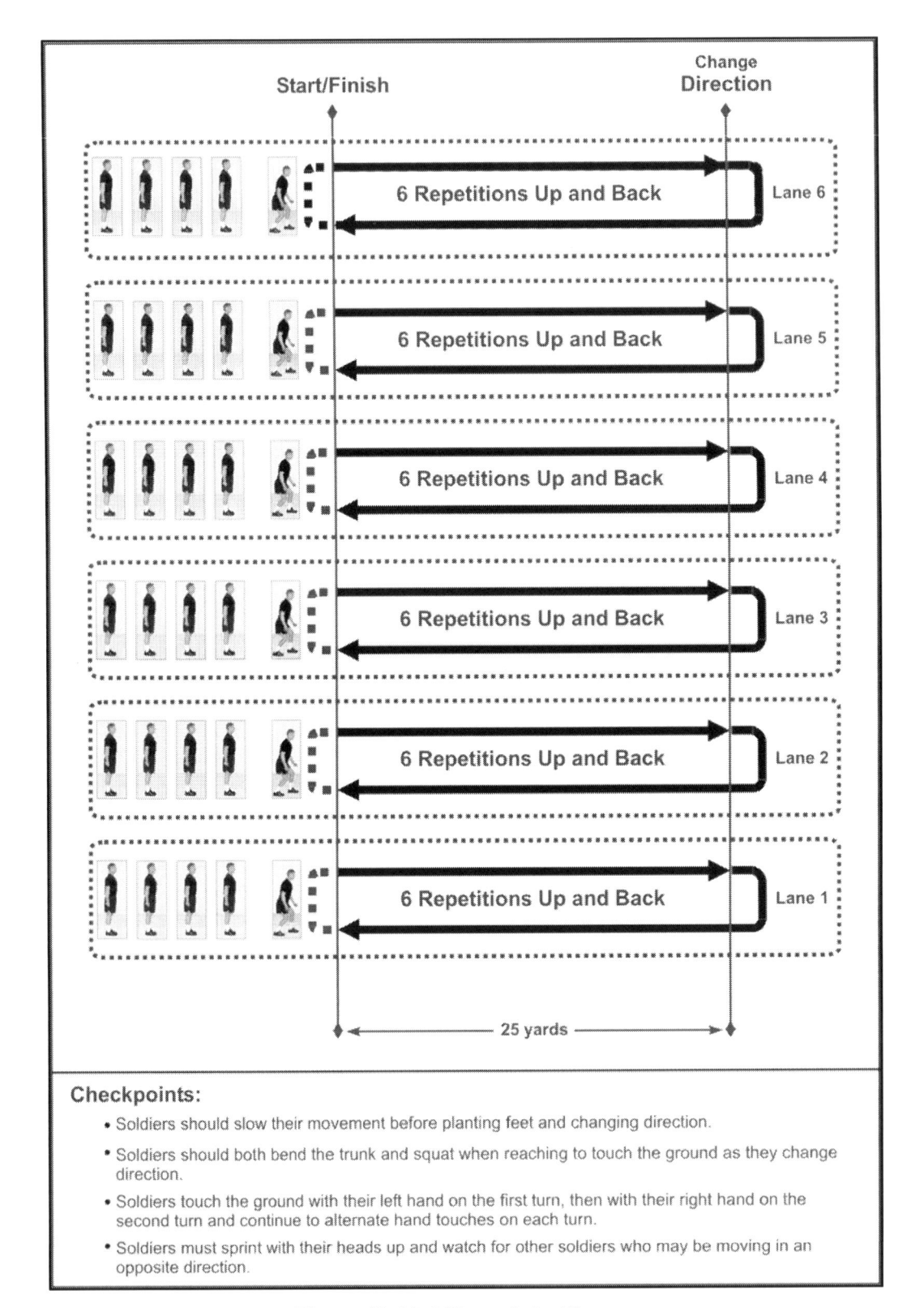

Figure 10-11. 300-yard shuttle run

HILL REPEATS

10-28. Hill repeats are an effective means of developing explosive leg strength, speed and anaerobic endurance. Both uphill and downhill running intervals are important. Uphill repeats build leg strength, power, and anaerobic endurance, while downhill repeats improve speed though rapid leg turn-over and releasing neural inhibitions. The intensity and duration of the repetitions will depend on the characteristics of the hill. The PRT leader designates the number of repetitions and signals the start of each group or individual. <u>Hill repeats should not be conducted under load</u>. Refer to Table 10-2 for endurance and mobility activities, prescriptions of intensity, duration, and volume within the toughening and sustaining phases. In addition, Chapter 5, Planning Considerations, provides the template for commanders and PRT leaders to implement endurance and mobility activities into their PRT programs.

UPHILL REPEATS

10-29. A short, steep hill is ideal for explosive uphill efforts of 15-20 seconds sprinting up 40-60 yards and 60-90 seconds walking back down for 6 to 10 repetitions. On uphill repeats, lean slightly forward without bending at the waist. On steep hills, the knees will need to rise higher than normal to permit a full stride. Refer to Table 10-2 for endurance and mobility activities, prescriptions of intensity, duration, and volume within the toughening and sustaining phases. In addition, Chapter 5, provides the template for commanders and PRT leaders to implement endurance and mobility activities into their PRT programs.

DOWNHILL REPEATS

10-30. Long, gentle slopes are best for improving speed through downhill repeats. Downhill repeats are performed at a high intensity of 15-20 seconds of downhill sprinting (near maximal effort) with rest intervals consisting of walking back up the hill for 60-90 seconds for 6 to 10 repetitions. It is important to maintain good form during HR, especially when running downhill. Refer to Table 10-2 for endurance and mobility activities, prescriptions of intensity, duration, and volume within the toughening and sustaining phases. In addition, Chapter 5, provides the template for commanders and PRT leaders to implement endurance and mobility activities into their PRT programs.

ABILITY GROUP RUN

10-31. The AGR trains Soldiers in groups of near-equal ability. Each ability group runs at a prescribed pace intense enough to produce a training effect for that group and each Soldier in it. Leaders should program these runs for specific lengths of time, not miles to be run. This training method provides a challenge for each ability group while controlling injuries. The PRT leader conducts a 1-mile run assessment to assign Soldiers in ability groups. Based on each Soldier's 1-mile run assessment time, the PRT leader assigns the Soldier to one of the groups shown in Table 10-3.

Table 10-3. Ability group assignment

Toughening Phase AGR Assignments	Sustaining Phase AGR Assignments
A Group, 7:15 and faster	A Group, 6:30 and faster
B Group, 7:16 to 8:15	B Group, 6:31 to 7:15
C Group, 8:16 to 10:15	C Group, 7:16 to 8:00
D Group, 10:16 and slower	D Group, 8:01 and slower

10-32. Some Soldiers may make the cut off times to qualify for an ability group but are unable to maintain the prescribed running pace listed in the PRT schedule. If this occurs, they may drop down to the slower group and progress later to the faster running group. Ability group runs must be conducted for the duration and intensity specified in the training schedules in Chapter 5, Planning Considerations. The frequency of AGRs is one or two times per week. Ability group runs, speed running, and foot marching (greater than 5 km) should not be conducted on the same or consecutive days. The running duration is determined by time, not distance. Soldiers

Endurance and Mobility Activities

should move to faster groups when they are ready because they progress at different rates. Those who have difficulty maintaining the specified pace within an ability group should be placed in a slower ability group. Supervision will prevent a constant shifting of Soldiers between groups due to lack of individual effort. See the training schedules in Chapter 5, Planning Considerations, for AGR times and pace. Routes used for sustained running in ability groups should be well lighted, free from hazards and traffic, and marked at ¼-mile intervals. Ability group leaders will ensure running is at the proper pace prescribed for their group by checking their split times at each ¼-mile marker along the route. Table 10-4 shows the appropriate ¼-mile split time based on the AGR pace.

Table 10-4. Quarter-mile split times based on AGR pace

Pace/Mile	1/4-Mile Split	Pace/Mile	1/4-Mile Split	Pace/Mile	1/4-Mile Split
6:00	1:30	8:15	2:03	10:30	2:38
6:15	1:34	8:30	2:07	10:45	2:42
6:30	1:37	8:45	2:11	11:00	2:45
6:45	1:42	9:00	2:15	11:15	2:49
7:00	1:45	9:15	2:19	11:30	2:53
7:15	1:48	9:30	2:23	11:45	2:57
7:30	1:52	9:45	2:27	12:00	3:00
7:45	1:56	10:00	2:30	12:15	3:04
8:00	2:00	10:15	2:34	12:30	3:07

10-33. Refer to Table 10-2 for endurance and mobility activities, prescriptions of intensity, duration, and volume within the toughening and sustaining phases. In addition, Chapter 5, Planning Considerations, provides the template for commanders and PRT leaders to implement endurance and mobility activities into their PRT programs.

UNIT FORMATION RUN

10-34. The UFR elicits intangible rewards gained from running with a group, such as esprit de corps, team building, and discipline. Unit formation runs are based on a time and/or distance that can be achieved with unit integrity and a display of unit cohesion. Unit formation runs are organized by squad, platoon, company, or battalion; not by ability. Keeping a large unit in step, with proper distance intervals and correct running form, offers intangible benefits that commander's desire. Commanders should not use UFRs as the foundation of their PRT program. They should be performed no more than once per quarter due to the limited training effect offered for the entire unit. The UFR begins with a gradual increase in intensity for the first three minutes or ¼ mile, then continues at a prescribed target pace for a specified time, and concludes with a gradual decrease in intensity for the last three minutes or ¼ mile. The gradual increase and gradual decrease quarter miles will be conducted at a pace two minutes slower than the target pace. The unit commander is responsible for establishing a pace achievable by all Soldiers in the unit. Refer to Table 10-2 for endurance and mobility activities, prescriptions of intensity, duration, and volume within the toughening and sustaining phases. In addition, Chapter 5, Planning Considerations, provides the template for commanders and PRT leaders to implement endurance and mobility activities into their PRT programs.

RELEASE RUN

10-35. The RR combines the benefits of formation running and individual performance at higher training intensities. Soldiers will run in formation for a specified time (no more than 15 minutes), then released to run as fast as they can back to the starting point. Upon completion of the release run, additional PRT activities may be conducted or recovery performed. Refer to Table 10-2 for endurance and mobility activities, prescriptions of intensity, duration, and volume within the toughening and sustaining phases. In addition, Chapter 5, Planning Considerations, provides the template for commanders and PRT leaders to implement endurance and mobility activities into their PRT programs.

TERRAIN RUN

10-36. The TR applies the train for combat proficiency concept to PRT. Running through local training areas, over hills, and around obstacles improves mobility, endurance, and the ability to stop, start, and change direction. Terrain running is designed to be conducted with small unit integrity. This type of running is best performed by squads and sections. Distances should generally be 1 mile for densely wooded areas and up to 2 miles on tank trails and open fields. Intensity is relative to the terrain. Physical readiness training leaders will form the unit and maintain an interval suitable for the terrain and environmental conditions. Soldiers should perform terrain running in ACUs and well-fitting boots. Soldiers may progress to performing terrain running with IBA, ACH, individual weapon, and under fighting load. Refer to Table 10-2 for endurance and mobility activities, prescriptions of intensity, duration, and volume within the toughening and sustaining phases. In addition, Chapter 5, Planning Considerations, provides the template for commanders and PRT leaders to implement endurance and mobility activities into their PRT programs.

FOOT MARCHES

10-37. Foot marching as a movement component of maneuver is a critical Soldier physical requirement. Regular foot marching helps to avoid the cumulative effects of lower injury trauma and prepares Soldiers to successfully move under load. See FM 21-18, *Foot Marches*, for specific instructions and guidance for the conduct of foot marches.

CONDITIONING OBSTACLE COURSE

10-38. Obstacle course running develops physical capacities, fundamental skills, and abilities that are important to Soldiers in combat operations. Soldiers must be able to crawl, creep, climb, walk, run, and jump. Furthermore, with individual body armor and fighting load, they must be able to perform all these tasks for long periods of time without exhaustion or injury, even after fatigue has set in. Refer to Appendix E for obstacle negotiation.

ENDURANCE TRAINING MACHINES

10-39. When using ETM there are four primary variables to consider: exercise mode, training frequency, exercise duration, and training intensity. Exercise prescription specifies training frequency, exercise duration, and training intensity. The mode of exercise (type of ETM) is determined by environmental constraints and/or training according to physical profile limitations (temporary/permanent). Each ETM contains specific instructions for proper use and adjustments for the Soldier to obtain optimal posture during endurance exercise (seat height on cycle ergometers or seat distance on rowing machines). If the ETM has no visible list of operating instructions, ask the PRT leader or AI for assistance (Figure 10-12).

EXERCISE MODE

10-40. Exercise mode refers to the specific activity performed by the Soldier: running, cycling, swimming, and the use of a variety of endurance training equipment. There are advantages to using endurance training equipment (environmental constraints, safety for Soldiers on physical profile, and isolation of specific muscle groups to be trained during rehabilitation and reconditioning). Consideration for use of specific types of equipment may be based on the Soldier's ability to participate in weight-bearing or non-weight-bearing activities. Weight-bearing activities include walking or running on a treadmill and using a stair climbing/stepping machine. Non-weight-bearing and limited weight-bearing activities include use of cycle ergometers (upright/recumbent) elliptical trainers, rowers, climbing machines, and cross-country ski machines. Use of limited or non-weight-bearing endurance training equipment is desirable for obtaining higher caloric expenditure through additional training sessions by overweight Soldiers trying to reduce body fat. Each of these modes typically provide the Soldier with a variety of individual exercise routines that monitor and display exercise duration, training intensity (heart rate/pace/watts), caloric expenditure, and distance completed (miles/km). See Figure 10-12 for examples of various types of endurance training equipment.

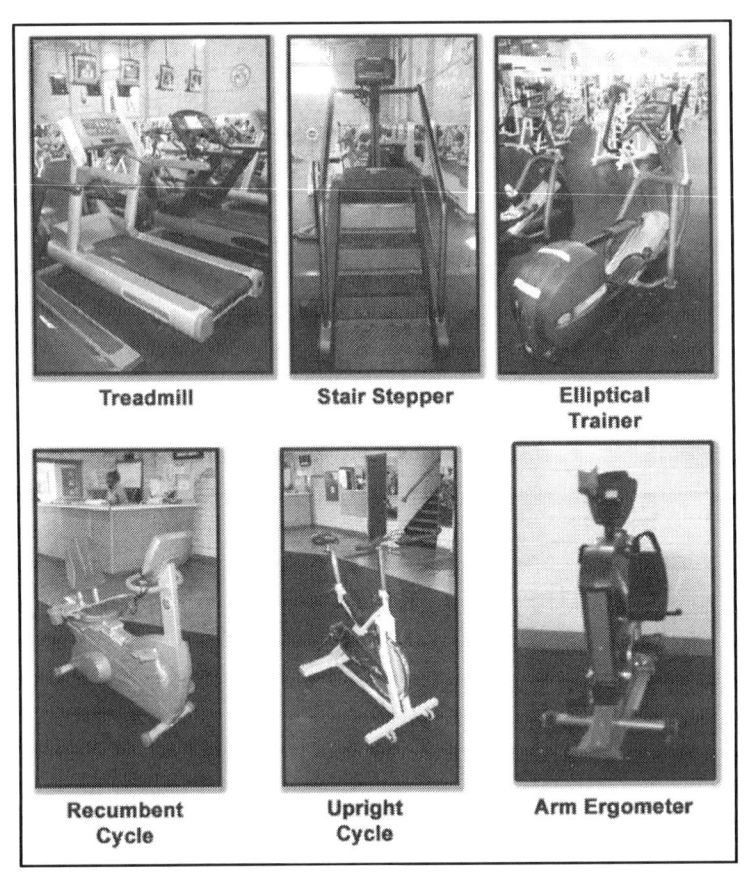

Figure 10-12. Endurance training machines

TRAINING FREQUENCY

10-41. Training frequency refers to the number of training sessions conducted per day or per week. Training frequency is determined by exercise duration and training intensity. Training sessions that involve high intensity or longer duration may necessitate less frequent training to allow for adequate recovery. Normal endurance training frequency is three to five exercise sessions per week.

EXERCISE DURATION

10-42. Exercise duration is 20 minutes or longer and varies from machine to machine, depending on the intensity of the exercise routine being performed (hill profile, speed, degree of incline, resistance). Most exercise sessions of high or moderate intensity should last 20 to 30 minutes. Endurance exercise sessions that address additional caloric expenditure for body fat reduction should be of low intensity and may last up to 60 minutes.

TRAINING INTENSITY

10-43. Training intensity is typically monitored and displayed on the exercise equipment control panel in terms of heart rate, pace (mph/kph, step rate), watts, kiloponds, caloric expenditure (kcals), or resistance.

> **Summary**
> The activities in this chapter develop the endurance and mobility demanded of WTBDs. A properly designed PRT running program strikes a balance between speed and sustained running to train the full spectrum of endurance. Endurance training equipment is used to accommodate environmental constraints and/or training according to Soldiers' physical profile limitations.

Appendix A
Army Physical Fitness Test (APFT)

The intent of the Army Physical Fitness Test (APFT) is to provide an assessment of the PRT program. Physical fitness testing is designed to ensure the maintenance of a base level of physical fitness essential for every Soldier, regardless of Army MOS or duty assignment. PRT programs must be developed to take this base level of conditioning and raise it to help meet or exceed mission-related physical performance tasks. Commanders must ensure that physical fitness testing does not form the foundation of unit or individual PRT programs. Temporary training periods solely devoted toward meeting APFT requirements are highly discouraged. See AR 350-1 for policy guidelines pertaining to the APFT.

APFT OVERVIEW

A-1. The APFT provides a measure of upper and lower body muscular endurance. It is a performance test that indicates a Soldier's ability to perform physically and handle his or her own body weight. Army Physical Fitness Test standards are adjusted for age and physiological differences between the genders.

FITNESS STANDARDS

A-2. The APFT consists of push-ups, sit-ups, and a 2-mile run—done in that order—on the same day. Soldiers are allowed a minimum of 10 minutes and a maximum of 20 minutes rest between events. All three events must be completed within two hours. The test period is defined as the period of time that elapses from the start to the finish of the three events (from the first push-up performed to the last Soldier crossing the finish line of the 2-mile run event).

A-3. In accordance with AR 350-1, all Soldiers must attain a score of at least 60 points on each event and an overall score of at least 180 points. Soldiers in BCT must attain 50 points in each event and an overall score of 150 points. The maximum score a Soldier can attain on the APFT is 300 points. The use of extended scale scoring IS NOT authorized.

A-4. Army Physical Fitness Test results will be recorded on DA Form 705 (sample and Army Knowledge Online (AKO) form reference located at chapter end). One scorecard will be maintained for each Soldier. The scorecard will be kept in a central location in the unit and will accompany the individual military personnel records jacket at the time of permanent change of station according to AR 350-1. Units and separate offices monitor the performance and progress of their Soldiers. Individual Soldiers are not authorized to administer the APFT to themselves to simply satisfy record test requirements. A minimum of four Soldiers are required to administer an APFT: OIC or NCOIC, an event supervisor(s), an event scorer, and support personnel. Another Soldier being tested or support personnel may act as the holder to secure the Soldier's ankles during the sit-up event.

A-5. Any piece of clothing not prescribed as a component of the IPFU, ACU or commander authorized civilian attire is not permitted for wear during the APFT. Neither are devices or equipment that offer any potential for unfair advantage during testing. Unless prescribed as part of the Soldier's medical profile, the wearing of the following items are not authorized: nasal strips, weight lifting gloves, back braces, elastic bandages, or braces. Electronic devices are also not authorized (MP3 players, radios, cell phones, and compact disc players). AR 670-1, *Wear and Appearance of Army Uniforms and Insignia*, specifies the components of the IPFU ensemble.

Appendix A

APFT ADMINISTRATION

SUPERVISION

A-6. The success of any physical fitness testing program depends on obtaining valid and accurate test results; therefore, the APFT must be administered properly to accurately evaluate individual Soldier and unit physical fitness. Supervision of the APFT is necessary to ensure the objectives of the physical fitness program are met. Proper supervision provides for standardization in the following:
- Test preparation.
- Control of performance factors.
- Training of test personnel.
- Test scoring.

PREPARATION

A-7. Preparation for the APFT should be directed at securing the most accurate evaluation of personnel participating in the test. Preparatory requirements include the following:
- Selecting and training of event supervisors, scorers, timers, demonstrators, and support personnel.
- Equipment inventory.
- Securing the test site.

PLANNING

A-8. The commander should ensure that testing is consistent with regard to events, scoring, clothing, equipment, and facilities. Testing should be planned to permit each Soldier to perform at his maximal level and should ensure the following:
- Soldiers are not tested when fatigued or ill.
- Soldiers do not participate in tiring duties before taking an APFT.
- Weather and environmental conditions do not inhibit physical performance.
- Risk analysis is conducted.

DUTIES OF TEST PERSONNEL

RESPONSIBILITIES

A-9. The Army Physical Fitness Test personnel must be familiar with all aspects of administration of the APFT. Supervision of Soldiers and laying out the test area are essential duties. The following test personnel are recommended to conduct an APFT: an OIC and/or NCOIC, an event supervisor(s), a timer, a back-up timer, an event scorer(s), a demonstrator(s), and support personnel. The minimum number of test personnel required to administer the APFT is four: an OIC/NCOIC, an event supervisor, an event scorer, and support personnel to hold the Soldiers' feet on the sit-up event.

A-10. The OIC and the NCOIC are responsible for the administration of the APFT. Responsibilities include:
- Preparation for push-up event (after reading instructions and before test start).
- Administration of the test.
- Conduct of recovery upon completion of the test.
- Procurement of all necessary equipment and supplies.
- Arrangement and layout of test area.
- Training of event supervisors, scorers, timer, back-up timer demonstrators, and support personnel.
- Ensure tests are properly administered and that events are explained, demonstrated, and scored according to standard.

- When test personnel required to administer the APFT are limited, the OIC/NCOIC may perform the duties of an event demonstrator and/or back-up timer.
- Reports results of test.

A-11. The event supervisors are responsible for administration of test events. Responsibility includes the following:
- Administers one or more test events.
- Ensures necessary equipment is on hand for each event(s).
- Reads APFT event instructions.
- Conducts APFT event demonstration.
- When test personnel required to administer the APFT are limited, the event supervisor(s) may perform the duties of the timer.
- Supervises event scoring to standard.
- Answers questions on scoring discrepancies and informs the OIC/NCOIC.

A-12. The event scorers are responsible for scoring events to standard. Responsibility includes the following:
- Enforces test standards.
- Records the correct number of repetitions in the raw score block on DA Form 705.
- Records initials in initials box on DA Form 705.
- Performs other duties assigned by the OIC or the NCOIC.
- Receives training conducted by the OIC/NCOIC to ensure scoring is to standard.

A-13. The demonstrators are responsible for demonstrating the push-up and sit-up events to standard. Responsibility includes the following:
- Assists the event supervisor by demonstrating push-ups and sit-ups to standard during the reading of event instructions.
- Performs other duties assigned by the OIC or the NCOIC.
- Receives training, conducted by the OIC/NCOIC, to ensure demonstration of push-ups and sit-ups are to standard.

A-14. Timers and back-up timers are responsible for properly timing the push-up, sit-up, and 2-mile run events.

A-15. Support personnel assist in preventing unsafe acts to ensure smooth operation of the APFT. The use of support personnel depends on local policy and unit standing operating procedures. For example, support personnel may perform the duties of the holder during the sit-up event. Medical support on site is not required unless specified by local policy. The OIC and/or the NCOIC should have a plan for medical support (if required).

TEST SITE

REQUIREMENTS

A-16. The OIC and the NCOIC should select a test site that is flat and free of debris. The test site should have the following:
- A site that is free of any significant hazards.
- A briefing area for the reading of event instructions.
- A preparation area (can be same as briefing area).
- A soft, flat, dry area for push-ups and sit-ups.
- A flat, measure 2-mile running course with a solid surface that is not more than 3 percent grade.

A-17. Sound judgment must be used in the selection of a 2-mile run course. There is no requirement to survey 2-mile run courses; however, selected test sites should be free of significant hazards such as traffic, slippery road surfaces, and areas where heavy pollution is present. Running tracks may be used to administer the 2-mile run event. If a 400-meter track is used, the OIC/NCOIC must add an additional 61 feet, 4 inches to the standard 8 laps to ensure the test's required 2-mile distance is covered. One lap on a 400-meter track is 92 inches shorter

Appendix A

than one lap on a 440-yard track. Eight laps on a 400-meter track is 736 inches shorter than eight laps (2 miles) on a 440-yard tack. Therefore, Soldiers running on a 400-meter track must run an additional 61 feet, 4 inches.

TEST PROCEDURES

A-18. The APFT test sequence is the push-up, sit-up, and 2-mile run (or an approved alternate aerobic event). The order of events cannot be changed. There are no exceptions to this sequence. Soldiers are allowed a minimum of 10 minutes and a maximum of 20 minutes to recover between events. The OIC or the NCOIC determines the recovery time. It is normally based on the number of Soldiers taking the test. If large numbers of Soldiers are being tested, staggered start times should be planned to allow for proper recovery between test events. Under no circumstances is the APFT valid if Soldiers cannot begin and end all three events in two hours or less. The following paragraphs describe procedures for APFT administration. On test day, the OIC or the NCOIC briefs Soldiers on the purpose and organization of the test. The OIC or the NCOIC explains test administration including the scorecard, scoring standards, and test sequence. In addition, the wearing of unauthorized items such as nasal strips, braces, elastic bandages, weight lifting gloves, electronic devices (MP3 players, radios, cell phones, and compact disc players) are addressed. Test instructions for the push-up, sit-up, and 2-mile run (or approved alternate aerobic event) are read prior to conducting preparation. After preparation is completed, the push-up event will begin. From the beginning of the push-up event to the completion of all remaining events, the total time elapsed cannot exceed two hours. Upon completion of all events, recovery will be conducted

A-19. The following instructions are **READ** aloud to all Soldiers taking the APFT.

> YOU ARE ABOUT TO TAKE THE ARMY PHYSICAL FITNESS TEST, A TEST THAT WILL MEASURE YOUR UPPER AND LOWER BODY MUSCULAR ENDURANCE. THE RESULTS OF THIS TEST WILL GIVE YOU AND YOUR COMMANDERS AN INDICATION OF YOUR STATE OF FITNESS AND WILL ACT AS A GUIDE IN DETERMINING YOUR PHYSICAL TRAINING NEEDS. LISTEN CLOSELY TO THE TEST INSTRUCTIONS, AND DO THE BEST YOU CAN ON EACH OF THE EVENTS.

A-20. If DA Form 705 has not been issued, scorecards will be handed out at this time. The OIC or the NCOIC will then instruct the Soldiers to fill in the appropriate spaces with the required personal data. The following instructions are **READ** aloud to all Soldiers taking the APFT:

> "IN THE APPROPRIATE SPACES, PRINT IN INK THE PERSONAL INFORMATION REQUIRED ON THE SCORECARD."

Note: The preceding remark is omitted if scorecards were issued prior to arrival at the test site.

A-21. Soldiers are then given time to complete the required information. Next, the OIC or the NCOIC explains procedures for scorecard use during testing. The following instructions are **READ** aloud to all Soldiers taking the APFT:

> "YOU ARE TO CARRY THIS CARD WITH YOU TO EACH EVENT. BEFORE YOU BEGIN, HAND THE CARD TO THE SCORER. AFTER YOU COMPLETE THE EVENT, THE SCORER WILL RECORD YOUR RAW SCORE, INITIAL THE CARD, AND RETURN IT TO YOU."

A-22. Now the OIC or the NCOIC explains how raw scores are converted to point scores. At this point in time, Soldiers are assigned to groups. The following instructions are **READ** aloud to all Soldiers taking the APFT:

> "EACH OF YOU WILL BE ASSIGNED TO A GROUP. STAY WITH YOUR TEST GROUP FOR THE ENTIRE TEST. WHAT ARE YOUR QUESTIONS ABOUT THE TEST AT THIS POINT?"

INSTRUCTIONS

A-23. The OIC, the NCOIC, or the event supervisor will read all three event instructions prior to the start of the test. Specific 2-mile run route instructions can be addressed at the 2-mile run event test site.

PUSH-UP

A-24. **The OIC, the NCOIC, or the event supervisor must read the following before beginning the push-up event** (Figure A-1).

Appendix A

> "THE PUSH-UP EVENT MEASURES THE ENDURANCE OF THE CHEST, SHOULDER, AND TRICEPS MUSCLES. ON THE COMMAND, 'GET SET', ASSUME THE FRONT-LEANING REST POSITION BY PLACING YOUR HANDS WHERE THEY ARE COMFORTABLE FOR YOU. YOUR FEET MAY BE TOGETHER OR UP TO 12 INCHES APART (MEASURED BETWEEN THE FEET). WHEN VIEWED FROM THE SIDE, YOUR BODY SHOULD FORM A GENERALLY STRAIGHT LINE FROM YOUR SHOULDERS TO YOUR ANKLES. ON THE COMMAND 'GO', BEGIN THE PUSH-UP BY BENDING YOUR ELBOWS AND LOWERING YOUR ENTIRE BODY AS A SINGLE UNIT UNTIL YOUR UPPER ARMS ARE AT LEAST PARALLEL TO THE GROUND. THEN, RETURN TO THE STARTING POSITION BY RAISING YOUR ENTIRE BODY UNTIL YOUR ARMS ARE FULLY EXTENDED. YOUR BODY MUST REMAIN RIGID IN A GENERALLY STRAIGHT LINE AND MOVE AS A UNIT WHILE PERFORMING EACH REPETITION. AT THE END OF EACH REPETITION, THE SCORER WILL STATE THE NUMBER OF REPETITIONS YOU HAVE COMPLETED CORRECTLY. IF YOU FAIL TO KEEP YOUR BODY GENERALLY STRAIGHT, TO LOWER YOUR WHOLE BODY UNTIL YOUR UPPER ARMS ARE AT LEAST PARALLEL TO THE GROUND, OR TO EXTEND YOUR ARMS COMPLETELY, THAT REPETITION WILL NOT COUNT, AND THE SCORER WILL REPEAT THE NUMBER OF THE LAST CORRECTLY PERFORMED REPETITION."
>
> "IF YOU FAIL TO PERFORM THE FIRST 10 PUSH-UPS CORRECTLY, THE SCORER WILL TELL YOU TO GO TO YOUR KNEES AND WILL EXPLAIN YOUR DEFICIENCIES. YOU WILL THEN BE SENT TO THE END OF THE LINE TO BE RETESTED. AFTER THE FIRST 10 PUSH-UPS HAVE BEEN PERFORMED AND COUNTED, NO RESTARTS ARE ALLOWED. THE TEST WILL CONTINUE, AND ANY INCORRECTLY PERFORMED PUSH-UPS WILL NOT BE COUNTED. AN ALTERED, FRONT-LEANING REST POSITION IS THE ONLY AUTHORIZED REST POSITION. THAT IS, YOU MAY SAG IN THE MIDDLE OR FLEX YOUR BACK. WHEN FLEXING YOUR BACK, YOU MAY BEND YOUR KNEES, BUT NOT TO SUCH AN EXTENT THAT YOU ARE SUPPORTING MOST OF YOUR BODY WEIGHT WITH YOUR LEGS. IF THIS OCCURS, YOUR PERFORMANCE WILL BE TERMINATED. YOU MUST RETURN TO, AND PAUSE IN, THE CORRECT STARTING POSITION BEFORE CONTINUING. IF YOU REST ON THE GROUND OR RAISE EITHER HAND OR FOOT FROM THE GROUND, YOUR PERFORMANCE WILL BE TERMINATED. YOU MAY REPOSITION YOUR HANDS AND/OR FEET DURING THE EVENT AS LONG AS THEY REMAIN IN CONTACT WITH THE GROUND AT ALL TIMES. CORRECT PERFORMANCE IS IMPORTANT. YOU WILL HAVE TWO MINUTES IN WHICH TO DO AS MANY PUSH-UPS AS YOU CAN. WATCH THIS DEMONSTRATION."

Figure A-1. Push-up event narrative

Figure A-2. Push-up additional checkpoints

A-25. During the push-up event, scorers sit or kneel 3 feet from the Soldier's left or right shoulder at a 45-degree angle (refer to Figure A-2). Additional checkpoints to explain and demonstrate for the push-up event are as follows:

- "Your chest may touch the ground during the push-up as long as the contact does not provide an advantage. You cannot bounce off the ground."
- "If a mat is used, your entire body must be on the mat. Sleeping mats are not authorized for use."
- "Your feet will not be braced during the push-up event."
- "You may do the push-up event on your fists."
- "You may not cross your feet while doing the push-up event."
- "You may not take any APFT event in bare feet."
- "You should not wear glasses while performing the push-up event."

A-26. In conclusion, the OIC/NCOIC, or the event supervisors, asks:

> **"WHAT ARE YOUR QUESTIONS ABOUT THIS EVENT?"**

SIT-UP

A-27. The OIC, the NCOIC, or the event supervisor, must **READ** the following before the sit-up event (Figure A-3).

"THE SIT-UP EVENT MEASURES THE ENDURANCE OF THE ABDOMINAL AND HIP-FLEXOR MUSCLES. ON THE COMMAND 'GET SET', ASSUME THE STARTING POSITION BY LYING ON YOUR BACK WITH YOUR KNEES BENT AT A 90-DEGREE ANGLE. YOUR FEET MAY BE TOGETHER OR UP TO 12 INCHES APART (MEASURED BETWEEN THE FEET). ANOTHER PERSON WILL HOLD YOUR ANKLES WITH THE HANDS ONLY. NO OTHER METHOD OF BRACING OR HOLDING THE FEET IS AUTHORIZED. THE HEEL IS THE ONLY PART OF YOUR FOOT THAT MUST STAY IN CONTACT WITH THE GROUND. YOUR FINGERS MUST BE INTERLOCKED BEHIND YOUR HEAD AND THE BACKS OF YOUR HANDS MUST TOUCH THE GROUND. YOUR ARMS AND ELBOWS NEED NOT TOUCH THE GROUND. ON THE COMMAND, 'GO', BEGIN RAISING YOUR UPPER BODY FORWARD TO, OR BEYOND, THE VERTICAL POSITION. THE VERTICAL POSITION MEANS THAT THE BASE OF YOUR NECK IS ABOVE THE BASE OF YOUR SPINE. AFTER YOU HAVE REACHED OR SURPASSED THE VERTICAL POSITION, LOWER YOUR BODY UNTIL THE BOTTOM OF YOUR SHOULDER BLADES TOUCH THE GROUND. YOUR HEAD, HANDS, ARMS OR ELBOWS DO NOT HAVE TO TOUCH THE GROUND. AT THE END OF EACH REPETITION, THE SCORER WILL STATE THE NUMBER OF SIT-UPS YOU HAVE CORRECTLY PERFORMED. A REPETITION WILL NOT COUNT IF YOU FAIL TO REACH THE VERTICAL POSITION, FAIL TO KEEP YOUR FINGERS INTERLOCKED BEHIND YOUR HEAD, ARCH OR BOW YOUR BACK AND RAISE YOUR BUTTOCKS OFF THE GROUND TO RAISE YOUR UPPER BODY, OR LET YOUR KNEES EXCEED A 90-DEGREE ANGLE. IF A REPETITION DOES NOT COUNT, THE SCORER WILL REPEAT THE NUMBER OF YOUR LAST CORRECTLY PERFORMED SIT-UP. IF YOU FAIL TO PERFORM THE FIRST 10 SIT-UPS CORRECTLY, THE SCORER WILL TELL YOU TO 'STOP' AND WILL EXPLAIN YOUR DEFICIENCIES. YOU WILL THEN BE SENT TO THE END OF THE LINE TO BE RE-TESTED. AFTER THE FIRST 10 SIT-UPS HAVE BEEN PERFORMED AND COUNTED, NO RESTARTS ARE ALLOWED. THE TEST WILL CONTINUE, AND ANY INCORRECTLY PERFORMED SIT-UPS WILL NOT BE COUNTED. THE UP POSITION IS THE ONLY AUTHORIZED REST POSITION.

"IF YOU STOP AND REST IN THE DOWN (STARTING) POSITION, THE EVENT WILL BE TERMINATED. AS LONG AS YOU MAKE A CONTINUOUS PHYSICAL EFFORT TO SIT UP, THE EVENT WILL NOT BE TERMINATED. YOU MAY NOT USE YOUR HANDS OR ANY OTHER MEANS TO PULL OR PUSH YOURSELF UP TO THE UP (REST) POSITION OR TO HOLD YOURSELF IN THE REST POSITION. IF YOU DO SO, YOUR PERFORMANCE IN THE EVENT WILL BE TERMINATED. CORRECT PERFORMANCE IS IMPORTANT. YOU WILL HAVE TWO MINUTES TO PERFORM AS MANY SIT-UPS AS YOU CAN. WATCH THIS DEMONSTRATION."

Figure A-3. Sit-up event narrative

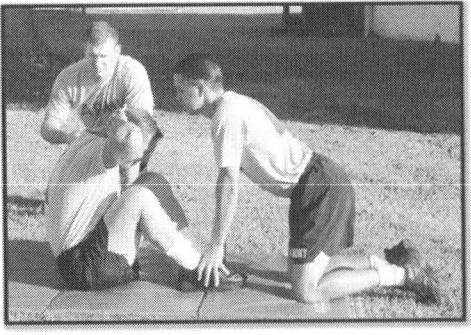

Figure A-4. Sit-up additional checkpoints

A-28. During the sit-up event, the scorer kneels or sits 3 feet from the Soldier's left or right hip. The scorer's head should be even with the Soldier's shoulder when he is in the vertical (up) position (refer to Figure A-4). Additional checkpoints to explain and demonstrate for the sit-up event are as follows:

- "If a mat is used, your entire body must be on the mat. Sleeping mats are not authorized for use."
- "You may not swing your arms or use your hands to pull yourself up or push off the ground to obtain the up position. If this occurs your performance in the event will be terminated."
- "You may wiggle to obtain the up position, but while in the up position, you may not use your elbows or any part of the arms to lock on to or brace against the legs. Your elbows can go either inside or outside the knees, but may not be used to hold yourself in the up position. If this occurs your performance in the event will be terminated."
- "During your performance of the sit-up, your fingers must be interlocked behind your head. As long as any of your fingers are overlapping to any degree, they are considered to be interlocked (Figure A-5). If they do not remain interlocked, that repetition will not count and the scorer will repeat the number of the last correct repetition performed."
- "Both heels must stay in contact with the ground (Figure A-5). If either foot breaks contact with the ground during a repetition, that repetition will not count and the scorer will repeat the number of the last correct repetition performed."

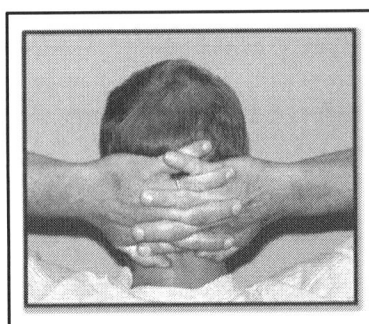

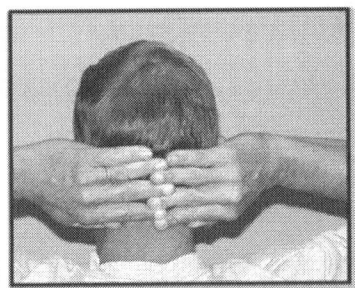

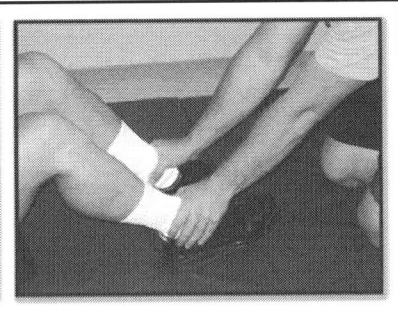

Figure A-5. Sit-up hand and feet position

A-29. In conclusion, the OIC/NCOIC, or the event supervisors, asks:

"WHAT ARE YOUR QUESTIONS ABOUT THIS EVENT?"

Appendix A

2-MILE RUN

A-30. The OIC, the NCOIC, or the event supervisor, must read the following before the 2-mile run event (Figure A-6).

> "THE 2-MILE RUN MEASURES YOUR AEROBIC FITNESS AND ENDURANCE OF THE LEG MUSCLES. YOU MUST COMPLETE THE RUN WITHOUT ANY PHYSICAL HELP. AT THE START, ALL SOLDIERS WILL LINE UP BEHIND THE STARTING LINE. ON THE COMMAND 'GO', THE CLOCK WILL START. YOU WILL BEGIN RUNNING AT YOUR OWN PACE. TO RUN THE REQUIRED TWO MILES, YOU MUST COMPLETE THE REQUIRED 2-MILE DISTANCE (DESCRIBE THE NUMBER OF LAPS, START AND FINISH POINTS, AND COURSE LAYOUT). YOU ARE BEING TESTED ON YOUR ABILITY TO COMPLETE THE TWO-MILE COURSE IN THE SHORTEST TIME POSSIBLE. ALTHOUGH WALKING IS AUTHORIZED, IT IS STRONGLY DISCOURAGED. IF YOU ARE PHYSICALLY HELPED IN ANY WAY (FOR EXAMPLE, PULLED, PUSHED, PICKED UP AND/OR CARRIED), OR LEAVE THE DESIGNATED RUNNING COURSE FOR ANY REASON, THE EVENT WILL BE TERMINATED. IT IS LEGAL TO PACE A SOLDIER DURING THE TWO-MILE RUN AS LONG AS THERE IS NO PHYSICAL CONTACT WITH THE PACED SOLDIER AND IT DOES NOT PHYSICALLY HINDER OTHER SOLDIERS TAKING THE TEST. THE PRACTICE OF RUNNING AHEAD OF, ALONG SIDE OF, OR BEHIND THE TESTED SOLDIER WHILE SERVING AS A PACER IS PERMITTED. CHEERING OR CALLING OUT THE ELAPSED TIME IS ALSO PERMITTED. THE NUMBER ON YOUR CHEST IS FOR IDENTIFICATION. YOU MUST MAKE SURE IT IS VISIBLE AT ALL TIMES. TURN IN YOUR NUMBER WHEN YOU FINISH THE RUN AND GO TO THE AREA DESIGNATED FOR RECOVERY. DO NOT STAY NEAR THE SCORERS OR THE FINISH LINE AS THIS MAY INTERFERE WITH TESTING."

Figure A-6. 2-mile run event narrative

A-31. In conclusion, the OIC/NCOIC, or the event supervisors, asks:

> "WHAT ARE YOUR QUESTIONS ABOUT THIS EVENT?"

APFT EQUIPMENT

A-32. The following equipment is required for administration of the APFT:
- Two stopwatches, clipboards, and black pens for each scorer.
- Run numbers and DA Forms 705 for each Soldier being tested.

APFT FACILITIES

A-33. The following facilities are required for administration of the APFT:

- Designated area for preparation and recovery.
- One test station (6 feet wide by 15 feet deep) for every 8 Soldiers participating in the push-up and sit-up events.
- A measured 2-mile run course.

APFT PERSONNEL

A-34. The following personnel are recommended for administration of the APFT:
- OIC and/or NCOIC.
- Event supervisor.
- One event scorer for every eight Soldiers being tested.
- Timer and back-up timer.
- Required support personnel.

APFT TIMER AND BACK-UP TIMER

A-35. The timer begins each push-up or sit-up event with the command, "GET SET." On the command, "GO," time starts on both the timer's and back-up timer's watches. The timer indicates time remaining at one minute (with the command, "ONE MINUTE REMAINING"), 30 seconds (with the command, "30 SECONDS REMAINING"), and counts down the remaining 10 seconds (with the command, "10, 9, 8, 7, 6, 5, 4, 3, 2, 1, STOP"). The timer begins the 2-mile run assessment with the command, "GET SET." The 2-mile run time starts on both the timer's and back-up timer's watches on the command, "GO." As Soldiers near the finish line, the timer calls out time in minutes and seconds (for example: "FOURTEEN-FIFTY-EIGHT, FOURTEEN-FIFTY-NINE, FIFTEEN MINUTES, FIFTEEN-O-ONE").

APFT SCORER

A-36. The scorer counts the correct number of repetitions out loud, repeats the last number of the correct repetitions when incorrect repetitions are performed, and verbally corrects push-up and sit-up performances. When Soldiers complete their APFT events, the scorer records the correct number of completed push-ups and sit-ups, records the 2-mile run time, and initials the DA Form 705. During the push-up event, scorers sit or kneel three feet from the Soldier's left or right shoulder at a 45-degree angle (refer to Figure A-2). A scorer's head should be even with the Soldier's shoulder when he is in the front-leaning rest position. During the sit-up event, the scorer kneels or sits three feet from the Soldier's left or right hip. The scorer's head should be even with the Soldier's shoulder when he is in the vertical (up) position (refer to Figure A-4). During the 2-mile run event, the scorer is at the finish line. When the scorer has entered the Soldier's 2-mile run time on the DA Form 705, he converts the raw scores into point scores for each event, enters the total on the DA Form 705, and initials each event on the scorecard. The scorer then returns all DA Forms 705 to the OIC or the NCOIC.

APFT FAILURES

A-37. Soldiers who fail to achieve the minimum passing score for their age and gender on any event are considered test failures. If a Soldier is ill or becomes injured during the APFT and fails to achieve the minimum passing score for their age and gender on any event, he is considered a test failure.

ALTERNATE AEROBIC EVENTS

A-38. Alternate aerobic events assess the cardio respiratory and muscular endurance of Soldiers with permanent medical profiles, or long-term temporary profiles that cannot perform the 2-mile run. The alternate aerobic APFT events are the following:
- 800-Yard-Swim Test.
- 6.2-Mile Stationary-Cycle Ergometer Test.
- 6.2-Mile Bicycle Test.
- 2.5-Mile Walk Test.

Appendix A

A-39. Required scores for alternate aerobic events are recorded in Table A-1.

Table A-1. Alternate aerobic event standards

ALTERNATE AEROBIC EVENT STANDARDS

EVENT	GENDER	AGE									
		17-21	22-26	27-31	32-36	37-41	42-46	47-51	52-56	57-61	62+
800-YARD SWIM	Men Women	20:00 21:00	20:30 21:30	21:00 22:00	21:30 22:30	22:00 23:00	22:30 23:30	23:00 24:00	24:00 25:00	24:30 25:30	25:00 26:00
6.2-MILE CYCLE ERGOMETER AND BICYCLE TEST	Men Women	24:00 25:00	24:30 25:30	25:00 26:00	25:30 26:30	26:00 27:00	27:00 28:00	28:00 30:00	30:00 32:00	31:00 33:00	32:00 34:00
2.5-MILE WALK	Men Women	34:00 37:00	34:30 37:30	35:00 38:00	35:30 38:30	36:00 39:00	36:30 39:30	37:00 40:00	37:30 40:30	38:00 41:00	38:30 41:30

A-40. Soldiers on permanent physical profile are given a DA Form 3349. This form annotates exercises and activities suitable for the profiled Soldier. The form also stipulates the events and/or alternate aerobic event the Soldier will do on the APFT. The Soldier must perform all regular APFT events his profile permits. Each Soldier must score a minimum of 60 points on each regular event taken to PASS. The profiled Soldier must complete the alternate aerobic event in a time equal to or less than the one listed in Table A-1. The Soldier must receive a minimum passing score in the alternate event taken to PASS the test. Soldiers profiled for two or more events must take the two-mile run or an alternate aerobic event to PASS the test. Soldiers who cannot perform the 2-mile run or an alternate aerobic event cannot be tested. There is no point score annotated on the DA Form 705 for the performance of alternate aerobic events. These events are scored as a GO or NO GO.

A-41. Soldiers with temporary physical profiles must take a regular three event APFT after the profile has expired. Soldiers with temporary profiles of long duration (more than three months) may take an alternate aerobic event as determined by the commander with input from health-care personnel. Once the profile has been lifted, the Soldier must be given twice the length of the profile (not to exceed 90 days) to train for the regular three event APFT. If a regularly scheduled APFT occurs during the profile period, the Soldier should be given a mandatory make-up date for the APFT.

800-YARD SWIM TEST

A-42. The 800-yard-swim test measures cardio respiratory (aerobic) fitness. Administrative and support requirements for this event are listed below.

EQUIPMENT

A-43. The timer and back-up timer each require a stopwatch and appropriate safety equipment. Event scorers require a clipboard and black pen.

FACILITIES

A-44. A swimming pool at least 25 yards long and three feet deep is required.

Personnel

A-45. One event supervisor, one scorer for every three Soldiers, one timer, one back-up timer, and support personnel to ensure proper control and safety. The event supervisor will not be an event scorer.

Instructions

A-46. The OIC, the NCOIC, or the event supervisor, must read the following before the 800-yard swim event (Figure A-7).

> "THE 800-YARD SWIM MEASURES YOUR LEVEL OF AEROBIC FITNESS. YOU WILL BEGIN IN THE WATER; NO DIVING IS ALLOWED. AT THE START, YOUR BODY MUST BE IN CONTACT WITH THE WALL OF THE POOL. ON THE COMMAND 'GO', THE CLOCK WILL START. YOU SHOULD THEN BEGIN SWIMMING AT YOUR OWN PACE, USING ANY STROKE OR COMBINATION OF STROKES YOU WISH. YOU MUST SWIM (STATE THE NUMBER) LAPS TO COMPLETE THIS DISTANCE. YOU MUST TOUCH THE WALL OF THE POOL AT EACH END OF THE POOL AS YOU TURN. ANY TYPE OF TURN IS AUTHORIZED. YOU WILL BE SCORED ON YOUR ABILITY TO COMPLETE THE SWIM IN A TIME EQUAL TO, OR LESS THAN, THAT LISTED FOR YOUR AGE AND GENDER. WALKING ON THE BOTTOM TO RECUPERATE IS AUTHORIZED. SWIMMING GOGGLES ARE PERMITTED, BUT NO OTHER EQUIPMENT IS AUTHORIZED."

Figure A-7. 800-yard swim test narrative

A-47. In conclusion, the OIC/NCOIC, or the event supervisors, asks:

> "WHAT ARE YOUR QUESTIONS ABOUT THIS EVENT?"

Administration

A-48. The OIC, the NCOIC, or the event supervisor, will read the instructions aloud and answer questions. The event supervisor will assign each Soldier to a lane and tell the Soldier to enter the water. He allows for a short acclimation and preparation period. The event supervisor must be alert to the safety of the Soldiers throughout the test.

Timing Techniques

A-49. When the timer gives the command, "GET SET" the Soldiers position themselves to begin the event. Time begins when the timer gives the command, "GO." The timer calls out times in minutes and seconds as Soldiers near the finish. Time is recorded by the scorer when the Soldier touches the end of the pool or crosses a predetermined line that establishes the 800-yard mark.

Scorer Duties

A-50. Scorers must observe the Soldiers assigned to them. They must ensure that each Soldier touches the bulkhead (wall) at every turn. The scorer records the time in the time block and circles the GO or NO GO. The 800-yard swim is entered in the alternate event block. Refer to Figure A-11 for scoring this event. If the pool

Appendix A

length is measured in meters, the scorer can convert the exact distance to yards. To convert meters to yards, multiply the number of meters by 39.37 and divide the product by 36.

6.2-MILE STATIONARY CYCLE ERGOMETER TEST

A-51. The 6.2-mile stationary cycle ergometer test measures cardio respiratory (aerobic) and leg muscle endurance. Administrative and support requirements for this event follow.

EQUIPMENT

A-52. The event supervisor requires two stopwatches (the timer and back-up timer each require a stopwatch), FM 7-22, Appendix A, and one stationary cycle ergometer. The ergometer must have mechanically adjustable resistance measured in kiloponds or newtons and must be available for training and testing. The seat and handlebars must be adjustable to accommodate Soldiers of different sizes. It should have an adjustable tension setting (resistance) and an odometer. The resistance is set by a tension strap on a weighted pendulum connected to the flywheel. The cycle ergometer must be calibrated prior to test administration. Event scorers require a clipboard and black pen.

FACILITIES

A-53. The test site can be any location (usually a gym) where there is an approved cycle ergometer. The test station should be two yards wide and four yards deep.

PERSONNEL

A-54. One event supervisor, one scorer for every three Soldiers tested, one timer, one back-up timer, and support personnel to ensure proper control and safety are required. The event supervisor will not be an event scorer.

INSTRUCTIONS

A-55. The OIC, the NCOIC, or the event supervisor, must read the following before the 6.2 cycle ergometer test event (Figure A-8).

> "THE 6.2-MILE STATIONARY-CYCLE ERGOMETER EVENT MEASURES YOUR CARDIO-RESPIRATORY FITNESS AND LEG MUSCLE ENDURANCE. THE ERGOMETER'S RESISTANCE MUST BE SET AT TWO KILOPOUNDS (20 NEWTONS). ON THE COMMAND, 'GO', THE CLOCK WILL START, AND YOU WILL BEGIN PEDALING AT YOUR OWN PACE WHILE MAINTAINING THE RESISTANCE INDICATOR AT TWO KILOPOUNDS. YOU WILL BE SCORED ON YOUR ABILITY TO COMPLETE 6.2 MILES (10 KILOMETERS), AS SHOWN ON THE ODOMETER IN A TIME EQUAL TO OR LESS THAN THAT LISTED FOR YOUR AGE AND GENDER."

Figure A-8. 6.2-mile stationary cycle ergometer test narrative

Army Physical Fitness Test (APFT)

A-56. In conclusion, the OIC/NCOIC, or the event supervisors, asks:

> **"WHAT ARE YOUR QUESTIONS ABOUT THIS EVENT?"**

ADMINISTRATION

A-57. The event supervisor will read the instructions aloud and answer questions. He will also allow each Soldier a short warm-up period and an opportunity to adjust handlebar and seat height.

TIMING TECHNIQUES

A-58. When the timer gives the command, "GET SET" the Soldiers will position themselves to begin the event. Time begins when the timer gives the command, "GO." The timer will call out times in minutes and seconds as Soldiers near the last two-tenths of the test distance. He calls out the time remaining every 30 seconds for the last two minutes of the allowable time and every second during the last ten seconds.

SCORER DUTIES

A-59. The scorer must observe that the ergometer is functioning correctly. He must then make sure that the ergometer's tension settings have been calibrated and are accurate, and that the resistance of the ergometer has been set at two kiloponds or 20 newtons. The scorer must observe the Soldiers throughout the event. He will have to make small adjustments to the resistance to ensure that a continuous resistance of exactly 2 kiloponds is maintained throughout the test. The scorer records the time in the time block and circles the GO or NO GO. The 6.2-mile stationary cycle ergometer is entered in the alternate event block. Refer to Figure A-11 for scoring of this event.

6.2-MILE BICYCLE TEST

A-60. The 6.2-mile bicycle test measures cardio respiratory (aerobic) and leg muscle endurance. Administrative and support requirements for this event are listed below.

EQUIPMENT

A-61. The event supervisor requires two stopwatches (the timer and back-up timer each require a watch) and FM 7-22, Appendix A. One-speed or multispeed bicycles are authorized for use. If a multispeed bike is used, the event supervisor and/or scorer will take measures to ensure that only one speed is used during the event. This can be accomplished by taping the gear shifters. The Soldier taking the event sets the speed by selecting the gear they wish to ride in. Event scorers require a clipboard, numbers, and black pen.

FACILITIES

A-62. A relatively flat course with a uniform surface and no obstacles must be used. The course must be clearly marked. Quarter-mile tracks are not authorized for use. The Soldiers being tested must be in view of the scorers at all times. The course should be free of walkers and runners.

PERSONNEL

A-63. One event supervisor, one scorer for every 10 Soldiers tested, one timer, one back-up timer and support personnel to ensure proper control and safety are required. The event supervisor should not be an event scorer.

Appendix A

INSTRUCTIONS

A-64. The OIC, the NCOIC, or the event supervisor, must read the following before the 6.2 mile bicycle test event (Figure A-9).

> "THE 6.2-MILE BICYCLE TEST MEASURES CARDIO RESPIRATORY FITNESS AND LEG MUSCLES ENDURANCE. YOU MUST COMPLETE THE 6.2-MILES WITHOUT ANY PHYSICAL HELP FROM OTHERS. YOU MUST KEEP YOUR BICYCLE IN ONE GEAR OF YOUR CHOOSING FOR THE ENTIRE TEST. CHANGING GEARS IS NOT PERMITTED AND WILL RESULT IN DISQUALIFICATION. TO BEGIN, YOU WILL LINE UP BEHIND THE STARTING LINE. ON THE COMMAND, 'GO,' THE CLOCK WILL START, AND YOU WILL BEGIN PEDALING AT YOUR OWN PACE. TO COMPLETE THE REQUIRED DISTANCE OF 6.2-MILES, YOU MUST COMPLETE (DESCRIBE THE NUMBER OF LAPS, START AND FINISH POINTS, AND COURSE LAYOUT). YOU WILL BE SCORED ON YOUR ABILITY TO COMPLETE THE DISTANCE OF 6.2-MILES (10 KILOMETERS) IN A TIME EQUAL TO OR LESS THAN THAT LISTED FOR YOUR AGE AND GENDER. IF YOU LEAVE THE DESIGNATED COURSE FOR ANY REASON, YOU WILL BE DISQUALIFIED."

Figure A-9. 6.2-mile bicycle test narrative

A-65. In conclusion, the OIC/NCOIC, or the event supervisors, asks:

> "WHAT ARE YOUR QUESTIONS ABOUT THIS EVENT?"

ADMINISTRATION

A-66. The OIC, the NCOIC, or the event supervisor, will read the instructions aloud and answer questions. He then assigns Soldiers to a scorer. Each scorer assigns each Soldier a number and records the Soldier's number on their scorecard in the comment block.

TIMING TECHNIQUES

A-67. The event supervisor is the timer. When the timer gives the command, "GET SET" the Soldiers will position themselves to begin the event. Time begins when the timer gives the command, "GO." The timer will call out times in minutes and seconds as Soldiers near the end of the 6.2-mile ride.

SCORER DUTIES

A-68. The scorer records the time in the time block and circles the GO or NO GO. The 6.2-mile-bicycle is entered in the alternate event block. Refer to Figure A-11 for scoring of this event.

2.5-MILE WALK TEST

A-69. The 2.5-mile-walk test measures cardio respiratory (aerobic) and leg muscle endurance. Administrative and support requirements for this event follow.

Army Physical Fitness Test (APFT)

EQUIPMENT

A-70. The event supervisor requires two stopwatches (the timer and back-up timer each require a stopwatch). Event scorers require a clipboard, FM 7-22, Appendix A, numbers, and a black pen.

FACILITIES

A-71. The event uses the same course as the 2-mile run, with the addition of ½ mile added to the 2-mile distance. The Soldiers being tested must be in view of the scorers at all time.

PERSONNEL

A-72. One event supervisor, one scorer for every three Soldiers tested, one timer, one back-up timer, and support personnel to ensure proper control and safety are required. The event supervisor will not be an event scorer.

INSTRUCTIONS

A-73. The OIC, the NCOIC, or the event supervisor, must read the following before the 2.5-mile walk test event (Figure A-10):

> "THE 2.5-MILE WALK MEASURES CARDIO RESPIRATORY FITNESS AND LEG-MUSCLE ENDURANCE. ON THE COMMAND, 'GO,' THE CLOCK WILL START, AND YOU WILL BEGIN WALKING AT YOUR OWN PACE. YOU MUST COMPLETE (DESCRIBE THE NUMBER OF LAPS, START AND FINISH POINTS, AND COURSE LAYOUT). ONE FOOT MUST BE IN CONTACT WITH THE GROUND AT ALL TIMES. IF YOU BREAK INTO A RUNNING STRIDE AT ANY TIME OR HAVE BOTH FEET OFF THE GROUND AT THE SAME TIME, YOUR PERFORMANCE IN THE EVENT WILL BE TERMINATED. YOU WILL BE SCORED ON YOUR ABILITY TO COMPLETE THE 2.5-MILE COURSE IN A TIME EQUAL TO OR LESS THAN THAT LISTED FOR YOUR AGE AND GENDER."

Figure A-10. 2.5-mile walk narrative

A-74. In conclusion, the OIC/NCOIC, or the event supervisors, asks:

> "WHAT ARE YOUR QUESTIONS ABOUT THIS EVENT?"

ADMINISTRATION

A-75. The OIC, the NCOIC, or the event supervisor, will read the instructions aloud and answer questions. He then assigns Soldiers to a scorer. Each scorer assigns each Soldier a number and records the Soldier's number on their scorecard in the comment block.

TIMING TECHNIQUES

A-76. When the timer gives the command, "GET SET" the Soldiers will position themselves to begin the event. Time begins when the timer gives the command, "GO." The timer will call out times in minutes and seconds as Soldiers near the end of the 2.5-mile walk.

Appendix A

SCORER DUTIES

A-77. Scorers must observe the Soldiers during the entire event and must ensure that the Soldiers maintain a walking stride. Soldiers that break into any type of running stride will be terminated from the event and will be a NO GO. The scorer records the time in the time block and circles the GO or NO GO. 2.5-mile walk is entered in the alternate event block. Refer to figure A-11 for scoring of this event.

DA FORM 705 SAMPLE

A-78. The following is a sample of the Army Physical Training Fitness Test Scorecard that leaders use to test the physical fitness of their units (Figure A-11, A through F). DA Form 705 can be downloaded from the AKO My Forms link. Forms can be filled out on screen or by hand. The samples provided are not to be used for scoring the APFT. Use the scorecard downloaded from AKO for scoring APFT events.

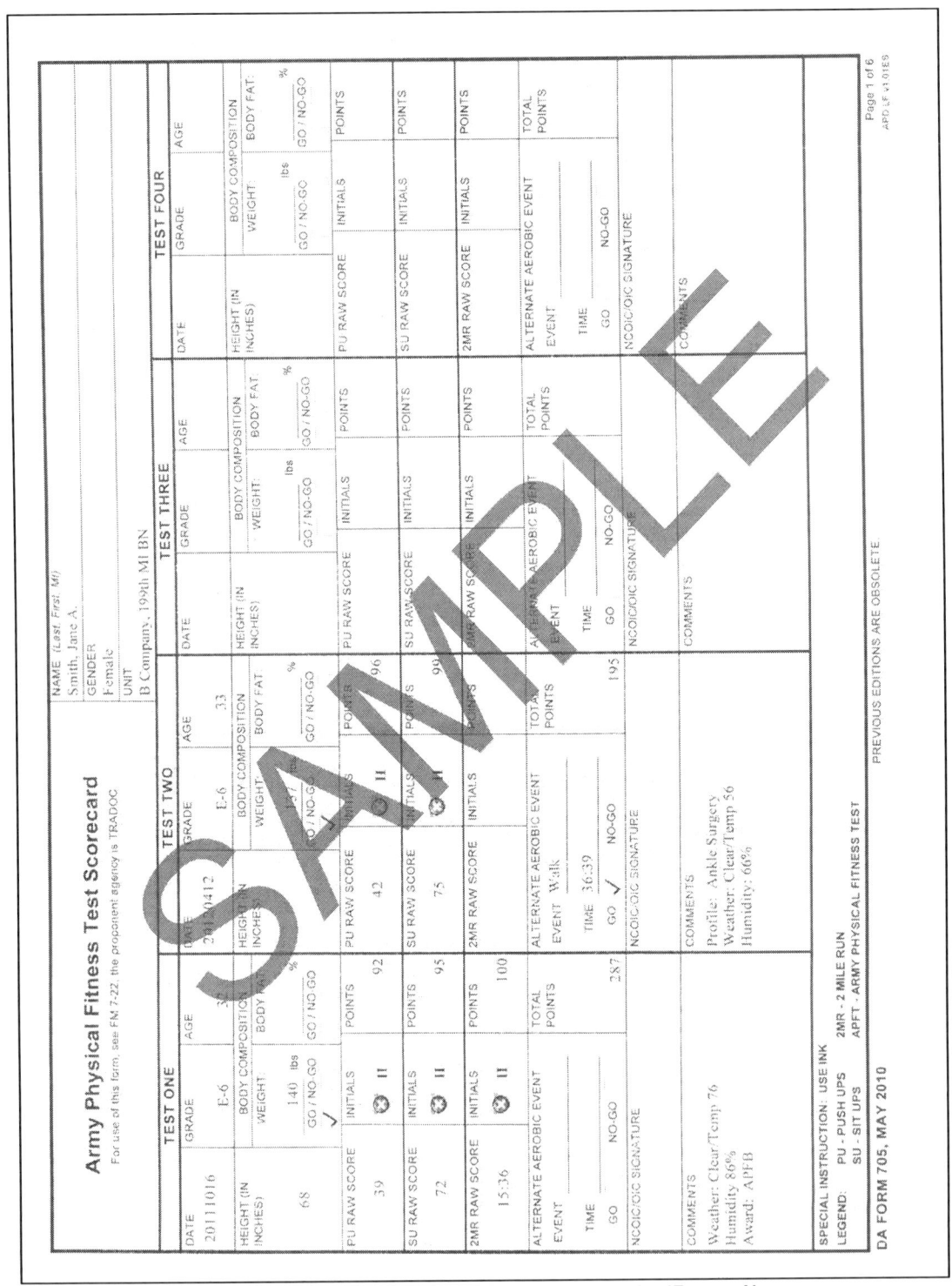

Figure A-11A. DA Form 705 sample (Page 1)

Appendix A

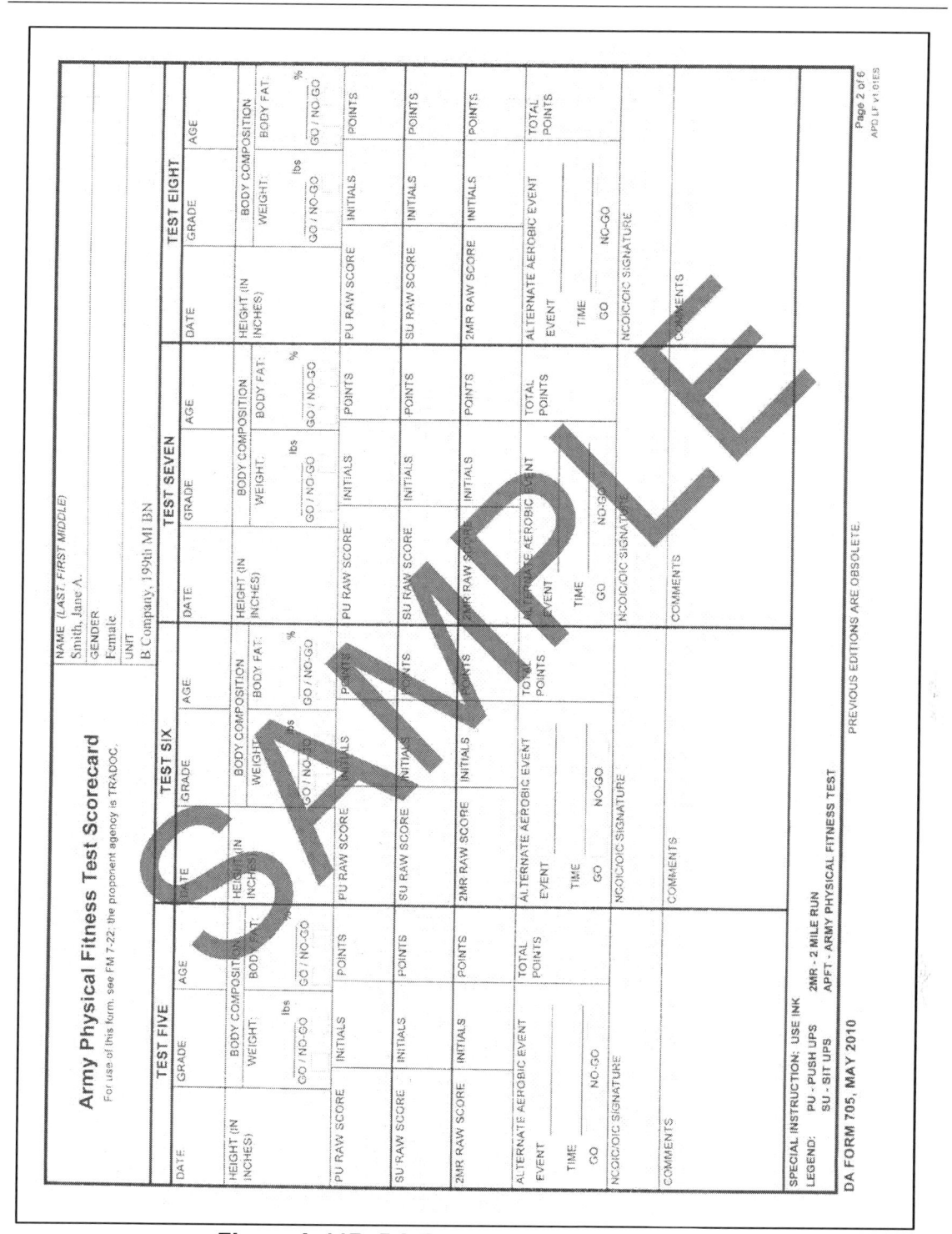

Figure A-11B. DA Form 705 sample (page 2)

Figure A-11C. DA Form 705 sample (page 3)

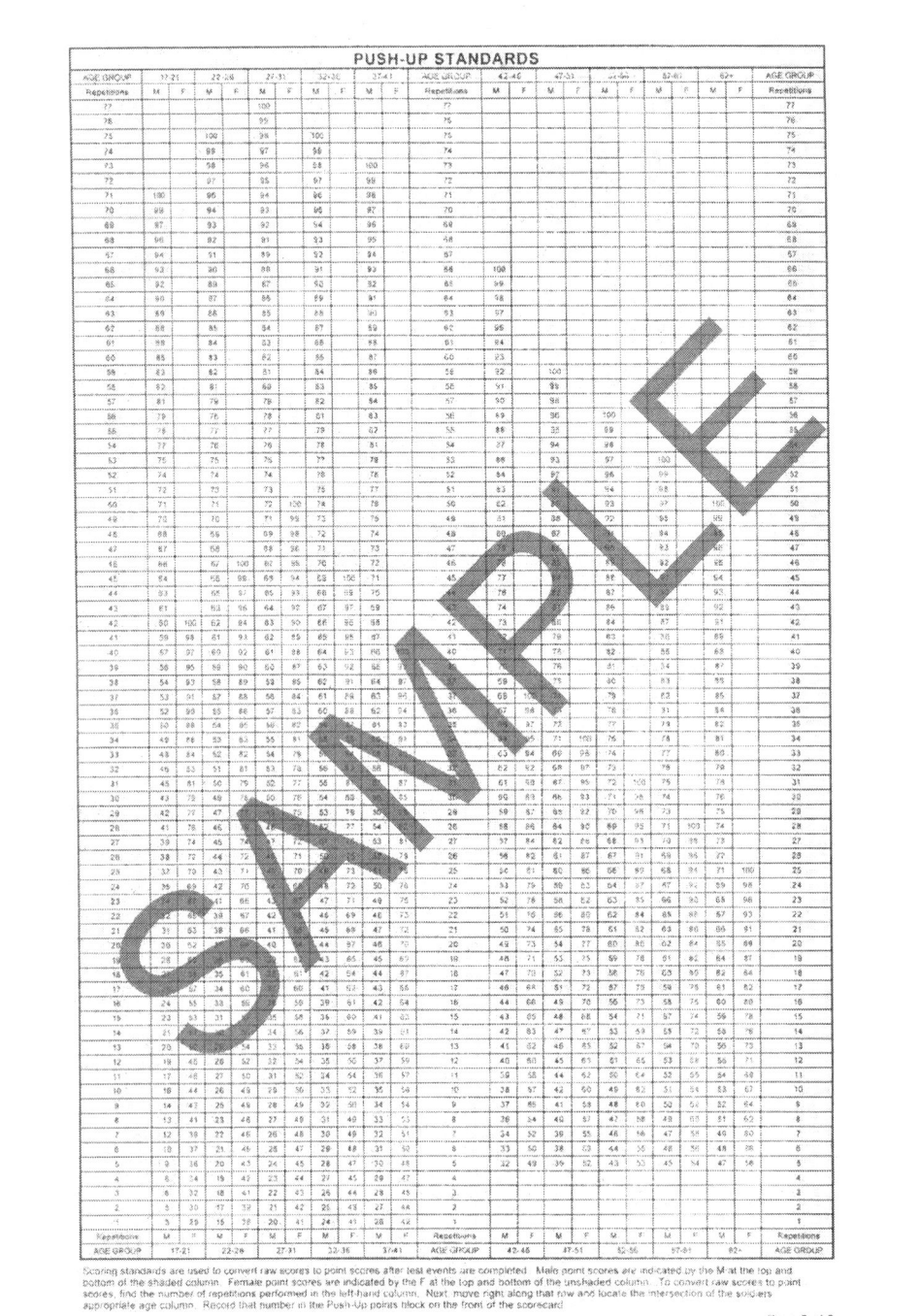

Appendix A

Figure A-11D. DA Form 705 sample (page 4)

Army Physical Fitness Test (APFT)

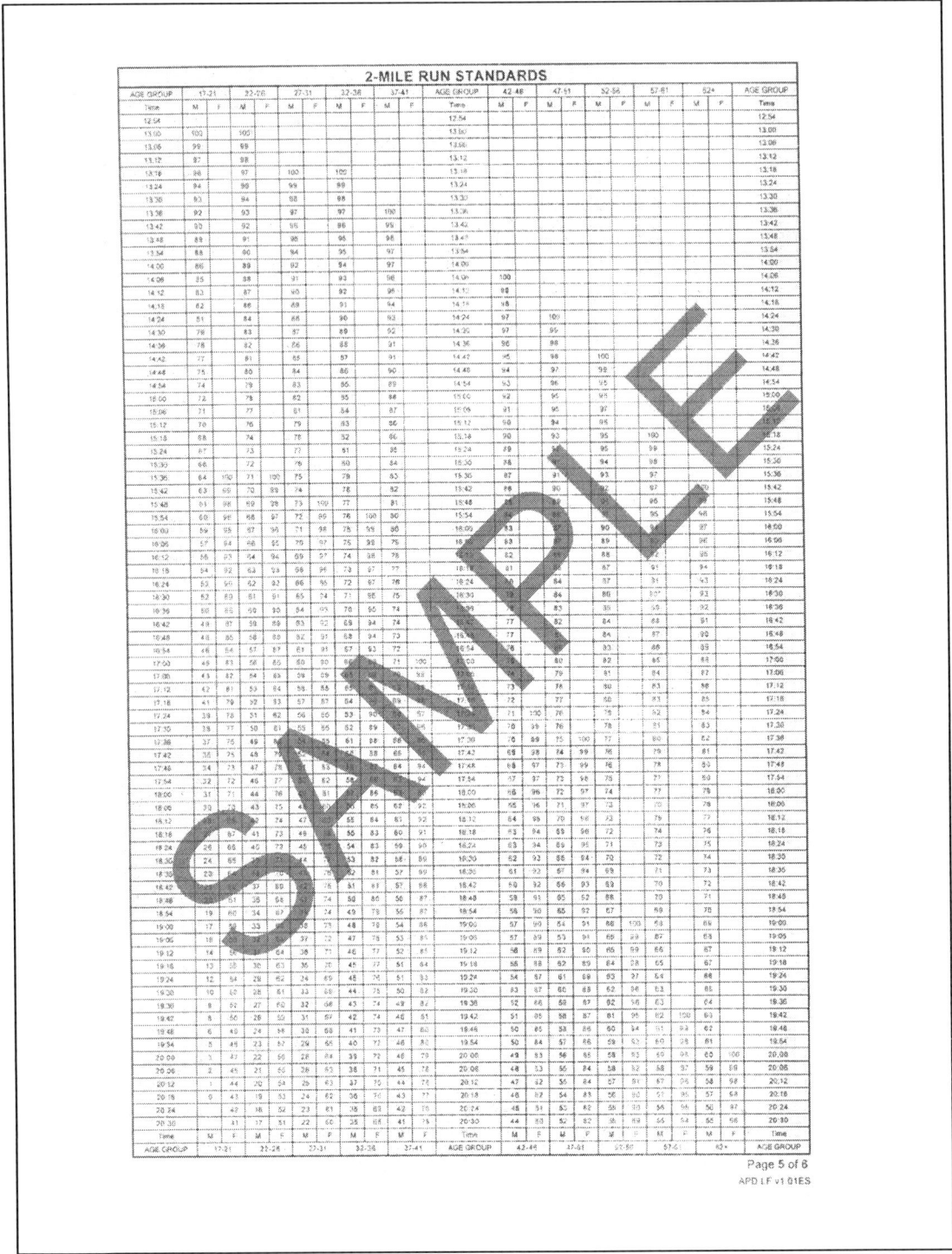

Figure A-11E. DA Form 705 sample (page 5)

2-MILE RUN STANDARDS

AGE GROUP	17-21		22-26		27-31		32-36		37-41		AGE GROUP	42-46		47-51		52-56		57-61		62+		AGE GROUP
Time	M	F	M	F	M	F	M	F	M	F	Time	M	F	M	F	M	F	M	F	M	F	Time
20:18	0	43	19	53	24	62	38	70	43	77	20:18	46	82	54	83	58	90	57	95	57	98	20:18
20:24		42	18	52	23	61	35	69	42	76	20:24	45	81	53	82	55	90	56	95	56	97	20:24
20:30		41	17	51	22	60	35	68	41	75	20:30	44	80	52	82	55	89	55	94	55	96	20:30
20:36		39	16	50	21	59	34	68	40	75	20:36	43	80	51	81	54	88	54	93	54	95	20:36
20:42		38	14	49	20	58	33	67	40	74	20:42	43	79	51	81	53	87	53	92	53	94	20:42
20:48		37	13	48	19	57	32	66	39	73	20:48	42	78	50	80	52	87	52	91	53	94	20:48
20:54		36	12	47	18	57	31	65	38	73	20:54	41	78	49	79	51	86	51	91	52	93	20:54
21:00		35	11	46	17	56	30	65	37	72	21:00	40	77	48	79	51	85	50	90	51	92	21:00
21:06		32	10	45	16	56	29	64	36	71	21:06	39	77	47	78	50	84	50	89	[illegible]	91	21:06
21:12		32	9	44	15	54	28	63	35	71	21:12	38	76	47	77	49	84	49	[illegible]	49	90	21:12
21:18		31	8	43	14	53	27	63	34	70	21:18	37	75	46	77	48	83	48	[illegible]	48	90	21:18
21:24		30	7	42	12	52	26	62	34	69	21:24	37	75	45	76	47	82	[illegible]	87	47	89	21:24
21:30		28	6	41	11	51	25	61	33	68	21:30	36	74	44	76	47	81	[illegible]	86	46	[illegible]	21:30
21:36		27	4	40	10	51	25	61	32	68	21:36	35	73	44	75	46	[illegible]	[illegible]	86	[illegible]	[illegible]	21:36
21:42		26	3	39	9	50	24	60	31	67	21:42	34	73	43	74	45	80	[illegible]	84	[illegible]	88	21:42
21:48		25	2	38	8	49	23	59	30	66	21:48	33	72	42	74	44	79	[illegible]	[illegible]	[illegible]	88	[illegible]
21:54		24	1	37	7	48	22	58	29	66	21:54	32	71	41	73	44	79	43	[illegible]	42	85	[illegible]
22:00		22	0	36	6	47	21	58	29	65	22:00	31	71	40	72	43	78	42	[illegible]	[illegible]	84	22:00
22:06		21		35	5	46	20	57	28	64	22:06	30	70	[illegible]	72	42	77	41	81	[illegible]	83	22:06
22:12		20		34	4	46	19	57	27	64	22:12	30	70	[illegible]	71	41	76	40	80	40	[illegible]	22:12
22:18		19		33	3	45	18	56	26	63	22:18	29	69	38	[illegible]	40	76	39	80	39	[illegible]	22:18
22:24		18		32	2	44	17	55	25	62	22:24	28	69	37	[illegible]	40	75	38	79	[illegible]	81	22:24
22:30		16		31	1	43	16	54	24	61	22:30	[illegible]	[illegible]	[illegible]	69	[illegible]	74	37	[illegible]	[illegible]	80	22:30
22:36		15		30	0	42	15	54	23	61	22:36	[illegible]	67	[illegible]	68	[illegible]	[illegible]	37	[illegible]	36	79	22:36
22:42		14		29		41	15	53	23	60	22:42	25	66	[illegible]	68	37	[illegible]	[illegible]	[illegible]	35	78	22:42
22:48		13		28		40	14	52	22	59	[illegible]	24	66	[illegible]	67	36	[illegible]	[illegible]	75	34	78	22:48
22:54		12		27		40	13	52	21	59	[illegible]	23	65	[illegible]	67	36	71	[illegible]	75	33	77	22:54
23:00		10		26		39	12	51	20	58	[illegible]	23	64	[illegible]	66	35	70	33	74	32	76	23:00
23:06		9		25		38	11	50	19	57	23:06	[illegible]	[illegible]	32	66	34	70	32	73	31	75	23:06
23:12		8		24		37	10	49	18	[illegible]	23:12	[illegible]	[illegible]	31	65	33	69	31	73	30	74	23:12
23:18		7		23		36	9	49	17	[illegible]	[illegible]	[illegible]	[illegible]	30	64	33	68	30	72	29	74	23:18
23:24		5		22		35	8	48	17	55	[illegible]	19	[illegible]	29	64	32	67	30	71	28	73	23:24
23:30		4		21		34	7	48	16	54	[illegible]	18	[illegible]	[illegible]	63	31	67	29	70	27	72	23:30
23:36		3		20		34	6	47	15	54	23:36	17	61	[illegible]	63	30	66	28	69	27	71	23:36
23:42		2		19		33	[illegible]	[illegible]	14	53	[illegible]	[illegible]	60	27	62	29	66	27	69	26	70	23:42
23:48		1		18		32	[illegible]	[illegible]	[illegible]	53	[illegible]	[illegible]	59	26	61	29	64	26	68	25	70	23:48
23:54		0		17		31	[illegible]	[illegible]	[illegible]	[illegible]	[illegible]	15	59	25	61	28	64	25	67	24	69	23:54
24:00				16		30	3	[illegible]	11	[illegible]	[illegible]	14	58	25	60	27	63	24	66	23	68	24:00
24:06				15		29	2	[illegible]	[illegible]	50	[illegible]	13	57	24	59	26	62	23	65	22	67	24:06
24:12				14		29	1	43	[illegible]	[illegible]	[illegible]	12	57	23	59	25	61	23	65	21	66	24:12
24:18				[illegible]		[illegible]	0	42	9	[illegible]	24:18	11	56	22	58	25	61	22	64	20	66	24:18
24:24				12		[illegible]		41	8	[illegible]	24:24	10	56	22	57	24	60	21	63	19	65	24:24
24:30				11		[illegible]		[illegible]	7	47	24:30	10	55	21	57	23	59	20	62	18	64	24:30
24:36				10		25		[illegible]	[illegible]	47	24:36	9	54	20	56	22	58	19	62	17	63	24:36
24:42				9		24		39	[illegible]	[illegible]	24:42	8	54	19	56	22	58	18	61	16	62	24:42
24:48				8		[illegible]		39	5	45	24:48	7	53	18	55	21	57	17	60	16	62	24:48
24:54				7		[illegible]		38	4	45	24:54	6	52	18	54	20	56	17	59	14	61	24:54
25:00				6		[illegible]		37	3	44	25:00	5	52	17	54	19	56	16	58	13	60	25:00
25:06				5		[illegible]		37	2	43	25:06	4	51	16	53	18	55	15	58	13	59	25:06
25:12				[illegible]		[illegible]		36	1	43	25:12	3	50	15	52	18	54	14	57	12	58	25:12
25:18				[illegible]		19		35	0	42	25:18	3	50	15	52	17	53	13	56	11	58	25:18
25:24				2		18		34		41	25:24	2	49	14	51	16	52	12	55	10	57	25:24
25:30				1		17		34		40	25:30	1	49	13	51	15	52	11	55	9	56	25:30
25:36				0		17		33		40	25:36	0	48	12	50	15	51	10	54	8	55	25:36
25:42						16		32		39	25:42		47	11	49	14	50	10	53	7	54	25:42
25:48						15		32		38	25:48		47	11	49	13	50	9	52	6	54	25:48
25:54						14		31		38	25:54		46	10	48	12	49	8	51	5	53	25:54
26:00						13		30		37	26:00		46	9	47	11	48	7	51	4	52	26:00
26:06						12		30		36	26:06		45	8	47	11	47	6	50	3	51	26:06
26:12						11		29		35	26:12		44	7	46	10	47	5	49	2	50	26:12
26:18						11		28		35	26:18		43	7	46	9	46	4	48	1	50	26:18
26:24						10		28		34	26:24		43	6	45	8	46	3	47	0	49	26:24
26:30						9		27		33	26:30		42	6	44	7	44	3	47	0	48	26:30
Time	M	F	M	F	M	F	M	F	M	F	Time	M	F	M	F	M	F	M	F	M	F	Time
AGE GROUP	17-21		22-26		27-31		32-36		37-41		AGE GROUP	42-46		47-51		52-56		57-61		62+		AGE GROUP

Scoring standards are used to convert raw scores to point scores after test events are completed. Male point scores are indicated by the M at the top and bottom of the shaded column. Female point scores are indicated by the F at the top and bottom of the unshaded column. To convert raw scores to point scores, find the number of repetitions performed in the left-hand column. Next, move right along that row and locate the intersection of the soldiers appropriate age column. In all cases, when a time falls between two point values, the lower point value is used. Record that number in the 2MR points block on the front of the scorecard.

Page 6 of 6

APD LF v1.01ES

SAMPLE

Figure A-11F. DA Form 705 sample (page 6)

Appendix B
Climbing Bars

This appendix discusses climbing bars required for physical readiness training.

LAYOUT

B-1. Figure B-1 shows the climbing bar layout required for toughening and sustaining phase PRT drills.

Figure B-1. Climbing bars

SPECIFICATIONS

B-2. The specifications for the climbing bars follow:
- Five posts.
  - Each of the five posts measures 6 inches square by 12 feet long.
  - Each post is sunk 3 feet into the ground.
- Two bars.
  - Each of the two bars is a threaded water pipe.
  - Each bar measures 1.5 inches outside diameter by 12 feet long.
  - Each bar has two 1-inch deep end caps.
  - The bars are through the 6 by 6s at 7.5 and 8 feet above the ground.
- The distance from inside post edge to inside post edge is about 62 inches (Figure B-2). This is to allow enough bar space to conduct all exercises safely.
- The step-ups (16 inches long) are cut from 4 by 4 inches by 8-foot posts and secured to the 6 by 6s with 3-inch screws that are countersunk.
- The step-ups on the outside 6 by 6 posts are 18 inches from the ground; the step-ups on the inside post are 24 inches above the ground (Figure B-3).

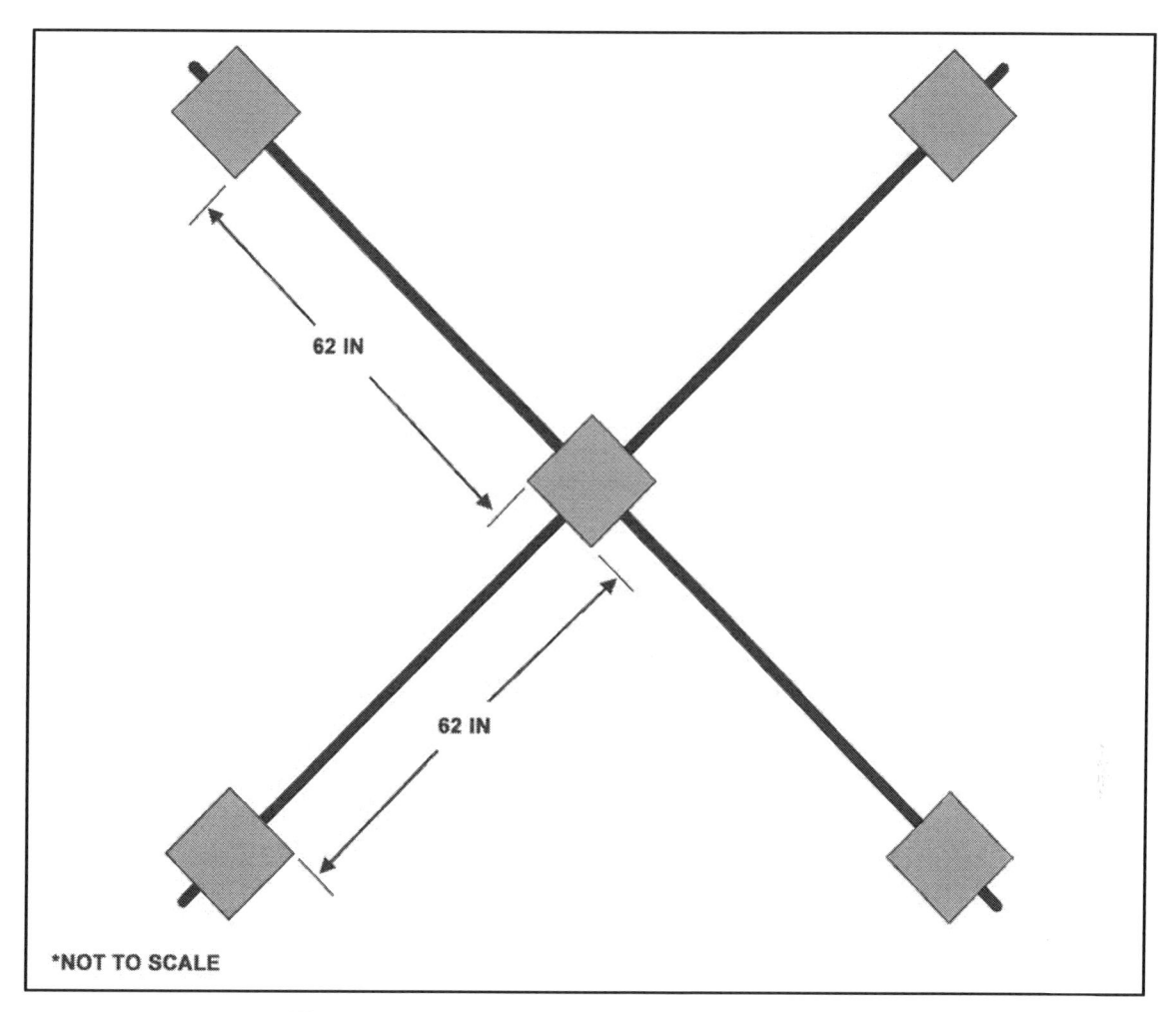

Figure B-2. Climbing bars, dimensions, top view

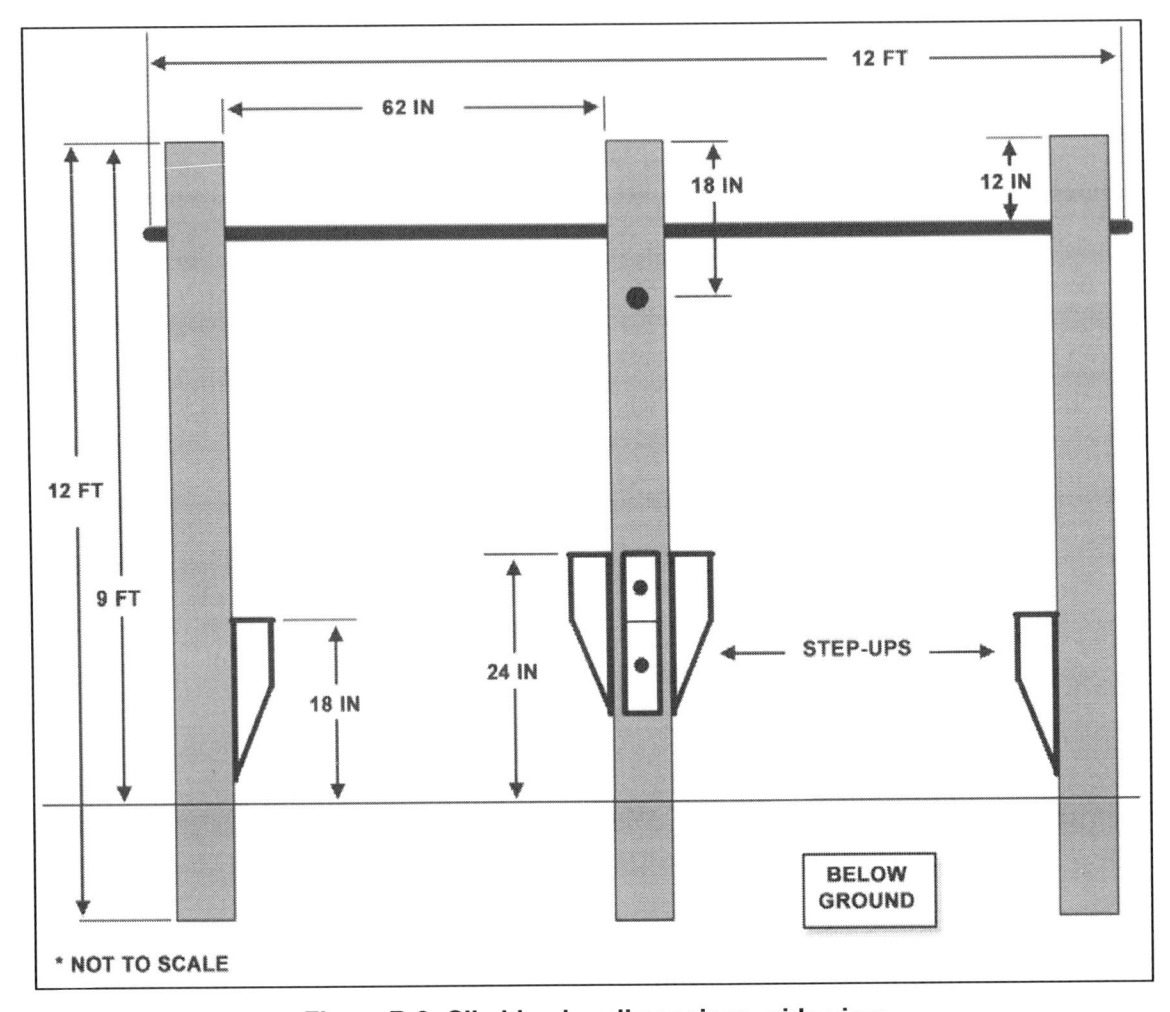

Figure B-3. Climbing bar dimensions, side view

PLANNING CONSIDERATIONS

B-3. The following planning considerations apply:
- Climbing bars provide adequate space and facilitate better command and control than traditional pull-up bars.
- Traditional pull-up bars are too narrow to safely and efficiently conduct the CLs.
- Use of multiple climbing bar "pods," as shown in Figure B-4, allow for efficient mass training.
- The CLs require one bar for every three Soldiers when performed as a single activity.
- Total ground surface area for four pods is only 625 square feet.
- Four pods will accommodate 16 stations of 3 Soldiers per station for a total of 48 Soldiers.
- Additional freestanding pods should be constructed to accommodate more Soldiers.

Appendix B

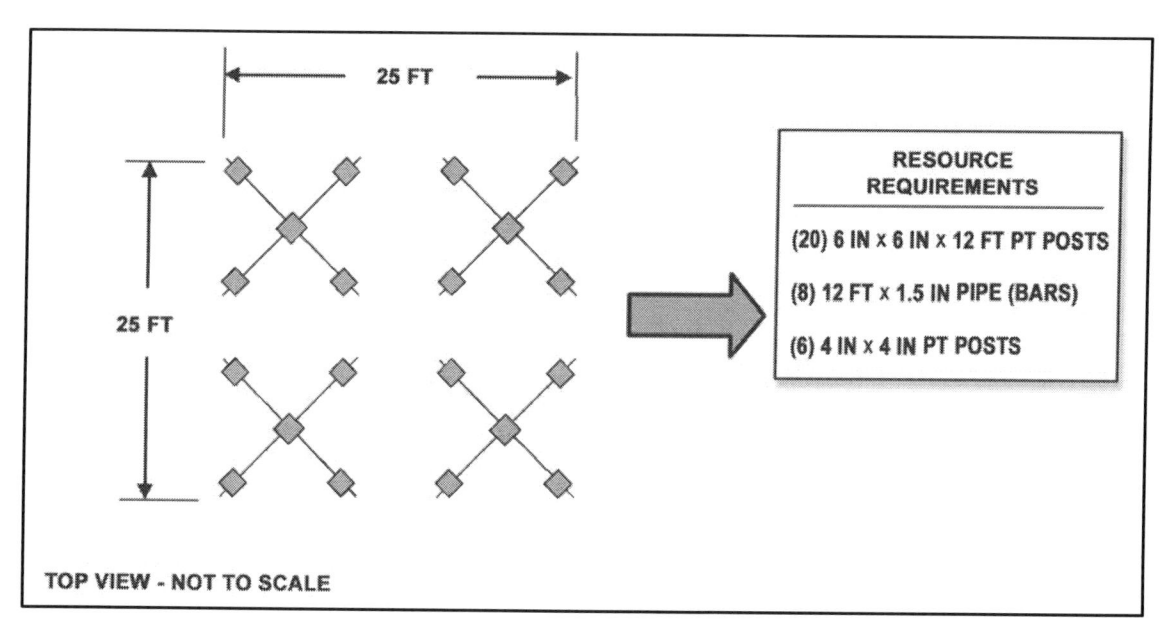

Figure B-4. Multiple climbing bar pods

Appendix C
Posture and Body Mechanics

"Good posture has many values for the Soldier. First, a Soldier is often judged by his appearance–the man with good posture looks like a good Soldier, he commands attention. Secondly, it is an accepted psychological fact that good posture is associated with good morale–a man with good posture feels better and is more positive. A man with poor posture cannot feel as positive, consequently he may develop a negative and discouraged attitude. Thirdly, good posture permits the body to function most efficiently."

FM 21-20, Physical Training (January 1946)

EFFECT OF POSTURE

C-1. Posture and body mechanics are critical factors for Soldier performance, allowing them to move efficiently with an ability to create great force and absorb heavy resistance. Posture is any position in which the body resides. It is further defined by the relationship of body segments to one another. Body mechanics is posture in motion. Though posture is often thought of as a stationary position, control of moving postures is perhaps even more important in task performance and injury control.

"In the training of anyone, nothing equals the importance of proper posture; it is the very foundation upon which the entire fabric of a successful course in physical training must be founded."

LTC Herman J. Koehler

C-2. When body segments are aligned properly, movement is efficient, and injury risk is minimized. When body segments are not aligned properly, movement is less efficient and risk of injury is increased. Consider a Soldier attempting to lift a heavy load from the ground with his legs straight and trunk twisted. Not only will the load seem heavier than if his knees were bent and his back straight, but he is at risk for injury. The back injury that occurs during an improper lift is an obvious example of the relationship between posture, body mechanics, performance, and health. Less obvious, but just as damaging, is the daily stress that takes its toll on the body when faulty postures are consistently assumed.

EFFECT OF GRAVITY

C-3. Gravity molds body tissues. The body adapts to the stresses placed upon it. Gravity exerts a constant influence. When body segments are not aligned properly, such as when the head is too far forward, gravity works to further pull the head forward, placing undue stress on the structures of the neck and upper back. Over time, the neck adapts to faulty posture and natural neck movements may become restricted. Another example of this effect is seen among those who allow their shoulders to round forward. Gravity compounds this effect, limiting overhead range-of-motion as shown in Figure C-1. By simply pulling the shoulders back as shown in Figure C-2, the arms are free to move fully overhead. To maintain this optimal position, Soldiers need to regularly stretch the chest muscles that are prone to tightness, and strengthen the upper back muscles that promote proper carriage of the shoulder girdle. More importantly, they need greater awareness of the manner in which they carry the shoulder girdle while performing everyday tasks. Rounding of the shoulders is a common postural problem with many Soldiers, perhaps from emphasizing pushing exercises at the expense of pulling motions.

Appendix C

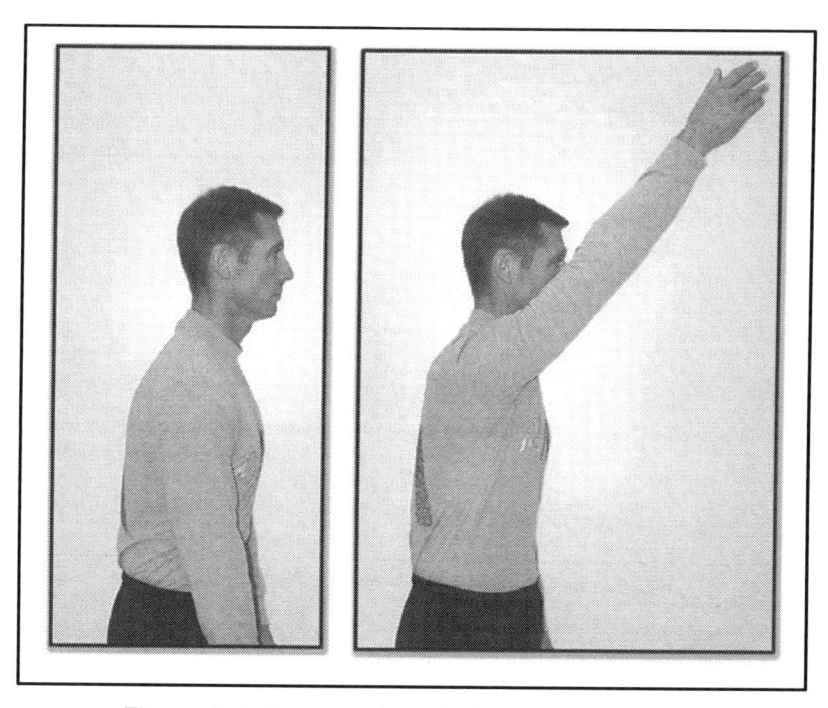

Figure C-1. Poor posture limits range of motion

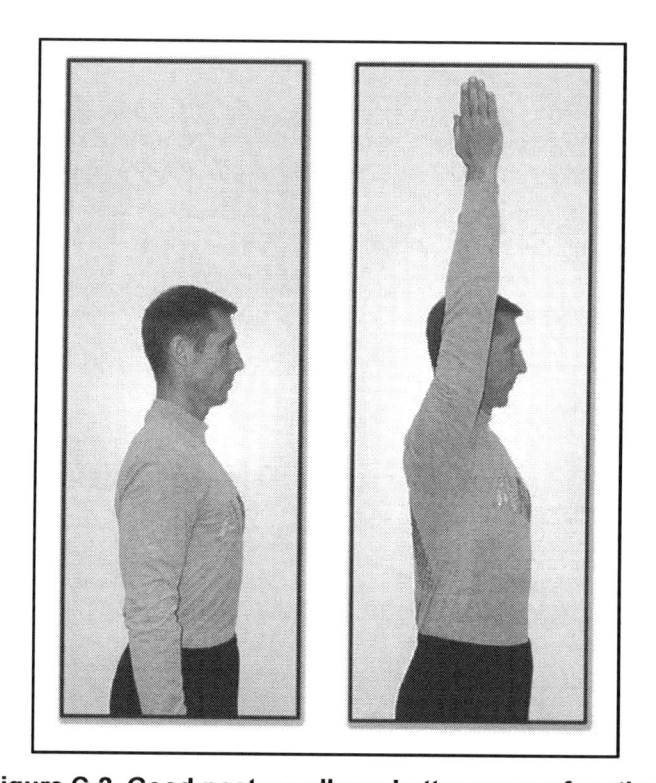

Figure C-2. Good posture allows better range of motion

EFFECT OF EXERCISE

C-4. Like gravity, exercise also molds body tissues. As previously noted, imbalanced exercise practices may adversely affect posture. When regularly performed with precision, the exercise drills and activities in this FM will enhance posture and improve body mechanics. For example, exercise 2 of preparation, rear lunge, provides an excellent stretch of the hip flexors, a muscle group that is prone to tightness. This tightness tilts the pelvis forward, creating an unbalanced base of support for the spine. (See Figure C-3.)

C-5. This exercise also extends the trunk and upper body, compensating for the many hours of flexion throughout the course of the day.

Figure C-3. Rear lunge

"We are all sculptors and painters, and our material is our own flesh, blood, and bones."
Henry David Thoreau

IMPROVING POSTURE

C-6. Improving posture must be built upon the desire to move correctly and efficiently at all times. Regardless of the amount of instruction given and exercise performed, Soldiers will habitually assume good postures only if they want to.

C-7. Good standing and sitting postures are characterized by vertical alignment of certain body segments. However, posture is not improved by forcefully holding the body in a position of ideal alignment. In fact, excessive effort to hold the body in a given posture will only serve to increase muscular tension and fatigue. Assuming naturally balanced postures shifts the weight of the body onto the bones, relieving muscles of the need to support weight bearing. Though the following recommendations are given in the form of a checklist, don't force the body to immediately conform to these ideals. Habits that have been reinforced over decades will take time to correct. Regular and precise performance of the PRT activities in this FM will enhance posture and body mechanics.

C-8. Checkpoints for sitting (Figure C-4):
- Center the head between the shoulders and keep the chin level.
- Draw the shoulders comfortably back; don't allow them to round forward.
- Carry the chest comfortably up and out.
- Maintain the inward curve of the lower back; don't allow it to roll outward or inward excessively.
- Use a firm support between the lower spine and the backrest of the seat or chair to assist in maintaining the proper position.
- Maintain 90-degree angles at the hips and knees with the feet flat on the floor.

Appendix C

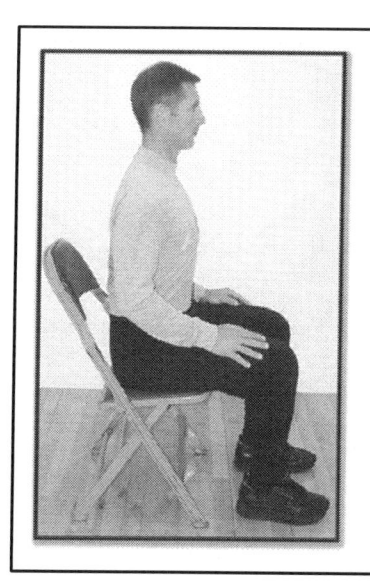

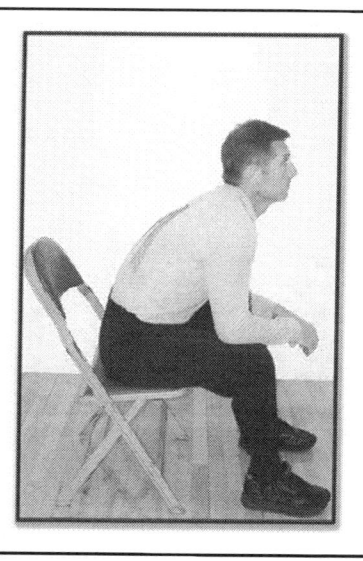

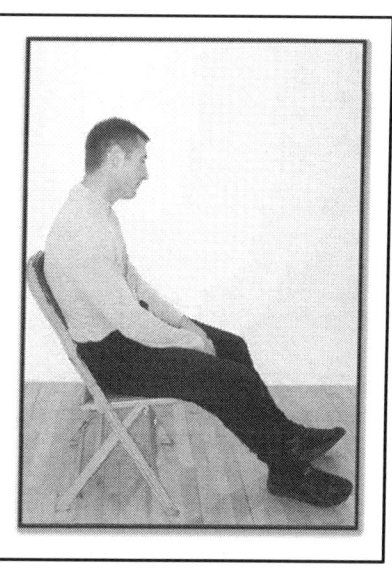

Figure C-4. Good (left) and poor (center and right) sitting posture

C-9. Checkpoints for standing (Figure C-5):
- Stand as tall as possible. The head should not be tilted or the shoulders raised.
- Center the head between the shoulders and keep the eyes and the chin level.
- Slightly draw the chin inward by pressing the neck back toward the collar.
- Moderately elevate the chest without strain. If the chest is raised properly, the abdomen flattens normally. Don't draw in the stomach to the extent that normal breathing is restricted.
- Relax the shoulders and let them fall evenly. If the shoulders round forward, draw them back slightly, without strain.
- Set the pelvis and hips level (refer to Figure C-10c).
- Keep the knees straight but not locked.
- Direct the feet forward without strain. Variations in skeletal alignment will prevent some individuals from assuming the feet-forward position.
- Distribute the weight evenly between the heels and the balls of the feet.

Posture and Body Mechanics

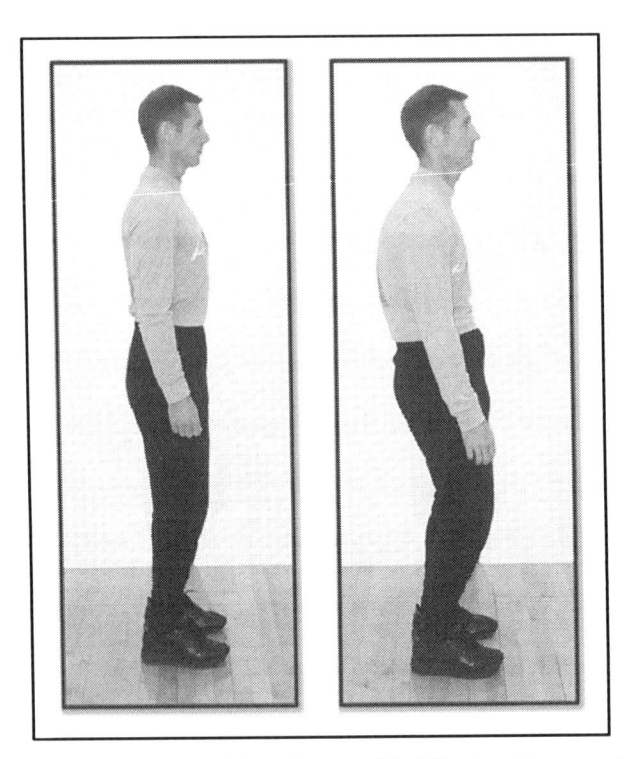

Figure C-5. Good (left) and poor (right) standing posture

COMPENSATING FOR THE EFFECTS OF COMMON POSTURES

C-10. Given the broad definition of posture (any position in which the body resides), the number of postures Soldiers may assume is infinite. However, Soldiers assume the same few postures throughout most of the duty day. The postures can be categorized as the flexed posture (associated with sitting, bending forward, lifting, and crouching); and the upright posture (associated with standing, walking, marching, and running). The body will eventually conform to accommodate these postures. Some muscles will become over-stretched and weak, while others will tighten and lose flexibility. The resulting muscle imbalances will hinder natural movement and increase the likelihood of injury. It is important to regularly compensate for time spent in these prolonged postures by performing exercises or activities that restore the optimal flexibility of muscles and joints:

- Performing extension compensates for flexion. The most common posture for many individuals is seated. This posture is associated with flexion of the spine. Unless great effort is made to sit straight (or a roll is used to maintain the inward curve of the low back), the trunk tends to assume a C-shape. The longer this flexed posture is assumed, the greater will be the effect on muscles around the trunk. The back muscles and ligaments become over-stretched and weak, while muscles on the other side of the trunk (for example, hip flexors) get tighter and pull the pelvis into an unbalanced position. The Soldier on the right in Figure C-6 is in a flexed position. Compensation for prolonged time in this position would occur if the Soldier assumed the prone position of extension demonstrated by the Soldier on the left. To prevent the imbalances associated with too much flexion, Soldiers should regularly perform extension exercises and activities such as those shown in Figure C-7.

Appendix C

Figure C-6. Soldiers in the flexed (right) and extended (left) postures

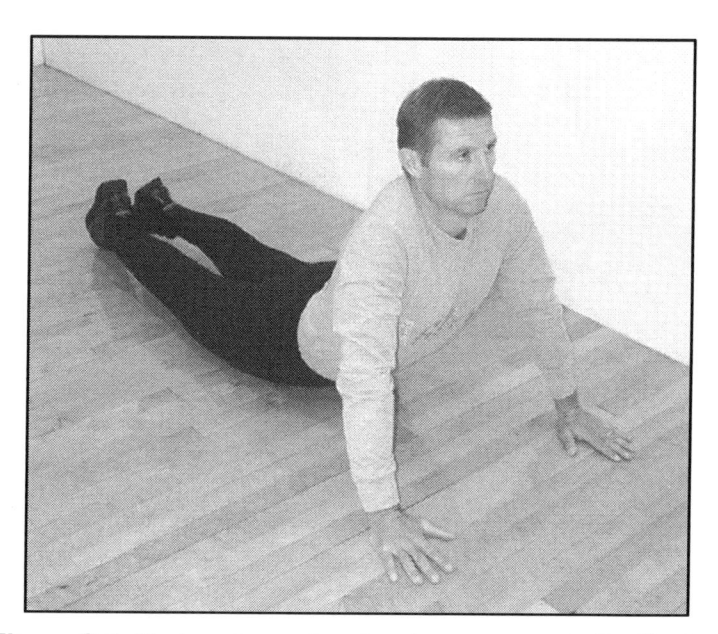

Figure C-7. Performing extension to compensate for flexion

- Performing decompression. This compensates for many of the compressive forces that act on the body throughout the day. Many Soldiers spend the majority of their day on their feet. The weight of the body and equipment creates a compressive effect on the spine and other weight-bearing joints. In fact, at the end of the day enough fluid will have been compressed out of the spinal discs that height measurements will usually indicate that Soldiers are noticeably shorter. Joints that are overly compressed may eventually compromise mobility. To compensate for compressive forces on the spine, it is useful to perform exercises or activities that decompress as shown in Figure C-8.

Posture and Body Mechanics

Figure C-8. Performing decompression to compensate for compression

BODY MECHANICS

C-11. Body mechanics (posture in motion) is the ability to control body movement. Many discussions of posture are limited to static positions such as sitting and standing. Good posture during movement is imperative for efficiency and injury control. Just as good posture requires balanced alignment of the body, so does exercise. Many Soldiers use awkward movements as they struggle to perform one last repetition. **When body mechanics are poor, the exercise serves little purpose and may do more harm than good.** The activities in the PRT system were designed to reinforce proper body mechanics. Of special importance to PRT leaders are the checkpoints given for each exercise. Adherence to these checkpoints ensures optimal execution of the exercise. Over time, skillful movements become second nature to the Soldier. When this occurs, physical readiness is enhanced and injury risk is minimized.

PREPARING THE BODY'S CORE

C-12. Muscles work to initiate and control movement. Because movement is more apparent than the lack of it, the focus is most often on the movement muscles create. Less obvious is the "braking" force that muscles apply to movement. Without this braking force, nearly all movement would be extremely sloppy and potentially dangerous. Around the body's core (trunk and pelvis), this braking action of the muscles becomes extremely important for two reasons. First, the spine and pelvis are the base of attachment for many muscles that power the arms and legs. Without a strong, stable base of support, using these muscles is like firing a cannon from a canoe. Second, the body's center of gravity is within the trunk area. Keeping it there leads to balanced, skillful movement. This is the job of the trunk muscles that do this primarily by putting on the brakes. The ability to maintain balanced postures is often referred to as stabilization. The load on the Soldiers shown in Figure C-9 demands strength and stability from the body's core.

Appendix C

Figure C-9. Soldiers moving under load

C-13. To promote stable postures during exercise, it is essential that Soldiers learn to prepare the trunk. A simple, two-part action prepares the trunk for exercise:

- Set the hips. This is also referred to as the neutral position of the pelvis. This position is found by first tilting the pelvis forward (buttocks goes back, belly goes forward, and the inward curve of the low back is increased), Figure C-10a. Second, tilt the pelvis backward (the buttocks and belly draw inward as far as possible, flattening the curve of the low back) (Figure C-10b). Then settle in between these two extremes (Figure C-10c).
- Tighten the abdominal muscles. Once the hips are set, tightening the abdominal muscles will ensure readiness of the muscles that control and protect the trunk. To contract the correct muscles, imagine drawing the gut straight inward as if preparing for a blow to the mid-section or trying to appear slimmer. Keep the hips set as the abdominals are tightened (Figure C-10c).

C-14. After setting the hips and tightening the abdominal muscles, the Soldier's posture should appear balanced and ready for exercise. The Soldier should not associate these two actions with a stiff, awkward posture. The goal is not to eliminate all movement from the trunk, but to simply control the natural motion that will occur.

Posture and Body Mechanics

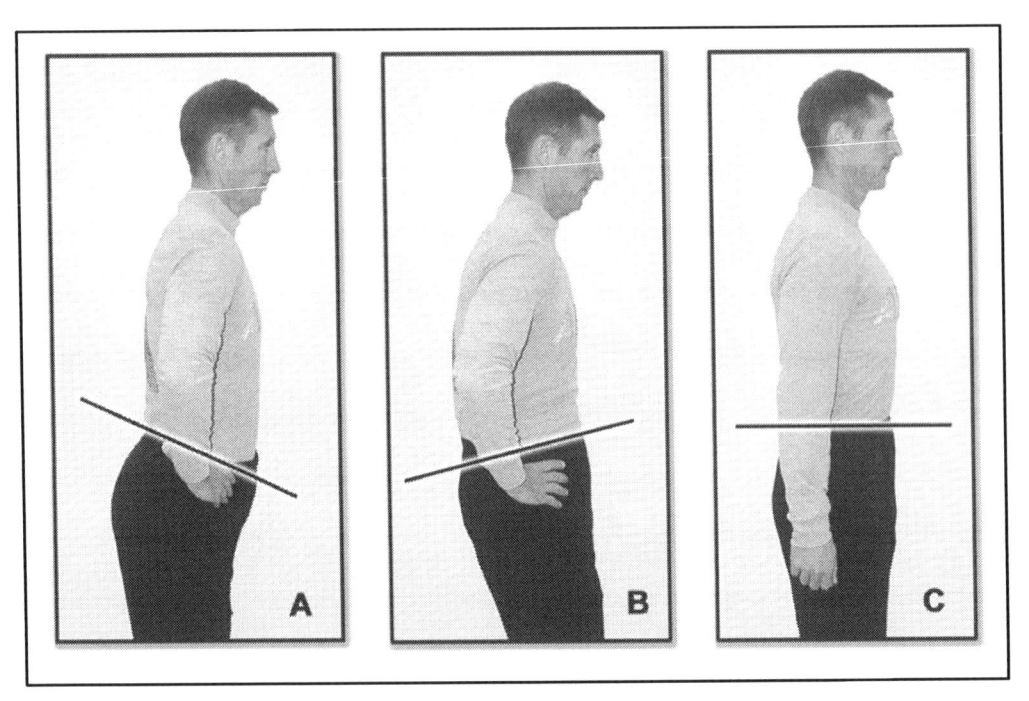

Figure C-10. Set the hips and tighten the abdominal muscles

POWER POSITION

C-15. Proper body mechanics are essential for the powerful movements required of Soldiers. From the power position (Figure C-11), the Soldier is ready to:
- Respond to or deliver aggression.
- Squat to lower or lift a heavy load.
- Accept a heavy load being passed from another individual.
- Sprint to cover.

Appendix C

Figure C-11. Power position

C-16. To assume the power position, first set the hips and tighten the abdominals as previously described. From the straddle stance, place one leg 6 to 8 inches behind the other and crouch so the hips go rearward and the trunk counterbalances by leaning slightly forward. The balls of the feet accept most of the body weight. The shoulder blades are pulled slightly back, but not forced. The chest is high, head is level and elbows and knees are comfortably bent (about 45-degrees).

LIFTING FROM THE GROUND

C-17. Power the lift with the legs, not the back (Figure C-12). Then continue to bend at the hips and knees to lower the body. To protect the back, keep the hips set and the abdominal muscles tight throughout the lift. Keep the load close to the body from start to finish. When Soldiers must turn under load, do so by pivoting the feet rather than twisting the trunk.

Posture and Body Mechanics

Figure C-12. Lifting from the ground

LIFTING OVERHEAD

C-18. Most of the power for pushing an object overhead comes from the legs. To transmit leg strength through the trunk and arms to the object being pushed, set the hips and tighten the abdominal muscles. Hands should be placed shoulder width apart with the upper arms aligned with the trunk. Squat slightly, then forcefully straighten the legs in a coordinated effort with the action of the arms (Figure C-13).

Figure C-13. Lifting overhead

Appendix C

PUSHING

C-19. Push with the hands in front of the shoulders and the upper arms close to the body. This technique creates a mechanical advantage that is lost the farther the hands and arms are from this position. Because this method is the most functional, push-ups performed in the CDs use this technique (Figure C-14).

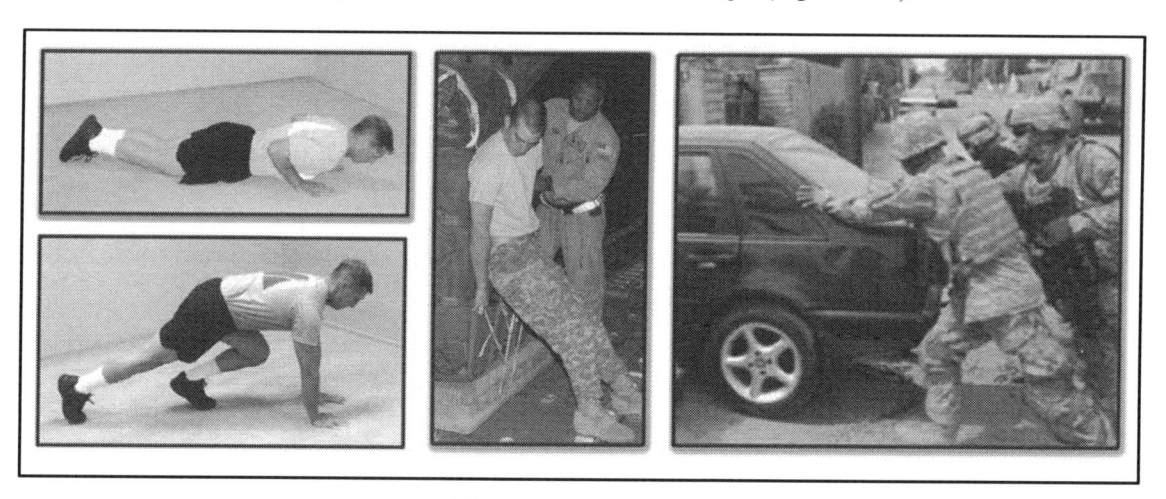

Figure C-14. Pushing

PULLING/CLIMBING

C-20. When pulling an object that is on the ground or horizontal to it, Soldiers must first assume the power position. Set the shoulder girdle by pulling the shoulder blades slightly to the rear. This is also important when pulling the body upward from an overhead grasp. Climbing will often require the legs to power the accent or gain leverage on support structures (Figure C-15). This will demand significant strength from the trunk muscles. The exercises in the CLs prepare Soldiers for these demands.

Figure C-15. Pulling/climbing

ROTATION

C-21. Prepare the body's trunk to control rotation. Coiling (rotating) the body, then quickly uncoiling is the primary source of power for many Soldier and athletic tasks such as throwing a punch or heaving an object onto a platform (Figure C-16). Each of these activities produces a torque on the spine and other joints that may cause injury if the forces are uncontrolled. Control comes from setting the hips, tightening the abdominals, and allowing the hips and knees to bend so as to absorb some of the stress of rotation.

Posture and Body Mechanics

Figure C-16. Rotation

JUMPING AND LANDING

C-22. Land softly with alignment of the shoulders, knees, and the balls of the feet. Land first on the balls of the feet with the heels touching down last. Bending of the hips and knees allows the legs to serve as coils that absorb the impact of the landing. The trunk should be straight but leaning forward so when it is viewed from the side, the shoulders, knees, and the balls of the feet are aligned (Figure C-17).

Figure C-17. Jumping and landing

LUNGING

C-23. Maintain the knee of the forward leg in vertical alignment with the ball of the foot. Do not allow the knee to go beyond the toes or to the right or left of the foot. Lunging is a component of many Soldier tasks. Figure C-18 shows Soldiers performing a proper lunge. Conditioning and kettlebell or dumbbell exercises that involve squat lunging prepare Soldiers for functional tasks such as this.

Appendix C

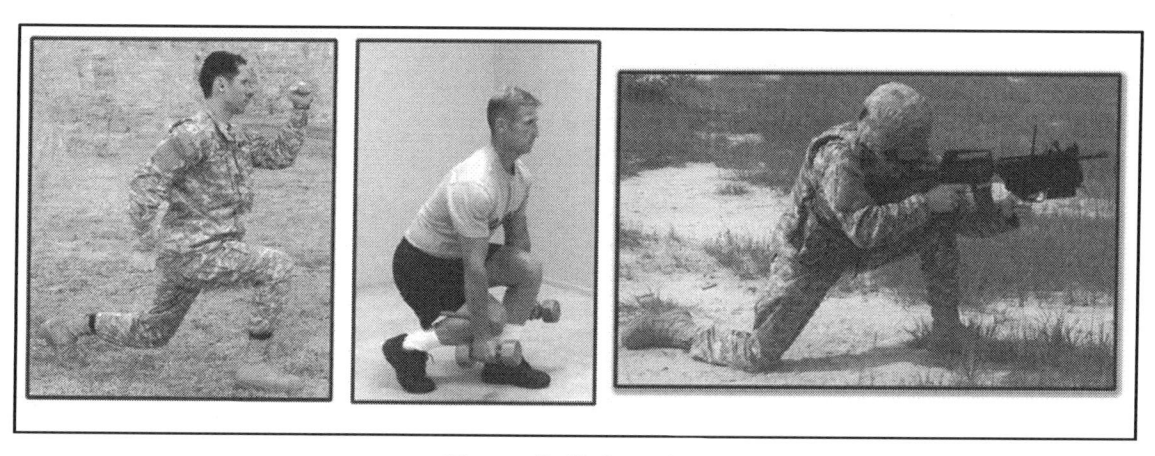

Figure C-18. Lunging

MARCHING

C-24. The head and trunk checkpoints for standing also apply to marching. Allow the arms to swing naturally, though crossing the midline of the body is excessive. Allow the hips to naturally rotate forward with each stride. Do not allow the knees to lock at any point in the walking cycle. Stride naturally, landing on the heel and pushing off with most of the weight toward the big toe. The feet remain directed forward. Do not strain to keep the feet directed forward, since variations in skeletal alignment will prevent some Soldiers from assuming the feet-forward position. Foot marching with a load on the back will require some forward lean of the trunk. Do not, however, allow the trunk and shoulders to round forward (Figure C-19).

Figure C-19. Marching and foot marching

RUNNING

C-25. Refer to Chapter 10, Endurance and Mobility Activities, for a discussion on running form.

CHANGING DIRECTION

C-26. Soldiers may be required to quickly change direction, while maintaining forward movement or to quickly reverse direction. To maintain forward movement, plant on the outside leg with plenty of bend in the hips and knees. The foot should turn slightly inward toward the change of direction. To reverse direction, as in the SR, reduce forward speed and crouch so the body is directed approximately 180 degrees from the forward direction. At the lowest point of the crouch, body weight should rest primarily on the leg closest to the new direction of travel, shifting momentum in that direction (Figure C-20).

Figure C-20. Changing direction

"If half of life is showing up and the other half is doing something when you get there, then the key to success is doing it well."

Stephen Van Camp, Chief of Doctrine, USAPFS

Appendix D
Environmental Considerations

Soldiers must be ready to perform physically demanding tasks in hot, cold, and high altitude environments. Acclimatization to these environments during the conduct of PRT and other daily training activities is essential to safely preparing Soldiers for physical success. When gradually exposed to these environments and the intensity and duration of activities are adjusted, Soldiers can safely acclimatize over time.

HEAT ACCLIMATIZATION

D-1. Soldiers need to acclimatize properly prior to conducting PRT in extremely hot environments. Heat acclimatization allows for those specific adaptations that aid in the reduction of physiological stress (heart rate, core temperature, and sweat adaptation). It also improves physical work capability in the heat and builds Soldier confidence. In hot environments Soldiers will safely acclimatize to the heat by conducting PRT sessions during the heat of the day at a lower intensity and volume. For example, PRT can be moved from early morning to late morning or from late morning to mid-afternoon. This allows for acclimatization by gradually progressing to a warmer/hotter environment. Consideration must also be given to wear of the IPFU ensemble (Figure D-2).

D-2. Heat acclimatization works on a principle of repeated bouts of heat exposure that are stressful enough to safely elevate core temperature and provoke the sweating mechanism. Limited physical activity accompanied by rest in hot environments will result in only partial acclimatization. Acclimatization requires a minimum daily heat exposure of two hours when combined with endurance and mobility, and strength and mobility training. Research has shown that repeated bouts of shorter duration exercise, like those found in speed running, allow for acclimatization more safely than sustained activity in the heat. Initially, Soldiers will train at a lower intensity and shorter duration, then safely progress, increasing physical exercise intensity, duration, and volume to achieve optimal acclimatization in warm/hot environments. In most cases Soldiers can acclimatize in approximately three weeks. Soldiers will maintain acclimatization for approximately one week with about 75 percent of acclimatization lost within three weeks once the Soldier no longer remains in that environment. Soldiers of low fitness levels or those susceptible to heat injuries may require additional days/weeks to fully acclimatize.

D-3. Soldiers must consume sufficient amounts of water to replace water lost due to sweat. Sweating rates greater than one quart per hour are not uncommon. Acclimatization increases sweating rates, which in turn increase water requirements. A risk to acclimatized Soldiers is dehydrating faster than their water intake. Dehydration reduces thermal regulatory advantages achieved through acclimatization and high levels of physical readiness.

HEAT INJURIES

D-4. Soldiers and PRT leaders must be aware of the signs and symptoms of heat injuries and their severities. They must know how to assess Soldiers who may be at risk and be ready to provide appropriate treatment immediately. If any of the below symptoms of heat cramps, heat exhaustion, or heatstroke are experienced, immediately stop physical activity and seek treatment and/or medical attention.

Appendix D

HEAT CRAMPS

Symptoms: Muscular twitching, cramping, muscular spasms in arms, legs, or abdomen.

Treatment: Monitor Soldier in a cool, shaded area, and give water and/or electrolyte sports drink. Call for medical attention if situation worsens.

HEAT EXHAUSTION (REQUIRES MEDICAL ATTENTION)

Symptoms: Excessive thirst, fatigue, lack of coordination, increased sweating, cool/wet skin, dizziness, and/or confusion.

Treatment: Monitor Soldier in a cool, shaded area, attempt to cool Soldier's head and body with cold water and give water and/or electrolyte sports drink and await medical attention.

HEATSTROKE (MEDICAL EMERGENCY, DIAL 911)

Symptoms: No sweating, hot/dry skin, rapid pulse, rapid breathing, seizure, dizziness and/or confusion, loss of consciousness.

Treatment: Monitor Soldier in a cool, shaded area, attempt to immediately cool Soldier's head and body with cold water or ice blanket and give water and/or electrolyte sports drink while awaiting medical attention.

HYPONATREMIA OR OVERHYDRATION (MEDICAL EMERGENCY, DIAL 911)

Symptoms: Confusion, weakness, nausea, and vomiting.

Treatment: Typically misdiagnosed and treated as dehydration. Monitor Soldier and follow treatment for heat exhaustion. If symptoms persist or become more severe with rehydration, replace salt loss and transport immediately to medical facility. DO NOT continue to have Soldier drink more water.

HYDRATION AND NUTRITION

D-5. Ensuring that Soldiers are properly hydrated and receive regular, adequate nutrition is a good way to prevent the onset of heat injuries. Water is the preferred hydration fluid before, during, and after physical training activities. Drink 13 to 20 ounces of cool water at least 30 minutes before beginning exercise (approximately 2 glasses of water). After exercise, drink to satisfy thirst, then drink a little more. Also avoid alcoholic beverages and soft drinks because they are not suitable for proper hydration and recovery. Sports drinks may be consumed, but are not required and contain a considerable number of additional calories. It is also possible to drink too much water. Be sure to limit intake to NO MORE THAN 1 ½ quarts per hour (48 oz) during heavy exertion. Remember, hydration is also important in the cold environment. Many times loss of water is not as noticeable when it is cool or cold.

D-6. Good nutrition practices helps ensure Soldiers have the needed vitamins and minerals for safe performance of exercise in hot environments. Sodium, potassium, and B complex vitamins are lost through sweat and exertion in the heat. It is important to replenish calories lost during exercise with foods containing these nutrients. Try to eat within an hour after exercise. This will assist in recovery as the body is still burning calories at an elevated rate.

COLD ACCLIMATIZATION

D-7. During exercise in the cold, the body usually produces enough heat to maintain its normal temperature. As Soldiers become fatigued, however, they slow down and their body produces less heat. Two types of cold injury conditions may occur due to prolonged exposure and/or loss of core temperature. Soldiers and PRT leaders must be aware of the signs and symptoms of cold injuries and their severities to assess Soldiers who may be at risk and to provide appropriate treatment immediately. If any of the following symptoms (frostbite or hypothermia) are experienced, immediately stop physical activity and seek treatment and/or medical attention.

COLD INJURIES

D-8. Soldiers participating in military training or deployments will often encounter cold stress that can impact successful mission accomplishment. Continued exposure in a cold environment degrades physical performance capabilities, significantly impacts morale, and eventually causes cold weather injuries. Cold environments include exposure to extremely low temperatures (Arctic regions), and cold-wet exposures (rain or water immersion) in warmer ambient temperatures. Cold-weather conditions impair many aspects of normal military functioning in the field, which in turn can influence Soldier health and performance.

Frostbite. When skin is exposed to temperatures/wind chill of 20 degrees Fahrenheit or below there is potential for freezing of skin tissue or frostbite (Figure D-1).

Symptoms: A white or grayish-yellow skin area; skin that feels unusually firm or waxy; numbness in body parts exposed to the cold such as the nose, ears, feet, hands, and skin.

Treatment: Keeping susceptible areas covered is the easiest way to prevent frostbite from occurring (Figure D-2). If any of the aforementioned symptoms are experienced, immediately stop physical activity and seek treatment and/or medical attention.

Hypothermia. This condition develops when the body cannot produce heat as fast as it is losing it (Figure D-1). When Soldiers experience prolonged exposure to cold temperatures or become wet or submerged in cool-water temperatures, they are susceptible to hypothermia.

Symptoms: Shivering, loss of judgment, slurred speech, drowsiness, and muscle weakness.

Treatment: Dressing in layers and wearing breathable undergarments that wick away moisture are helpful in preventing hypothermia. If a Soldier has the symptoms listed above, attempt to make him warmer and request medical attention.

Wind Chill Chart

Wind Speed (in MPH)	Actual Thermometer Readings (F)							
	50	40	30	20	10	0	-10	20
	Equivalent Chill Temperature (F)							
Calm	50	40	30	20	10	0	-10	-20
5	48	37	27	16	6	-5	-15	-26
10	40	28	16	3	-9	-21	-33	-46
15	36	22	9	-5	-18	-32	-45	-58
20	32	18	4	-10	-25	-39	-53	-67
25	30	15	0	-15	-29	-44	-59	-74
30	28	13	-2	-18	-33	-48	-63	-79
35	27	11	-4	-20	-35	-51	-67	-82
40	26	10	-6	-22	-37	-53	-69	-85
	Little Danger for Properly Clothed Soldiers			Increased Danger: Exposed Skin May Freeze (1 min)			Great Danger: Exposed Skin May Freeze (30 sec)	

Figure D-1. Wind chill chart

Appendix D

Endurance and Mobility

	Temperature			
	60 or +	50 - 59	40 - 49	39 & below
Uniform Items				
S/S Shirt	X			
Shorts	X	X	X	X
L/S Shirt		X	X	X
Outer-garment Shirt			X	X
Outer-garment Pants				X
Gloves w/ Inserts				X
Watch Cap				X

Strength and Mobility

	Temperature			
	60 or +	50 - 59	40 - 49	39 & below
Uniform Items				
S/S Shirt	X	X		
Shorts	X	X	X	X
L/S Shirt		X	X	X
Outer-garment Shirt			X	X
Outer-garment Pants			X	X
Gloves w/ Inserts				X
Watch Cap				X

Figure D-2. Clothing recommendations for PRT

ALTITUDE ACCLIMATIZATION

D-9. Soldiers may be deployed to theaters of operation that are at altitudes in excess of 3000 feet above sea level. Altitude acclimatization allows Soldiers to decrease their susceptibility to altitude illness and achieve optimal physical and cognitive performance for the altitude to which they are acclimatized. Altitude acclimatization has no negative side effects and will not harm health or physical performance upon return to low altitude. However, Soldiers with good aerobic endurance may acclimatize sooner and perform better than those of low fitness levels. Refer to the following website for more detailed discussion on altitude acclimatization.

http://www.usariem.army.mil/Pages/download/altitudeacclimatizationguide.pdf

AIR POLLUTION

D-10. Avoid exercising near heavily traveled streets and highways during peak traffic hours. If possible, avoid exposure to pollutants before and during exercise (including tobacco). In areas of high smog concentrations, train early in the day or later in the evening.

SUNLIGHT/SUNBURN

D-11. Use a waterproof or sweat proof sun block (SPF 15 or higher) when exercising in warm weather to avoid sunburn. Follow the instructions on the bottle for proper use.

D-12. For more information related to environmental considerations, see the following websites for more detailed information.

U.S. Army Public Health Command

http://phc.amedd.army.mil/pages/default.aspx

U.S. Army Research Institute of Environmental Medicine

http://www.usariem.army.mil/
http://www.usariem.army.mil/Pages/download/heatacclimatizationguide.pdf
http://www.usariem.army.mil/Pages/download/altitudeacclimatizationguide.pdf

Appendix E

Obstacle Negotiations

Obstacle course running develops physical capacities and fundamental skills and abilities that are important to Soldiers in combat operations (Figure E-1). Soldiers must be able to crawl, creep, climb, walk, run, and jump in order to accomplish certain missions. They must be able to do all these things while carrying full field equipment for long periods of time without exhaustion or injury, even after fatigue has set in. This chapter focuses on obstacle negotiation and obstacle courses used in PRT.

Figure E-1. Obstacles in combat

Appendix E

OBSTACLE COURSES

E-1. Conditioning and CFOC confidence obstacle courses as prescribed in this chapter must comply with installation safety requirements. Considerable time and effort must be expended to teach Soldiers how to correctly negotiate conditioning and confidence obstacles. Soldiers are required to receive instruction for each obstacle negotiated, have each obstacle demonstrated to standard by a PRT leader or AI, and be allowed to practice obstacle negotiation prior to course negotiation. Soldiers will wear ACUs and boots. Conditioning obstacle courses may be run for time. Confidence obstacle courses incorporate complex obstacles that involve height and will not be run for time.

- Conditioning obstacle course. The CDOC has low obstacles that must be negotiated quickly. Running the course challenges the Soldier's basic motor skills and physical condition. After Soldiers receive instruction and practice negotiation skills, they may run the course against time.
- Confidence obstacle course. The CFOC has higher and more difficult obstacles than the conditioning course. It gives Soldiers confidence in their mental and physical abilities and cultivates their spirit and daring. Soldiers are encouraged but not forced to negotiate each obstacle. Unlike conditioning courses, confidence courses are not run against time.

E-2. Physical readiness training leaders will ensure that AIs are positioned at each conditioning and confidence obstacle to ensure proper negotiation and Soldier safety. Physical readiness training leaders are required to perform risk management procedures as specified by their installation. One of the objectives of PRT is to develop Soldiers who are proficient in military physical skills (running, jumping, climbing, and carrying). Fast and skillful execution of these skills may mean the difference between the success and failure of combat missions.

RUNNING

E-3. Running is used to develop endurance. Soldiers should be exposed to running in the following situations:
- On roads.
- Over rough ground.
- Up and down hills.
- Across country.
- Over low obstacles.

JUMPING

E-4. In vertical and long jumping, the takeoff foot is planted firmly. The spring comes from the extension of the take-off foot leg as the other leg reaches for the high or far side of the obstacle (like a ditch). The arms are forcibly raised forward and upward to assist in propelling the body. Landing may be on one or both feet, depending upon the length of the jump. When jumping downward from a height, the jumper should aim his feet at the desired landing spot and jump with the knees slightly bent, feet together, and the trunk inclined slightly forward. As the feet strike the ground, the shock is absorbed by bending the knees into a full squatting position. If the height is too great or the ground too hard to absorb the shock, the jumper should execute a forward or side roll to absorb some of the shock.

DODGING

E-5. In combat situations it is often necessary to change directions quickly. To dodge while running, the lead foot (the left foot if the direction is to the right; the right foot if the direction is to the left) is firmly planted on the ground. The opposite foot is moved in the new direction. The knees are flexed slightly during the movement and the center of gravity is low. The head and trunk are quickly turned in the new direction at the instant of directional change.

CLIMBING AND SURMOUNTING

E-6. The Soldier should know how to effectively climb and surmount various types of obstacles.

VERTICAL CLIMBING A ROPE OR POLE

E-7. Whether climbing a rope or pole, the techniques are similar. The hands grasp the rope or pole overhead with the palms toward the face. The body is pulled upward with the arms and shoulders assisted by the feet (which grip and assist by pushing downward). If shoulder-girdle strength and body coordination are not adequate to permit alternating hands, the arms act together in pulling upward. For rope climbing technique, refer to paragraph E-37.

CLIMBING OVER A WALL

E-8. When going over a wall, the body is as close to the top as possible to maintain a low silhouette. (In combat operations, it is important to offer as small a target as possible. When preparing to go over a wall, the rifle is slung across the back so the hands are free.) Chinning and creeping are the most common methods used for surmounting a wall of moderate height.

CHINNING

E-9. Approach the wall at a walk or slow run. Jump upward and grasp the top of the wall and chin upward until it is possible to change into a push-up. Place the chest on the wall and kick vigorously upward and over with both legs. A creeping motion with the toes against the wall will help the upward progress of the chinning and pushing up.

CREEPING

E-10. Approach the wall at either a walk or slow run. Jump upward and grasp the top of the wall. Make contact with both knees and start a creeping motion upward. As the knees reach their limit of upward motion, place both feet against the wall and continue with a walking-creeping motion until one leg can be thrown over the top of the wall. Make sure a creeping walk is used.

RUN, JUMP, AND VAULT

E-11. Approach the wall at a run, jump forward and upward at it, and place one foot against it as high up as possible. Use the foot in contact with the wall to help push the body upward while grasping the top of the wall with the hands. Pull the body up with the arms, assisted by pressure from the foot against the wall, and swing the legs over, propelling the body over the wall.

HOOK AND SWING

E-12. Approach the wall at a run and jump forward and upward. Hook one elbow over the wall, locking the arm in place by pulling up until the top of the wall is underneath the armpit. Grasp the top of the wall with the other hand. Draw the leg that is closer to the wall up toward the abdomen as far as possible. Then swing the outside leg over the wall. The body is carried over with a rolling motion. Soldiers who are unable to draw up the leg as described can use a variation of this leg action. While hanging with both legs fully extended, start a swinging motion with the legs together. When the legs have enough momentum, swing the outside leg over the wall with a vigorous kick; then follow with the body.

DROPPING

E-13. Execute all drops from the wall in the same manner. Place one hand against the far side of the wall while the other hand grasps the top. From this position, roll over the wall and vault away from it with the legs swinging clear. As the body passes over the wall and drops, face the wall. This keeps the rifle and other equipment clear. Balance is maintained by retaining a grasp on the top of the wall as long as possible.

CLIMBING LADDERS AND CARGO NETS

E-14. Rope ladders, stationary vertical ladders, and cargo nets require the same general climbing technique. Grasp the side supports firmly in the hands about shoulder height and place the feet on a rung, which will cause

Appendix E

the body to be extended. To move up, obtain a higher grasp and move the opposite leg up a rung. The body is elevated as the knee straightens.

TRAVERSING HORIZONTAL OBJECTS BY HAND

E-15. Traversing horizontal objects puts stress on the arms and the shoulder-girdle area when the feet are suspended in the air and the arms and shoulders support body weight.

TRAVERSING HORIZONTAL ROPES OR PIPES

E-16. The hands grasp the horizontal support overhead with the palms facing. To propel the body forward, one hand is released and moved forward to secure a new grasp. At the same time, the opposite side of the body is swung forward. The other hand is then released and moved forward as the Soldier continues to move.

TRAVERSING HORIZONTAL LADDERS

E-17. The movement is the same as used in traversing a rope or pipe. The hands, however, are placed on the rungs palms forward. Otherwise, the technique is the same.

VAULTING

E-18. Vaulting is employed to overcome low barriers or fences. The object to be surmounted is approached at an angle. The hand on the side next to the obstacle is placed on top of the obstacle, then with a straight-arm movement the body weight is pushed upward. At the same time, the leg on the side next to the obstacle is thrown upward and over the top, followed by the other leg. In landing, the weight comes down on the leading leg first, followed by regaining the balance on both legs. The free arm serves as a balance. A direct (front) approach can also be used, at which time both legs go over the object together.

BALANCING

E-19. Balancing the body while walking or running on a narrow object, as when crossing obstacles, is a skill that requires practice and confidence. Balance is required in negotiating a log placed across a stream, or in crossing any narrow beam or rail. To perform this skill, place the feet on the object to be crossed, hold the arms to the side at shoulder level, then fix the eyes on the object approximately 5 yards in front of the feet. Walk the object by placing first one foot and then the other in the center of the object, slowly moving forward using the arms to aid in maintaining balance.

CRAWLING

E-20. Crawling in combat situations is a useful skill. Crawling may be in the high or low stance.

HIGH CRAWL

E-21. The Soldier moves on his hands and knees, moving one hand and the opposite knee and then continuing to move the hands in alternation with the knees.

LOW CRAWL

E-22. The Soldier is in the prone position. Pulling with both arms and pushing with one leg, accomplish forward movement. The other leg is dragged behind. The legs are alternated frequently to avoid fatigue.

THROWING

E-23. Throwing may be from the kneeling or standing position. The object to be thrown is held in the throwing hand with the throwing arm is bent at the elbow. The hand is then moved to the rear until it is behind the ear. The body is turned so that the lead foot and balance (other) arm point toward the target. The balance arm is used to sight over and align the throwing hand with the target. When properly aligned, the elbow is move rapidly forward until it is at a point just in front of the body where the arm is straightened and the wrist "snapped." This whip motion propels the object to the target. Underhand throws get momentum by the thrower bending his knees and swinging the throwing arm to the rear. As the knees are straightened, the arm is forcefully swung forward from the shoulder and the object is released.

FALLING

E-24. Injuries may be avoided if Soldiers are taught to fall properly by using body momentum to their advantage instead of resisting it. If enough momentum is present, as in falling while running or in jumping from a height, the Soldier can extend his hands to catch his weight while ducking his head and forward rolling onto his feet. The key to falling without injury from the standing position is relaxation and rolling the body to take the momentum of the fall on the outside of one leg, hip, and buttock.

CONDITIONING OBSTACLE COURSES

E-25. Conditioning obstacle courses are typically not standardized because of varying topographical conditions; however, individual obstacles within the conditioning course are standardized for both construction and negotiation. Commanders should use ingenuity in constructing courses, making good use of streams, hills, trees, rocks, and other natural obstacles. Since conditioning courses are run against time, they should not be made dangerous.

E-26. Conditioning courses should be developed based on the following guidelines:
- Courses should be horseshoe-shaped with route signs and the finish close to the start.
- Total course distance ranges from 300 to 450 yards.
- Courses contain 15 to 25 obstacles placed 20 to 30 yards apart.
- Obstacles are arranged so that those that exercise the same muscle groups are separated and not performed consecutively.
- Obstacles must be solidly built with no sharp points or corners and landing pits filled with sawdust or ground tires.
- Lanes will be wide enough for 6 to 8 Soldiers to run the course at the same time and avoid congestion.
- Courses will be built and marked so Soldiers cannot sidestep the obstacles or detour around them.
- To minimize the possibility of falls and injures due to fatigue, the last two or three obstacles should not be too difficult or involve high climbing.

OBSTACLES FOR JUMPING

E-27. These types of obstacles include ditches Soldiers can clear with one leap, trenches they can jump in or out of, and hurdles (Figure E-2).

Appendix E

Figure E-2. Jumping obstacles

OBSTACLES FOR DODGING

E-28. These obstacles include mazes or lanes for change of direction. The maze is constructed from posts set in the ground at irregular intervals. The spaces between the posts are narrow so Soldiers must pick their way carefully through and around them. Lane guides are built to guide Soldiers in dodging and change of direction (Figure E-3).

Figure E-3. Dodging obstacles

OBSTACLES FOR VERTICAL CLIMBING AND SURMOUNTING

E-29. These obstacles (Figure E-4) include climbing ropes 1 ½-inches in diameter (plain or knotted), cargo nets, walls (7 or 8 feet high) or vertical poles (6 to 8 inches in diameter and 15 feet high).

Figure E-4. Climbing obstacles

OBSTACLES FOR HORIZONTAL TRAVERSING

E-30. These obstacles include ladders, ropes, pipes or beams positioned 8 to 10 feet off the ground. These obstacles may be traversed using the arms only or a combination of arms and legs (Figure E-5).

Figure E-5. Horizontal traversing obstacles

OBSTACLES FOR CRAWLING

E-31. These obstacles (Figure E-6) include large pipe sections (tunnels 4 feet in diameter and 8 feet long); low rails (8 inch diameter log, 8 feet long, and 2 feet off the ground); and wire (all wire lanes will be 10 feet wide, 30 feet long, and 2 feet off the ground).

Figure E-6. Crawling obstacles

OBSTACLES FOR VAULTING

E-32. These obstacles (Figure E-7) include fences or low walls (3 to 3 ½ feet high).

Figure E-7. Vaulting obstacles

OBSTACLES FOR BALANCING

E-33. These obstacles (Figure E-8) include beams, logs, or planks that span water obstacles or dry ditches (2 feet deep).

Obstacle Negotiations

Figure E-8. Balancing obstacles

NEGOTIATION STANDARDS FOR CONDITIONING COURSES

E-34. The following paragraphs describe a variety of negotiation standards for successful completion of obstacle courses.

LANES TO GUIDE CHANGE OF DIRECTION

E-35. To successfully negotiate laned obstacles Soldiers must enter and exit the change of direction lanes while running, using the following technique. To change direction while running, plant the lead foot (left foot if the direction is to the right; right foot if the direction is to the left) firmly on the ground. Then, move the opposite foot in the new direction. The knees are flexed slightly and the center of gravity is low. Turn the head and trunk quickly in the new direction at the instant of the directional change.

DITCH

E-36. To successfully negotiate this obstacle the Soldier must jump over the ditch while running and use the following technique. When jumping over a ditch, the takeoff foot is planted firmly and the spring comes from the extension of this leg as the other leg reaches for the opposite side of the ditch. Raise the arms forcibly forward and upward to assist in propelling the body. The landing may be on one or both feet, depending on the length of the jump.

CLIMBING ROPE

E-37. The following technique is used to successfully negotiate this obstacle. To initiate the climbing action, grasp the rope with the hands, palms toward the face. Grapevine the rope by wrapping it around the lower leg, crossing the instep. With the opposite leg, anchor the rope by placing the bottom of the foot on the instep. Stand up pushing down with the legs while reaching overhead grasping the rope at a higher point. Draw the knees toward the chest while allowing the rope to slide between the knees and feet. Repeat the following sequence to continue climbing:

- Anchor the feet.
- Stand up pushing down with the legs.
- Reach overhead and re-grasp the rope.
- Draw the knees toward the chest.
- Re-anchor the feet on a higher point on the rope.
- Repeat sequence until reaching the top of the rope.

Appendix E

Logs

E-38. To successfully negotiate this obstacle, walk or run the log using the following technique. Place the feet on the log, hold the arms at the sides at shoulder level, and fix the eyes on the log approximately five yards in front of the feet. Walk or run the log by placing first one foot then the other in the center of the log, moving forward using the arms to maintain balance.

Horizontal Ladder

E-39. To successfully negotiate this obstacle, traverse the ladder using the following technique. Grasp the first rung overhead with the palms facing forward and suspend the body. To propel the body forward, release one hand and move forward to secure a new grasp. At the same time, swing the opposite side of the body forward. Release the other hand and move it forward to re-grasp another rung. Continue this technique grasping each and every rung until reaching the last rung. Suspend the body from the last rung, then drop to the ground.

Alternate High Stepping

E-40. To successfully negotiate this obstacle, enter and exit the maze while running, using the following technique. Run on the balls of the feet and raise the knees up high with each step while crossing over the obstacles and placing each foot in adjacent grid squares.

Horizontal Rope

E-41. To successfully negotiate this obstacle, traverse the rope using the following technique. Reach up and grasp the rope with both hands and swing the legs up to assume the position used when climbing a vertical rope. Leading with the head, traverse the rope horizontally by pulling with the arms. The feet and legs are used to secure the position on the rope and may also be used to assist in the movement as in the vertical rope climb. To complete negotiation of this obstacle, one hand must touch the post securing the end anchor point.

Wire

E-42. To successfully negotiate this obstacle, enter and exit using the low crawl technique. Start in the prone position. To move forward, pull with both arms and push with one leg. The other leg is dragged behind. The legs are alternated frequently to avoid fatigue. Continue this technique until the body has cleared the low wire.

Cargo Net

E-43. To successfully negotiate this obstacle, approach the net while running. Leap to grasp the rope rungs overhead and step up on the lower rope rungs. The Soldier may use either of the following methods to climb the cargo net: The first technique performs alternating arm and leg movements (reach up with the right arm to grasp a higher rung while simultaneously stepping up with the left leg). The second technique would be to grasp and step with the same side arm and leg, ascending the rope in a crawling fashion. Continue this technique to the top of the net, then propel the body over the platform and descend the net on the other side using a similar technique.

Fence

E-44. To successfully negotiate this obstacle, use the vaulting technique. Approach the fence at an angle with the hand on the side, next to the fence, placed on top of the fence. With a straight-arm movement, the Soldier pushes his body weight upward. At the same time, his leg on the side next to the fence is thrown upward and over the top, followed by his other leg. When landing, his weight comes down on his landing leg first, followed by regaining his balance on both legs. His free arm serves to balance him. A direct front approach can also be used, at which time both legs go over the fence together.

Trench

E-45. To successfully negotiate this obstacle, use the following technique. Jump downward into the trench, aiming the feet at the desired landing spot with the knees slightly bent, feet slightly apart, and trunk inclined slightly forward. As the feet strike the ground, the Soldier absorbs the shock by bending his knees to a squatting position. If the height is too great or the ground too hard to absorb the shock, he should land with his feet together and execute a forward or side roll to absorb some of the shock. To exit the trench, he uses one of the following techniques: Approach the trench wall at a run, jump forward and upward at it, and place one foot against the trench wall as high as possible. He uses the foot that is in contact with the wall to help push his body upward while grasping the top of the trench with his hands. He pulls his body up with his arms, assisted by the pressure of his foot against the wall and swings his legs over to propel himself out of the trench. Using the second technique, the Soldier approaches the trench wall at a run and jumps forward and upward. He hooks one elbow over the top of the trench, locking his arm in place by pulling up until the top of the trench is under his armpit. He grasps the top of the trench with his other hand. He draws his leg that is closest to the trench wall up toward his abdomen as far as possible, then swings his outside legs over the top of the trench. His body is then carried over with a rolling motion. Soldiers who are unable to draw up the leg as described can use a variation of this leg action. While hanging with both legs fully extended, he starts a swinging motion with his legs together. When his legs have enough momentum, he swings his outside leg over the trench wall with a vigorous kick, then follows with his body to exit the trench.

Low Rails

E-46. To successfully negotiate this obstacle, use the low crawl technique to move under the low rails.

Planks And Beams

E-47. To successfully negotiate this obstacle, use the same technique listed to traverse the logs.

Wall

E-48. To successfully negotiate this obstacle, use either of the following techniques to surmount the wall. Run, jump, and vault. When using this method, the Soldier approaches the wall at a run, jumps forward and upward at it, and places one foot against the wall as high as possible. He uses his foot in contact with the wall to help push his body upward while grasping the top of the wall with his hands. He pulls his body up with his arms, assisted by the pressure of his foot against the wall, and swings his legs over to propel himself over the wall. The second technique is the hook and swing. The Soldier approaches the wall at a run and jumps forward and upward. He hooks one elbow over the wall, locks his arm in place by pulling up until the top of the wall is under his armpit. He grasps the top of the wall with his other hand. He draws his leg that is closest to the wall up toward his abdomen as far as possible, then swings the outside leg over the wall. The body is then carried over with a rolling motion. A variation of this leg action can be used by Soldiers who are unable to draw up the leg as described. While hanging with both legs fully extended, he starts a swinging motion with his legs together. When his legs have enough momentum, he swings the outside leg over the wall with a vigorous kick, then follows with his body. To drop from the wall to the ground, he places one hand against the far side of the wall while his other hand grasps the top. From this position, he rolls over the wall and vaults away from it with his legs swinging clear. As his body passes over the wall and drops, he faces the wall. He maintains his balance by retaining his grasp on the top of the wall as long as possible and then dropping to his feet.

Low Wall

E-49. To successfully negotiate this obstacle, use the vaulting technique. The Soldier must approach the fence at an angle, his hand on the side next to the fence is placed on top of the fence, then with a straight-arm movement, he pushes his body weight upward. At the same time, his leg on the side next to the fence is thrown upward and over the top, followed by his other leg. In landing, his weight comes down on his landing leg first, followed by regaining his balance on both legs. His free arm also serves as a balance. A direct front approach can also be used, at which time both legs go over the fence together.

Hurdle

E-50. To successfully negotiate this obstacle, leap over the hurdle one leg at a time or step on the hurdle with one leg and leap down from the hurdle with the other or both legs to the ground.

Appendix E

PLATFORM

E-51. To successfully negotiate this obstacle, the Soldier surmounts the platform by using the support beams to step up and pull himself to the top. When jumping down from the platform to the ground, perform the same technique used for jumping downward from a height, as in negotiating a trench.

TUNNEL

E-52. To successfully negotiate this obstacle, two crawling methods may be used; the high crawl and low crawl. The Soldier performs the high crawl technique on his hands and knees. He propels himself forward by moving one hand forward while simultaneously moving his opposite knee forward. He continues moving on his hands and knees in an alternating fashion. The low crawl technique starts in the prone position. To move forward, he pulls with both arms and pushes with one leg. His other leg is dragged behind. Both legs are alternated frequently to avoid fatigue. The Soldier continues this technique until he exits the tunnel.

CONDUCTING THE CONDITIONING OBSTACLE COURSE

E-53. Before Soldiers run the CDOC in its entirety, they should be taken to each obstacle and instructed in the proper negotiation techniques previously mentioned. In each case the techniques should be explained in detail with emphasis on avoidance of injury. All Soldiers should be given the opportunity to practice on each obstacle until they become proficient at negotiation. Before the course is run against time, several practice runs should be run at a slower pace. During such practice runs, PRT leaders and AIs observe their performance and make appropriate corrections. Soldiers should never be permitted to run CDOCs for time until they have mastered all obstacles thoroughly. The best method of timing Soldiers on the obstacle course is to have the timer stand at the finish line and call out minutes and seconds as each Soldier crosses the finish line. If Soldiers fail to negotiate an obstacle, a predetermined penalty (5 to 10 seconds) should be assessed.

CONFIDENCE OBSTACLE COURSES

E-54. Confidence obstacle courses challenge Soldiers' strength, endurance, and mobility while instilling self-confidence and promoting teamwork. Soldiers do not negotiate these obstacles at high speed or against time. Obstacles vary in difficulty. Some stand very high. Safety nets and crash pads are provided for these high obstacles. Soldiers may skip any obstacle they are unwilling to attempt. PRT leaders and AIs should encourage, but not force Soldiers to attempt every obstacle. Fearful Soldiers should be encouraged to negotiate the easier obstacles before attempting the higher more difficult ones. Some of the higher, more difficult obstacles may be negotiated as a group effort, with stronger Soldiers assisting those unable to negotiate the obstacles by themselves. Gradually, as their confidence and negotiation skills improve, the weaker Soldiers will be able to successfully negotiate all obstacles individually. PRT leaders and AIs should be available to assist Soldiers in proper obstacle negotiation throughout the course. At no time are PRT leaders or AIs to make obstacles more difficult by shaking ropes, rolling logs, and so forth. This practice destroys confidence and greatly jeopardizes safety. Confidence obstacle courses must be constructed according to Folio Number 1, "Training Facilities," Corps of Engineers, Drawing number 28-13-95. Contact the installation Directorate of Public Works for blueprints. The Army's standardized CFOC consists of 22 obstacles that are grouped into color-coded quadrants with five or six obstacles in each. Negotiation becomes more difficult beginning with the black quadrant followed by the blue quadrant, white quadrant, and red quadrant. All Soldiers begin CFOC negotiation in the black quadrant. Soldiers progress to the more difficult quadrants (blue, white, and red) when they become proficient and successfully negotiate obstacles in previous quadrants.

BLACK QUADRANT

E-55. The black quadrant consists of the following obstacles.

HIGH STEP OVER

E-56. Soldiers step over each bar: they either alternate legs or use the same leg each time while making an effort not to use their hands. (Shorter Soldiers may be required to use hands). Soldiers must be spaced so as to prevent kicking each other.

Obstacle Negotiations

Low Wire

E-57. Soldiers move forward on their backs while at the same time raising the wire with their hands so their bodies will clear the wire. They continuing moving forward in this manner until they reach end of the obstacle.

Swing, Stop, and Jump

E-58. Soldiers gain momentum with a short run, grasp the rope, and swing their bodies forward to the top of the wall. They release the rope while standing on the wall and jump to the ground.

Six Vaults

E-59. Soldiers vault over each log using one or both hands.

Easy Balancer

E-60. Soldiers walk up one incline log and down the one on the other side to the ground. Running is not encouraged (Figure E-9).

Figure E-9. Black quadrant CFOC

BLUE QUADRANT

E-61. The blue quadrant consists of the following obstacles.

BELLY BUSTER

E-62. Soldiers vault, jump, or climb over a moving log.

REVERSE CLIMB

E-63. Soldiers approach the underside of the climbing ladder, climb up to and over the top of the ladder, then climb down the opposite side.

WEAVER

E-64. Soldiers move from one end of the obstacle to the other by weaving their bodies under one bar and over the next.

HIP-HIP

E-65. Soldiers step over each bar by either alternating legs or using the same leg each time while making an effort not to use their hands.

BALANCING LOGS

E-66. Soldiers walk or run along logs while maintaining their balance.

ISLAND HOPPERS

E-67. Soldiers jump from one log to another until obstacle is negotiated from near to far side (Figure E-10).

Obstacle Negotiations

Figure E-10. Blue quadrant CFOC

Appendix E

WHITE QUADRANT

E-68. The white quadrant consists of the following obstacles.

TOUGH NUT

E-69. Soldiers step over each "X" in each lane.

SLIDE FOR LIFE

E-70. Soldiers climb the tower, mount the center of the platform, grasp the rope firmly with their hands, and perform a heel hook. Soldiers begin traversing down the rope by moving hand-over-hand and reaching with the legs. Soldiers brake by use of the hands, legs, and feet. Soldiers traverse the rope to a marked release point. Soldiers dismount the rope by removing their legs from the rope, hanging with their arms fully extended, then drop to the ground landing on their feet. If during negotiation a Soldier's legs come off the rope, he should attempt to heel hook and lock his legs back on the rope. Soldiers must be instructed on proper technique for landing in the net if they should fall from the obstacle. Soldiers need to draw their knees toward their chest, tuck their chin, then attempt to land on their back or side. Only one Soldier is allowed on the rope at one time. This obstacle is dangerous if the rope is wet. This obstacle requires one instructor on the platform and one instructor at the release point. A safety net will extend from below the platform to the release point.

LOW BELLY OVER

E-71. Soldiers mount the low log and jump onto the high log, grasping with both hands the high log's top, keeping the belly area in contact with it. Soldiers swing their legs over the log, then lower themselves to the ground.

BELLY CRAWL

E-72. Soldiers move forward under the wire on their stomachs to the end of the wire obstacle.

DIRTY NAME

E-73. Soldiers mount the low log and jump onto the high log. Soldiers swing their legs over the top log, then lower themselves to the ground.

TARZAN

E-74. Soldiers mount the lowest log and maintain balance while walking the length of it. Soldiers then mount the higher log and maintain balance until they reach the horizontal ladder. Soldiers then step onto the foot blocks and grasp the first rung of the ladder. They begin traversing the ladder by releasing one hand at a time and swinging forward, grasping a more distant rung each time. Upon reaching the last rung, Soldiers hang with their arms fully extended and drop to the ground landing on their feet (Figure E-11).

Obstacle Negotiations

Figure E-11. White quadrant CFOC

RED QUADRANT

E-75. The red quadrant consists of the following obstacles.

INCLINING WALL

E-76. Soldiers approach the underside of the wall, jump up and grasp the top, and pull themselves over the top. Soldiers slide or jump down the incline to the ground.

SKYSCRAPER

E-77. A team of Soldiers (4+) jumps or climbs to the first floor and either climb the corner posts or help one another to higher floors. All climbing from the second to the fourth floor is accomplished only on sides containing safety nets. Crash pads are positioned on the non-climbing sides of the obstacle. The top of the obstacle is off limits and will not be negotiated. Only one team should be on the obstacle at a given time. Soldiers descend from floor to floor individually or as a team. They should not jump to the ground from above the first floor and must be instructed on proper technique for landing in the net if they should fall from the obstacle. Soldiers need to draw their knees towards their chest, tuck their chin, and attempt to land on their back or side.

Confidence Climb

E-78. Soldiers climb the vertical ladder to the second rung from the top, climb over the rung, and descend the other side. Only one Soldier is allowed on the obstacle at a time. An instructor is harnessed in at the top of the obstacle to assist Soldiers with obstacle negotiation. Soldiers must be instructed on proper technique for landing on the crash pad if they should fall from the obstacle. They must draw their knees toward their chest, tuck their chin, and attempt to land on their back or side. Pads will be placed at the base of the obstacle on both climbing sides.

Belly Robber

E-79. Soldiers step on the lower log and assume a prone position on the horizontal logs. They crawl over the logs to the opposite end of the obstacle. Rope gaskets must be attached to the ends of the logs to keep the hands from being pinched and to ensure logs cannot fall from the perpendicular cradle.

Tough One

E-80. Soldiers climb the cargo net up and over at the low end of the obstacle (13 feet). They move across the top of the logs, climb the ladder, and go over the log at the high end (33 feet). An instructor is harnessed in at the high end of the obstacle to assist climbers with obstacle negotiation. Soldiers then climb down the cargo net to the ground. The net will extend from below the log walk. Crash pads are positioned at the base of each cargo net. Soldiers must be instructed on proper technique for landing in the net if they should fall from the obstacle. They must draw their knees towards their chest, tuck their chin, and attempt to land on their back or side (Figure E-12).

Figure E-12. Red quadrant CFOC

CONFIDENCE OBSTACLE COURSE CONSTRUCTION AND SAFETY

E-81. The following paragraphs discuss course sketches that describe in detail CFOC construction and safety requirements.

COURSE SKETCHES

E-82. The following course sketches supplement the Department of the Army Engineer Drawings 28-13-95, Confidence Course Layout Plan. They serve as the minimum construction/safety standards for CFOCs.

E-83. Criteria for safety and structural inspections are specified in the obstacle risk assessment and according to the material manufacturer's directions. Re-inspection must include a review of the risk assessment, an analysis and assessment of accidents/injuries sustained since the obstacle was put into (or back into) service, following repairs, major renovations, or modifications.

Appendix E

E-84. CFOC safety precautions include:
- Inspection of structural integrity and safety devices prior to use.
- Current risk assessment updated prior to each day's training and updated as conditions change.
- Instructor training and certification on operation of obstacles prior to conduct of course.
- Preparation exercises before commencing course and recovery exercises upon completion.
- Muscular strength/muscle failure physical training that should not be conducted within 12 hours prior to the CFOC.
- Landing/fall areas under obstacles raked and refilled as needed before each use.
- Puddles of water under obstacles filled to preclude a false sense of security.
- Training that is postponed/modified when obstacles are slippery due to inclement weather.
- Instructors who instruct and demonstrate obstacle negotiation before allowing Soldiers to negotiate the CFOC.
- A sign posted at each obstacle detailing exact procedures to be used for proper negotiation.
- A maintenance and inspection log that is maintained for each CFOC. The log should include:
  - A detailed checklist for course and obstacle inspection.
  - A record of all course inspections and maintenance deficiencies.
  - A list of any uncorrected deficiencies remaining on the course and countermeasures in place.

E-85. Detailed obstacle illustrations are provided for:
- Tough One.
- Slide for Life.
- Confidence Climb.
- Skyscraper.
- Belly Robber.
- Tarzan.
- Low Belly Over.
- Dirty Name.
- Tough Nut.
- Belly Crawl.
- Inclining Wall.
- High Step Over.
- Swing, Stop, and Jump.
- Six Vaults.
- Easy Balancer.
- Belly Buster.
- Low Wire.
- Hip-Hip.
- Reverse Climb.
- Weaver.
- Balancing Logs.
- Island Hopper.

E-86. Safety equipment (nets, pads, ground covering) must be procured from reliable sources, inspected and tested frequently, and replaced before failure/deterioration. Figures E-13 through E-34 display differing obstacle constructions in use today.

Obstacle Negotiations

Purpose of obstacle is to give Soldiers confidence in their mental and physical abilities while cultivating personal courage. This obstacle is not timed.

Execution of obstacle: Soldier mounts and climbs net on lowest end (13 ft) of obstacle. Soldier goes over or between logs at top of rope, net or pole. Soldier moves across log walkway, climbs ladder to the high end (33 ft), then climbs down the cargo net to the ground.

Safety: Instructors conduct inspection and provide orientation and demonstration on apparatus. At a minimum, all ropes and wood surfaces are inspected prior to use for rips, tears or worn/insecure surfaces. Distance between rungs on log ladder should not exceed 36 in. Safety padding sufficient to break a fall should be emplaced at bottom of high (33 ft) cargo net. Instructor should be positioned at the top of the wooden ladder to observe/assist Soldiers over top log at high point and onto cargo net; instructor is to be secured with safety belt or harness to horizontal log to prevent instructor from being pulled off by Soldier negotiating apparatus.

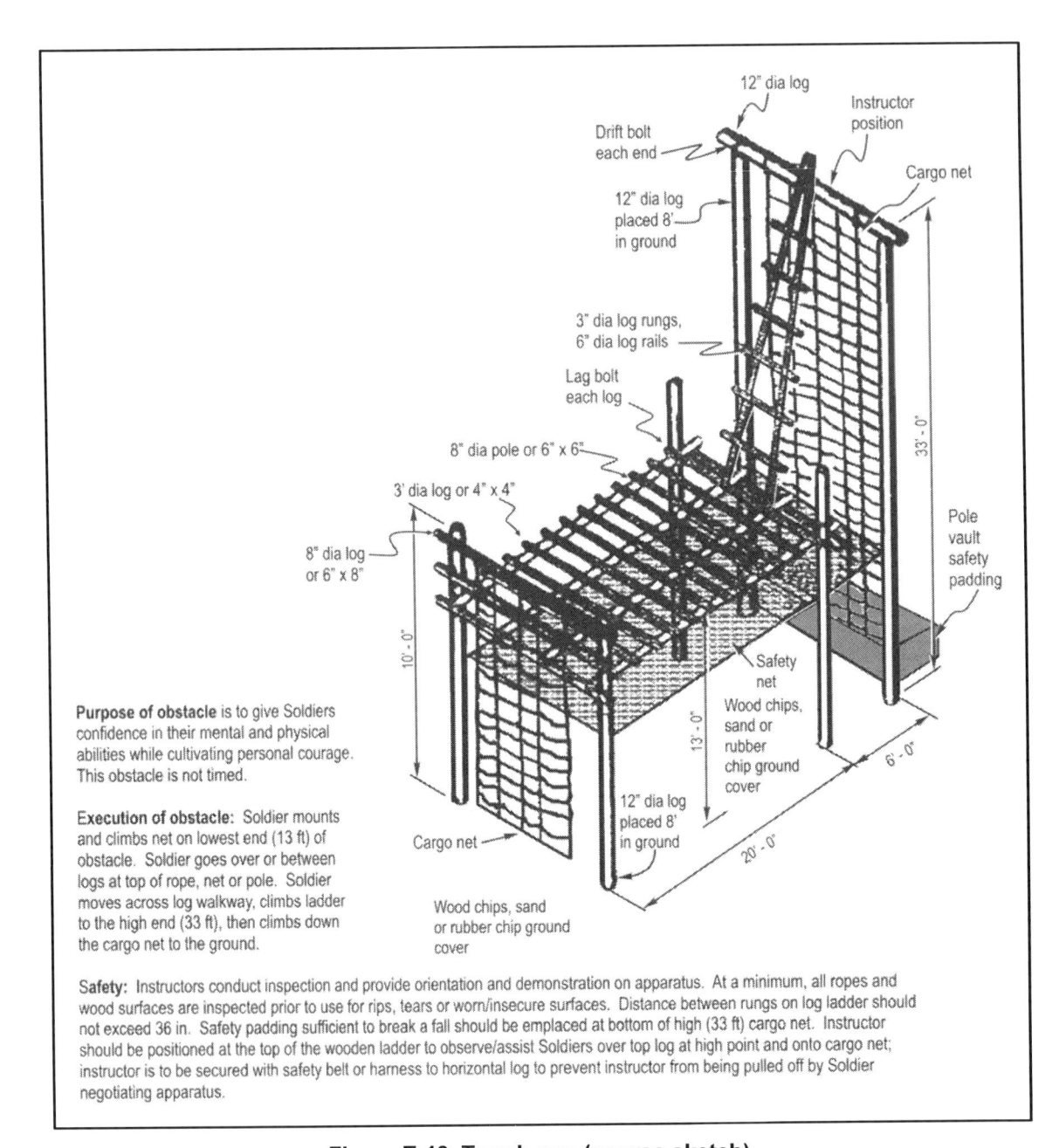

Figure E-13. Tough one (course sketch)

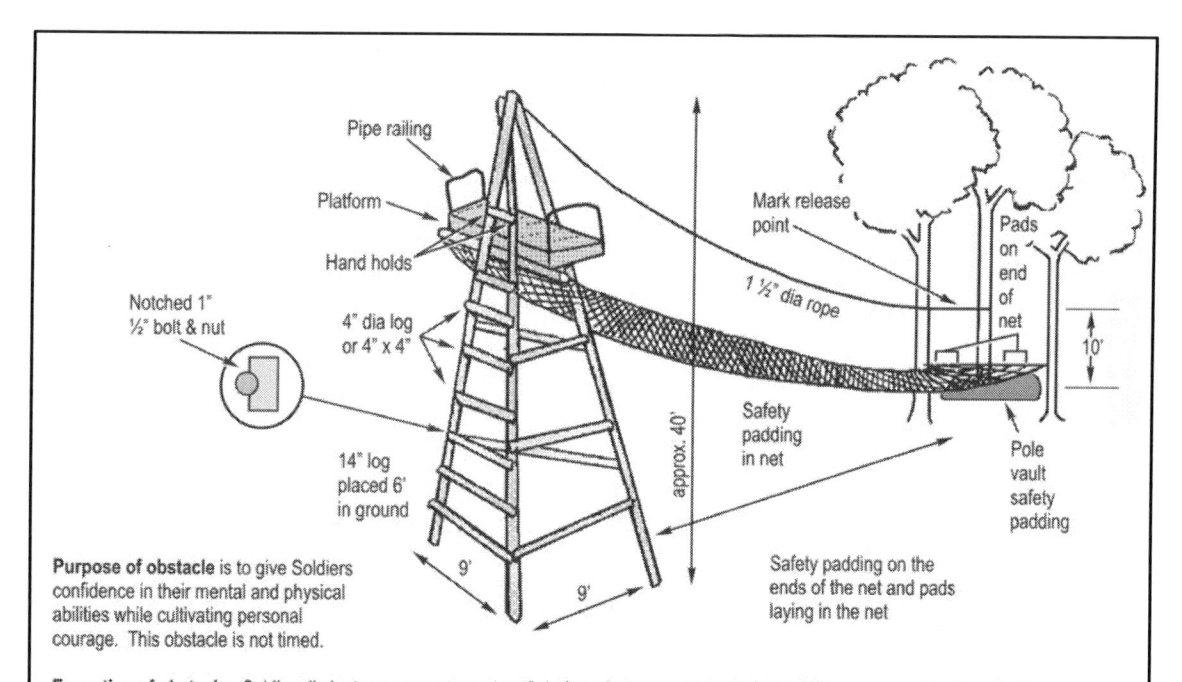

Appendix E

Purpose of obstacle is to give Soldiers confidence in their mental and physical abilities while cultivating personal courage. This obstacle is not timed.

Execution of obstacle: Soldier climbs tower, mounts center of platform (instructor available to assist), grasps rope firmly and swing legs upward. Soldier holds rope with legs to distribute weight between legs and arms. Braking the slide with feet and legs, Soldier proceeds down the rope. Soldiers must be warned that they could get rope burns on their hands if improperly executed. This obstacle can be dangerous when the rope is slippery. Soldiers leave the rope at a clearly marked point of release. Only one Soldier at a time is allowed on the rope.

This obstacle requires two instructors – one on the platform and the other on the ground.

Safety: Instructors conduct inspection and provide orientation and demonstration on apparatus. At a minimum, all ropes, nets and wood surfaces are inspected prior to use for rips, tears or worn/unsecured surfaces. Spacing between the rungs on the log ladder should not exceed 36 inches. Rope will be 1.5 inch diameter with no knots in the vicinity of the mounting point. A safety net is attached so that a Soldier falling from any portion of the rope will land in the net before striking any part of the tower. Padding placed in the net will reduce likelihood of hands/fingers being twisted in the net. Safety padding sufficient to break a fall should be emplaced at the drop off point. Instructor is positioned on the tower platform to assist Soldiers mounting the rope; instructor is to be secured to tower to prevent instructor from being pulled off by Soldier negotiating apparatus. Padding is emplaced at the bottom end of the net (nearest release point) to prevent Soldier from injury on tightened portion of net. This obstacle is dangerous when rope becomes wet/slippery and should not be used. Gloves should not be worn on this apparatus.

Figure E-14. Slide for life (course sketch)

Obstacle Negotiations

Purpose of obstacle is to give Soldiers confidence in their mental and physical abilities while cultivating personal courage. This obstacle is not timed.

Execution of obstacle: Soldier climbs vertical ladder. Soldier goes up to second rung from top, climbs over, and climbs down other side of ladder. Soldier does not climb over top rung. Only one Soldier at a time is allowed.

Safety: Instructors conduct inspection and provide orientation and demonstration on apparatus. At a minimum, all surfaces and cables are inspected prior to use for breaks, splinters, tears or worn/unsecured surfaces. Safety padding sufficient to break a fall is emplaced at each side on bottom of ladder/tower (inclined ladders are removed to prevent falling soldier from striking cross members). Instructor is positioned on the tower to assist Soldiers climbing to other side; instructor is to be secured to tower to prevent instructor from being pulled off by Soldier negotiating apparatus. This obstacle is dangerous when beams become slippery and should not be used. Gloves should not be worn on this apparatus.

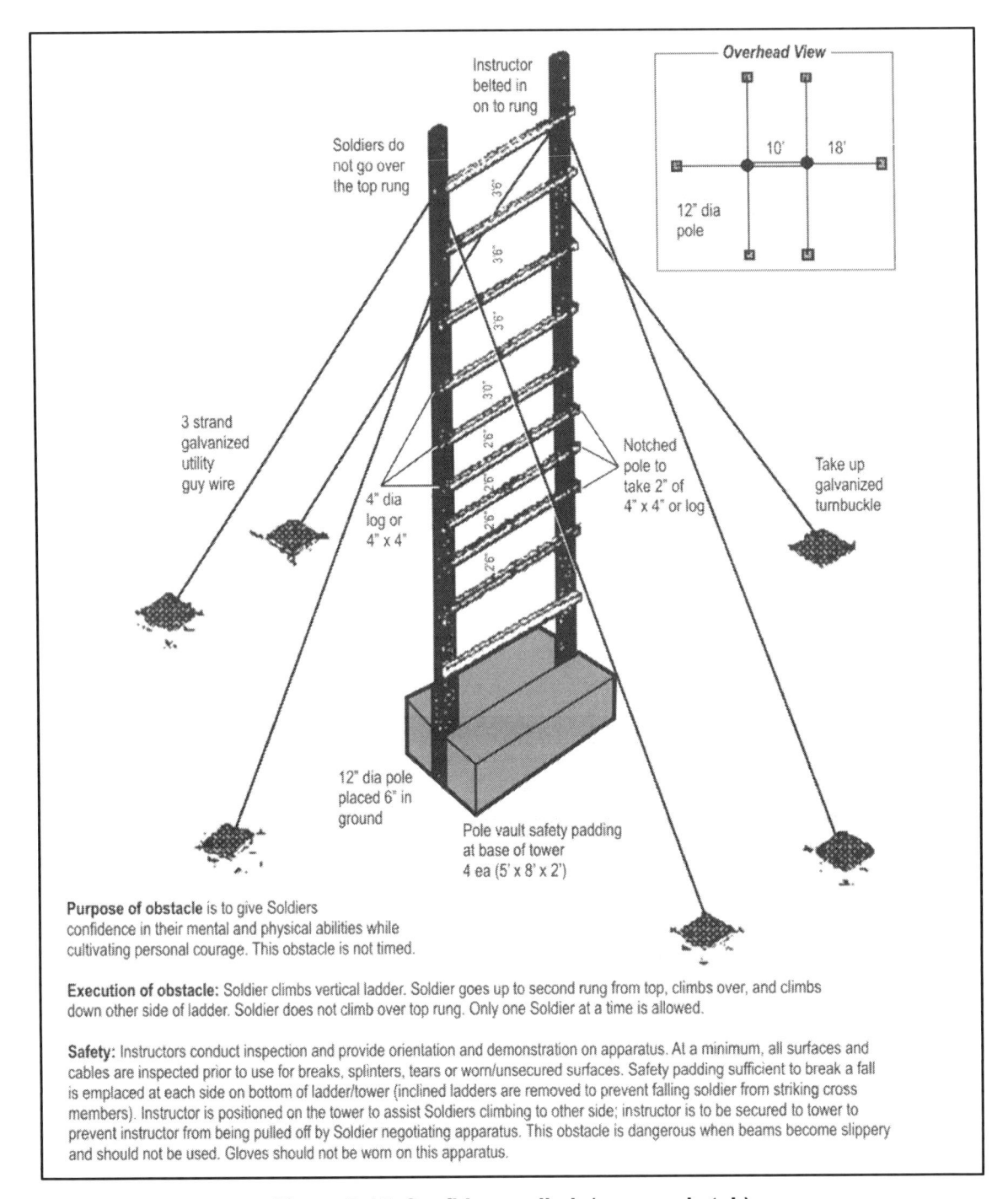

Figure E-15. Confidence climb (course sketch)

Appendix E

Purpose of obstacle is to give Soldiers confidence in their mental and physical abilities while cultivating personal courage and developing teamwork. This obstacle is not timed.

Execution of obstacle: Team of Soldiers (4+) jump or climb to first floor and either climb corner posts or help one another to higher floors. Subsequent climbing is done on side of tower over net (if available). They descend to the ground as a team as well. The top level roof is off limits/not used. One team at a time should be on the obstacle. Soldiers should never jump to the ground from above the first level.

Safety: Instructors conduct inspection and provide orientation and demonstration on apparatus. At a minimum, all surfaces and any supporting cables are inspected prior to use for breaks, splinters, tears or worn/unsecured surfaces. Safety padding sufficient to break a fall is emplaced on the ground under the climbing side(s) of the tower. This obstacle is dangerous when slippery and should not be used. Gloves should not worn on the apparatus.

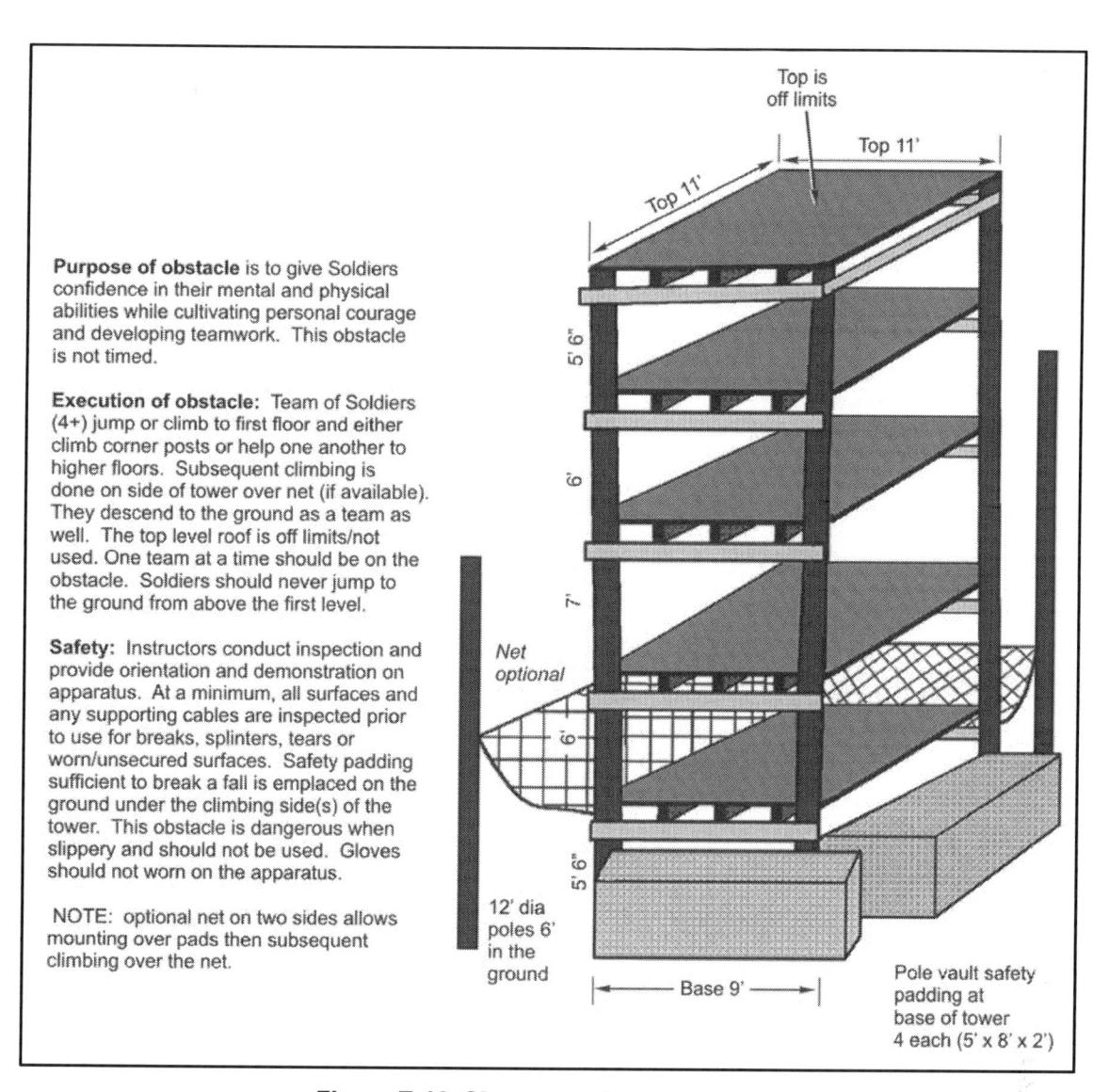

NOTE: optional net on two sides allows mounting over pads then subsequent climbing over the net.

Figure E-16. Skyscraper (course sketch)

Obstacle Negotiations

Purpose of obstacle is to give Soldiers confidence in physical abilities while cultivating toughness.

Execution of obstacle: Soldiers step on lower log and take prone, stomach down position on the horizontal logs. Soldiers crawl over logs to opposite end of obstacle, then dismount feet first.

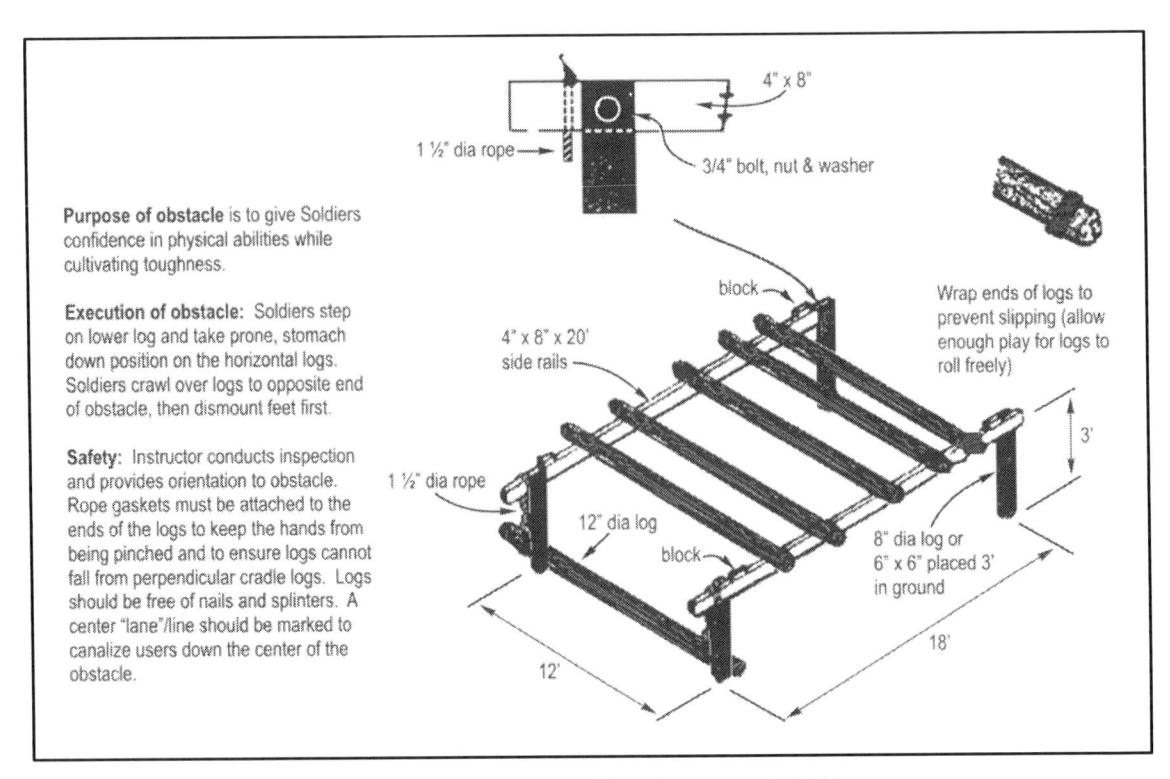

Safety: Instructor conducts inspection and provides orientation to obstacle. Rope gaskets must be attached to the ends of the logs to keep the hands from being pinched and to ensure logs cannot fall from perpendicular cradle logs. Logs should be free of nails and splinters. A center "lane"/line should be marked to canalize users down the center of the obstacle.

Figure E-17. Belly robber (course sketch)

Purpose of obstacle is to give Soldiers confidence in physical abilities to include balance and upper body strength.

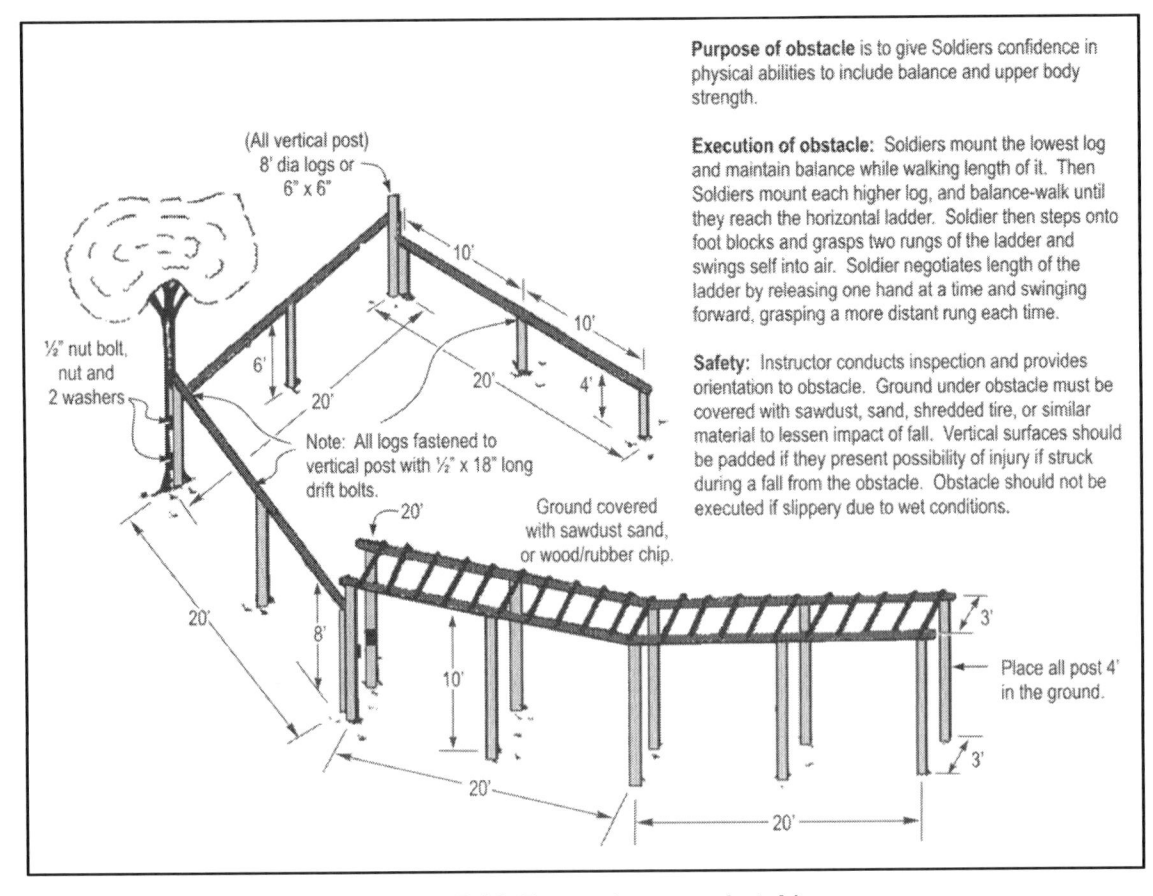

Execution of obstacle: Soldiers mount the lowest log and maintain balance while walking length of it. Then Soldiers mount each higher log, and balance-walk until they reach the horizontal ladder. Soldier then steps onto foot blocks and grasps two rungs of the ladder and swings self into air. Soldier negotiates length of the ladder by releasing one hand at a time and swinging forward, grasping a more distant rung each time.

Safety: Instructor conducts inspection and provides orientation to obstacle. Ground under obstacle must be covered with sawdust, sand, shredded tire, or similar material to lessen impact of fall. Vertical surfaces should be padded if they present possibility of injury if struck during a fall from the obstacle. Obstacle should not be executed if slippery due to wet conditions.

Figure E-18. Tarzan (course sketch)

Appendix E

Purpose of obstacle is to give Soldiers confidence in physical abilities to include balance and upper body strength.

Execution of obstacle: Soldiers mount the low log and jump onto high log. They grasp over the top of the log with both arms, keeping the belly area in contact with it. They swing their legs over the log, then lower themselves to the ground.

Safety: Instructor conducts inspection and obstacle must be covered with sawdust, sand, shredded tire, or similar material to lessen impact of fall. Vertical surfaces should be padded if they present possibility of injury if struck during a fall from the obstacle. Obstacle should not be executed when slippery due to wet conditions. Spotters should be used.

½" x 12" dia long drift bolts at each end

8" dia log or 6" x 6"

3'
7'
2' 8"
2' 6"
2'
3'
3'

Piece of fire hose or other traction material

Ground covered with sawdust, sand, or wood/rubber chips

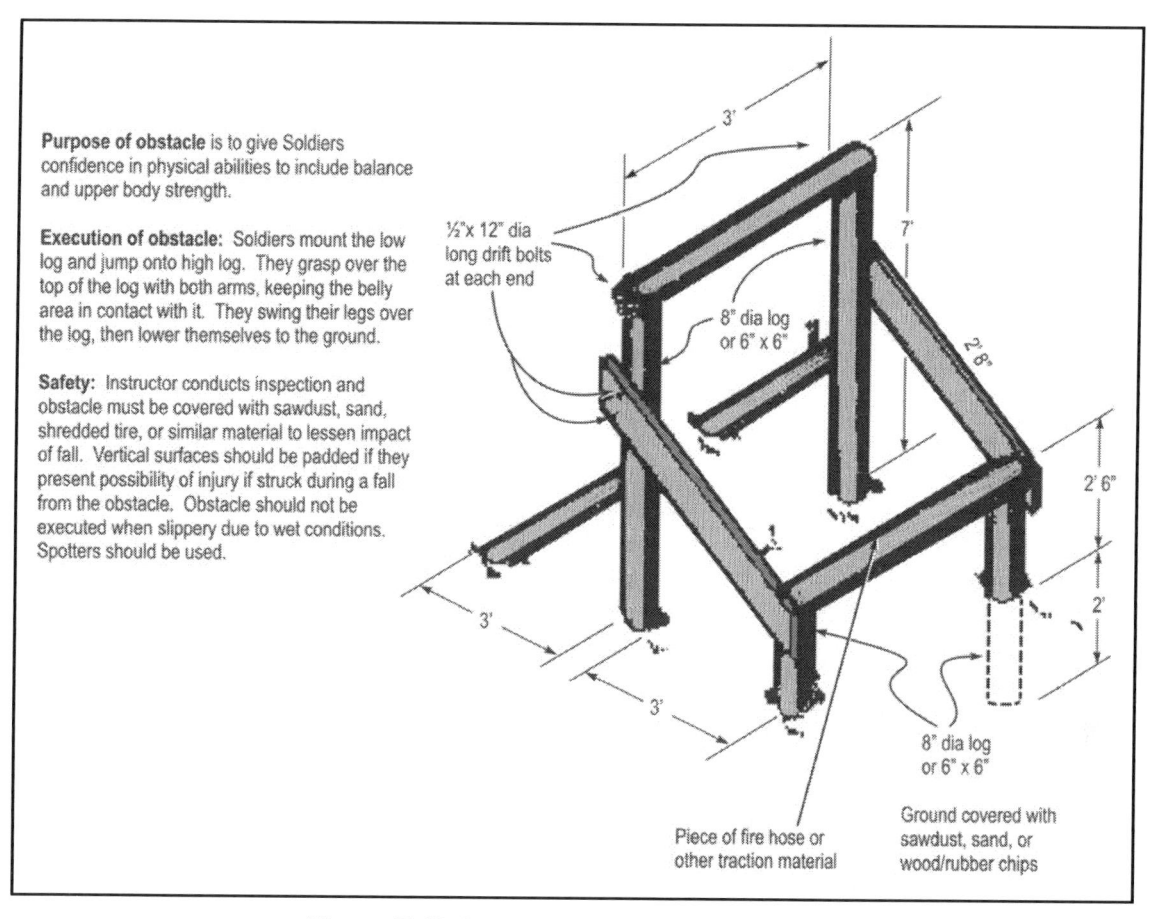

8" dia log or 6" x 6"

Figure E-19. Low belly over (course sketch)

Obstacle Negotiations

Purpose of obstacle is to give Soldiers confidence in physical abilities to include balance and upper body strength.

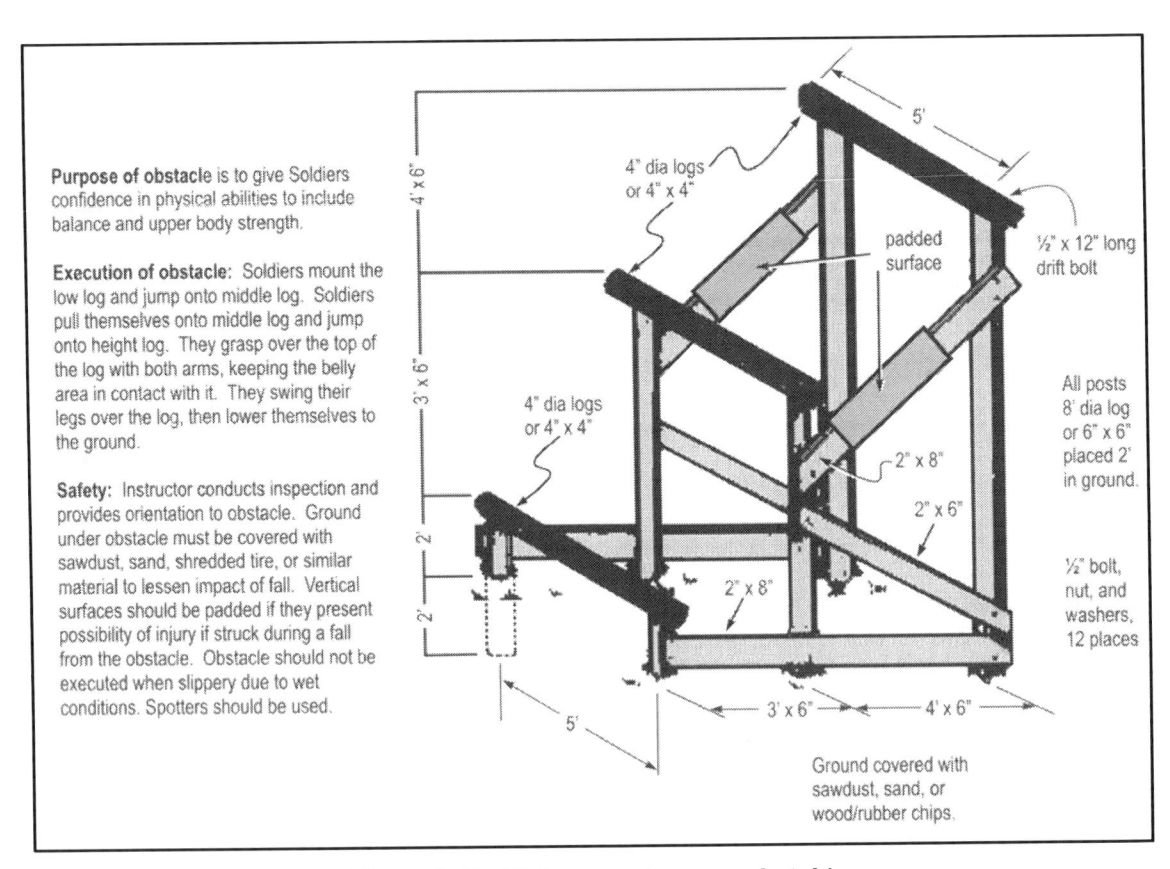

Execution of obstacle: Soldiers mount the low log and jump onto middle log. Soldiers pull themselves onto middle log and jump onto height log. They grasp over the top of the log with both arms, keeping the belly area in contact with it. They swing their legs over the log, then lower themselves to the ground.

Safety: Instructor conducts inspection and provides orientation to obstacle. Ground under obstacle must be covered with sawdust, sand, shredded tire, or similar material to lessen impact of fall. Vertical surfaces should be padded if they present possibility of injury if struck during a fall from the obstacle. Obstacle should not be executed when slippery due to wet conditions. Spotters should be used.

Figure E-20. Dirty name (course sketch)

Appendix E

Purpose of obstacle is to give Soldiers confidence in physical abilities.

Execution of obstacle: Soldiers step over each "X" in each lane.

Safety: Instructor conducts inspection and provides orientation to obstacle. Ensure obstacle does not have sharp edges or splinters.

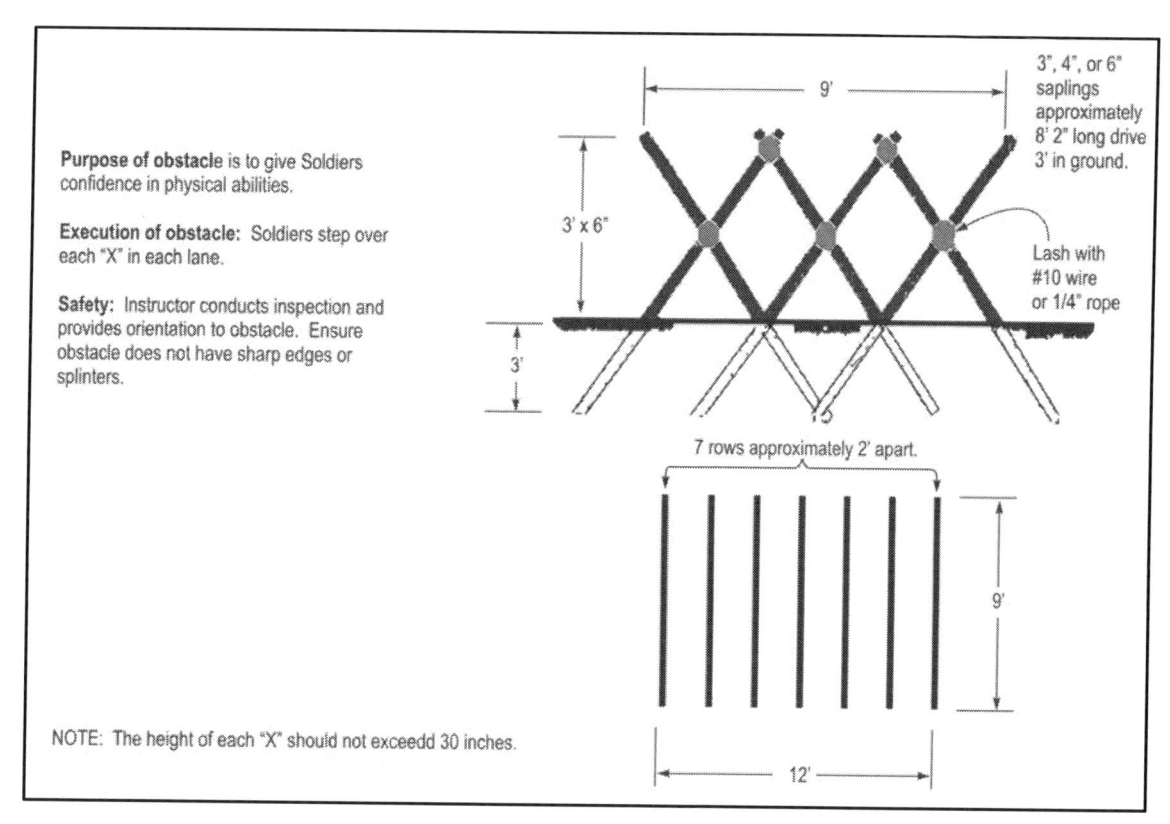

Dimensions shown: 9' wide, 3' x 6", 3' in ground. 3", 4", or 6" saplings approximately 8' 2" long drive 3' in ground. Lash with #10 wire or 1/4" rope.

7 rows approximately 2' apart. 9' by 12'.

NOTE: The height of each "X" should not exceedd 30 inches.

Figure E-21. Tough nut (course sketch)

Purpose of obstacle is to give Soldiers confidence in physical abilities.

Execution of obstacle: Soldiers move forward under wire, on their stomachs, to the end of the wire obstacle.

Safety: Instructor conducts inspection and provides orientation to obstacle. Wire should be 16" above ground. Crawling surface should be sand or sawdust, free of sharp objects. Direction of negotiating crawl may be reversed from time to time to maintain more level crawling surface.

Ground covered with sawdust or sand.

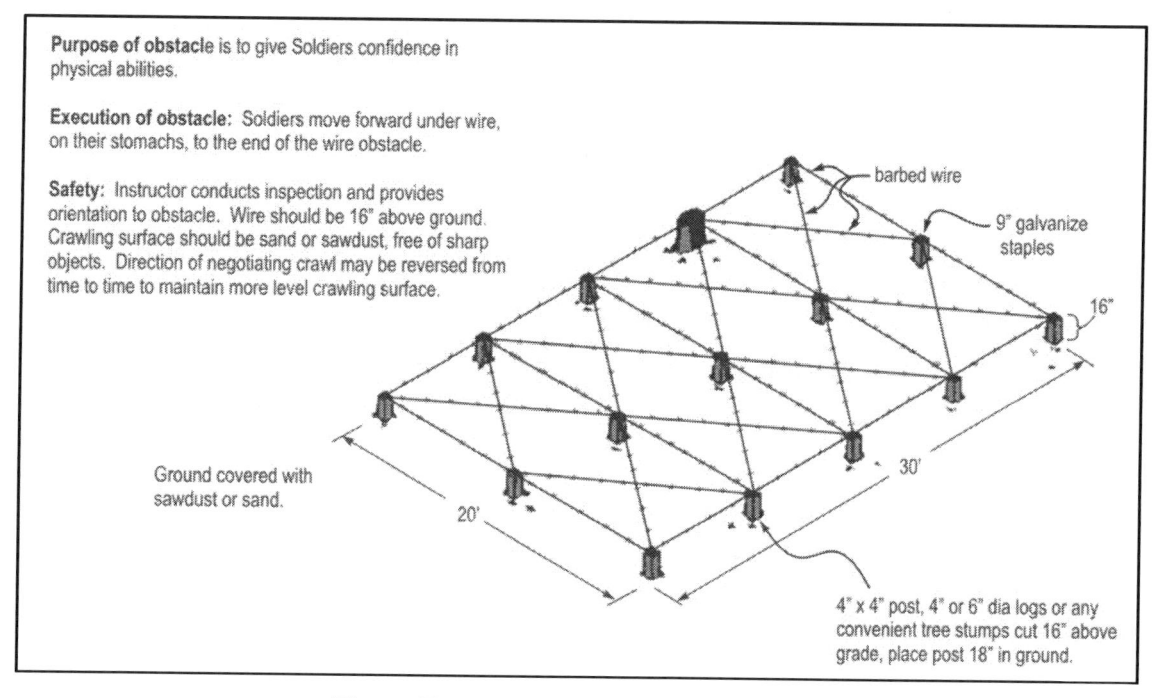

Dimensions: 20' x 30', wire 16" above ground. barbed wire, 9" galvanize staples. 4" x 4" post, 4" or 6" dia logs or any convenient tree stumps cut 16" above grade, place post 18" in ground.

Figure E-22. Belly crawl (course sketch)

Obstacle Negotiations

Purpose of obstacle is to give Soldiers confidence in physical abilities.

Execution of obstacle: Soldiers approach the underside of the wall, jump up and grasp the top and pull themselves over the top. Soldiers slide or jump down the incline to the ground.

Safety: Instructor conducts inspection and provides orientation to obstacle. Wire should be 16" above ground. Crawling surface should be sand or sawdust, free of sharp objects. Direction of negotiating crawl may be reversed from time to time to maintain more level crawling surface.

Remove bark on face of logs. Secure planks with 20d nails.

15'
2" x 6" ties, 3' apart
6' 6"
2" planking 16' long
8" dia logs or 6" x 6" placed 2' in ground.
5' 6"
Ground covered with sawdust, sand, or wood/rubber chips.

Figure E-23. Inclining wall (course sketch)

Purpose of obstacle is to give Soldiers confidence in physical abilities.

Execution of obstacle: Soldiers step over each bar; they either alternate legs or use the same leg each time while making an effort not to use their hands).

Safety: Instructor conducts inspection and provides orientation to obstacle. Wood surface must be free of nails and splinters. Soldiers must be spaced so as to prevent kicking each other.

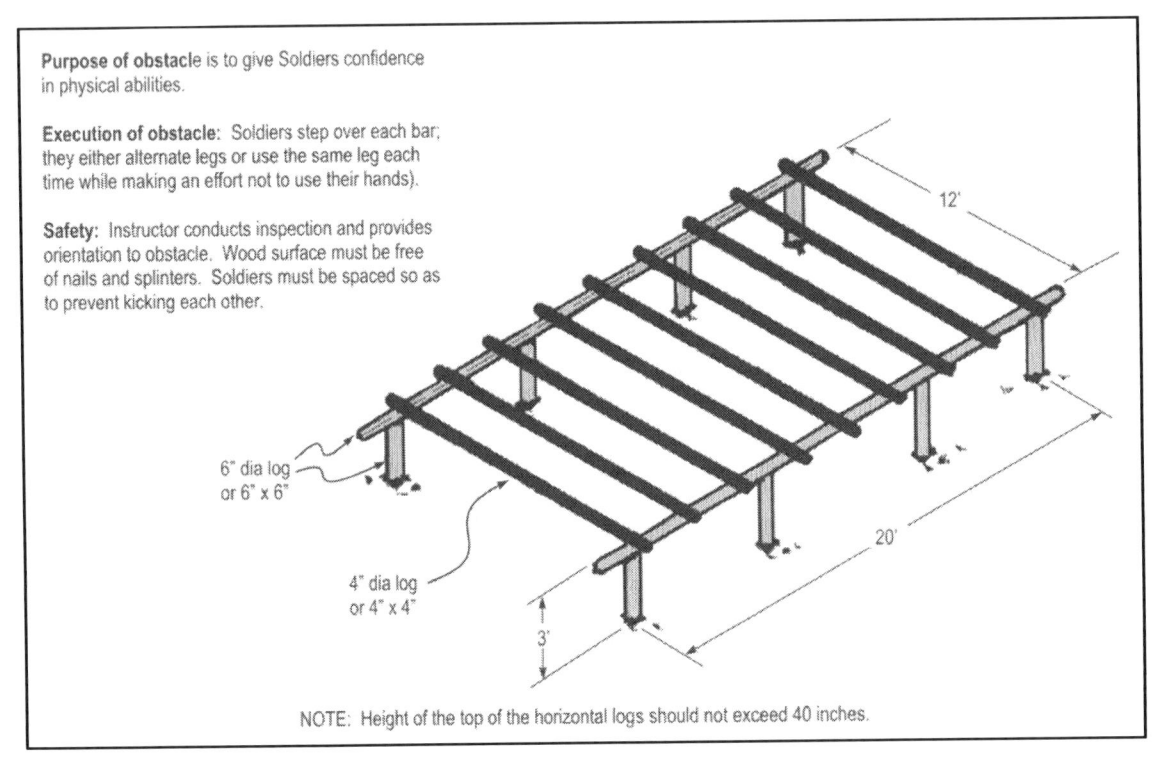

6" dia log or 6" x 6"
4" dia log or 4" x 4"
12'
20'
3'

NOTE: Height of the top of the horizontal logs should not exceed 40 inches.

Figure E-24. High step over (course sketch)

Appendix E

Purpose of obstacle is to give Soldiers confidence in physical abilities.

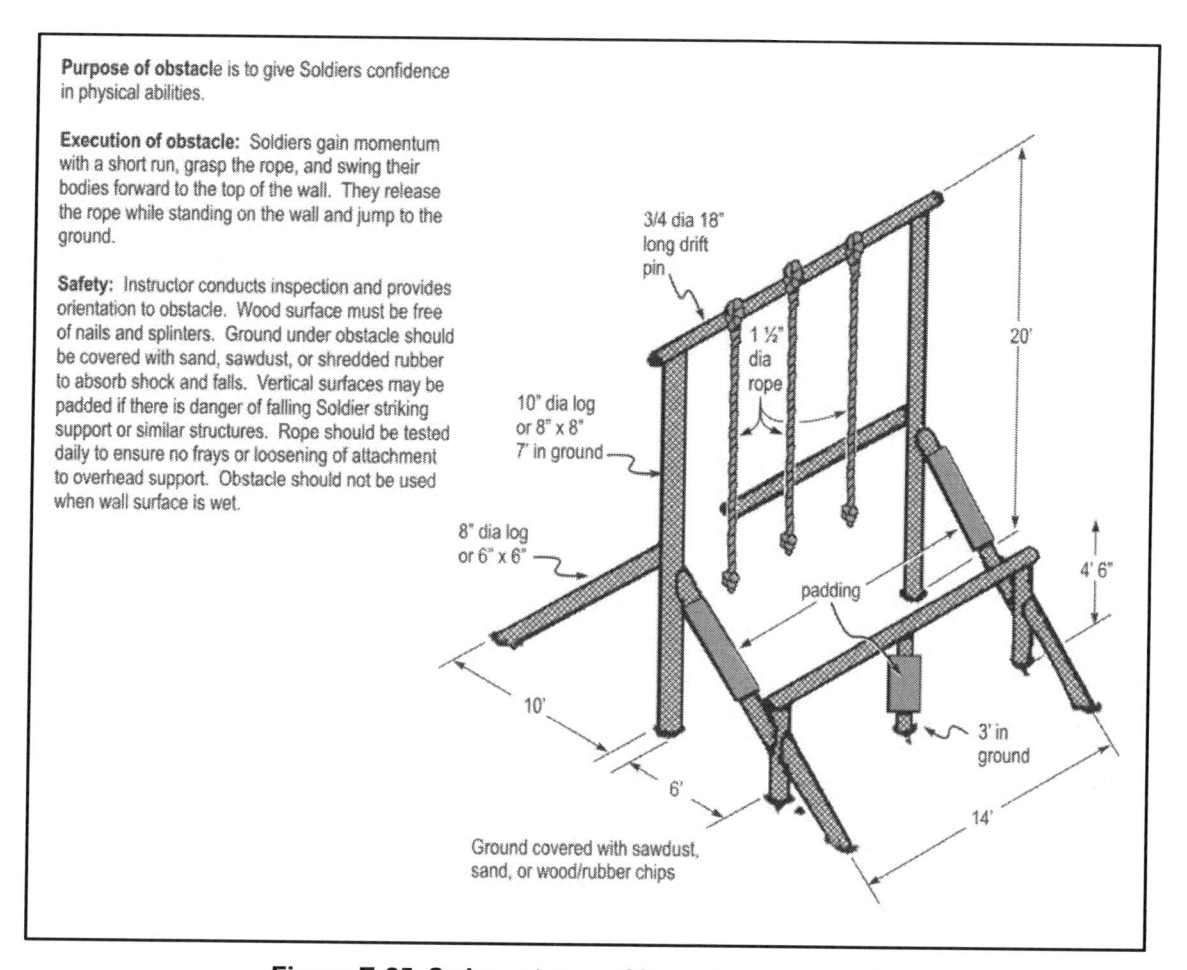

Execution of obstacle: Soldiers gain momentum with a short run, grasp the rope, and swing their bodies forward to the top of the wall. They release the rope while standing on the wall and jump to the ground.

Safety: Instructor conducts inspection and provides orientation to obstacle. Wood surface must be free of nails and splinters. Ground under obstacle should be covered with sand, sawdust, or shredded rubber to absorb shock and falls. Vertical surfaces may be padded if there is danger of falling Soldier striking support or similar structures. Rope should be tested daily to ensure no frays or loosening of attachment to overhead support. Obstacle should not be used when wall surface is wet.

Figure E-25. Swing, stop, and jump (course sketch)

Obstacle Negotiations

Purpose of obstacle is to give Soldiers confidence in physical abilities.

Execution of obstacle: Soldiers step over each bar; they either alternate legs or use the same leg each time while making an effort not to use their hands).

Safety: Instructor conducts inspection and provides orientation to obstacle. Wood surface must be free of nails and splinters. Soldiers must be spaced so as to prevent kicking each other.

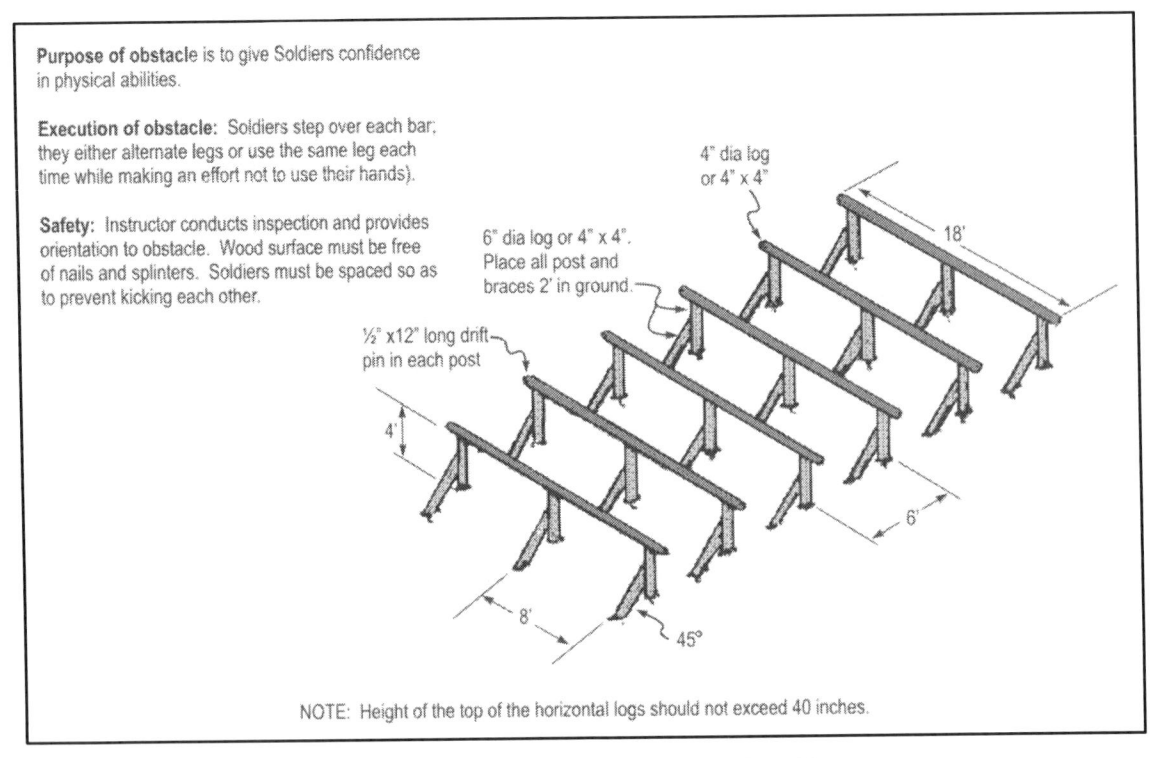

NOTE: Height of the top of the horizontal logs should not exceed 40 inches.

Figure E-26. Six vaults (course sketch)

Purpose of obstacle is to give Soldiers confidence in physical abilities.

Execution of obstacle: Soldiers walk up one inclined log and down the one on the other side to the ground. (No Running).

Safety: Instructor conducts inspection and provides orientation to obstacle. Wood surface must be free of nails and splinters. Ground should be covered with sand, sawdust, or shredded rubber. Notches can be cut into the logs to assist with traction.

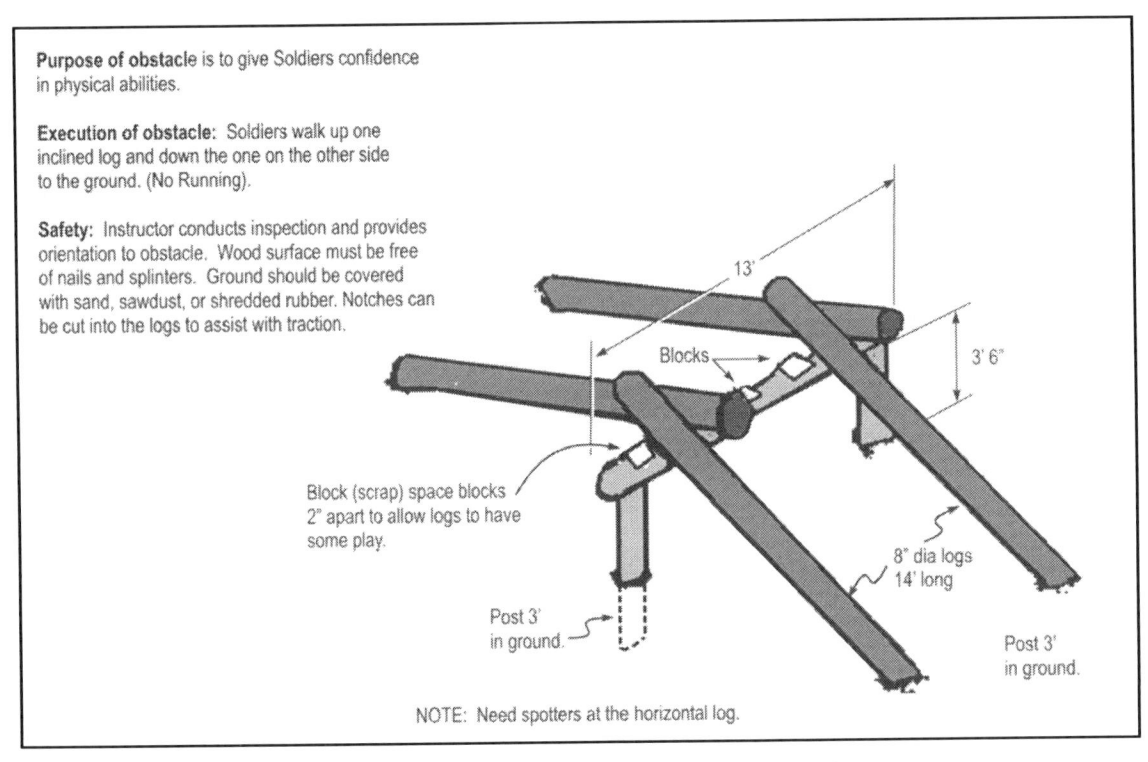

NOTE: Need spotters at the horizontal log.

Figure E-27. Easy balancer (course sketch)

Appendix E

Purpose of obstacle is to give Soldiers confidence in physical abilities.

Execution of obstacle: Soldiers vault, jump or climb over log.

Safety: Instructor conducts inspection and provides orientation to obstacle. Soldiers must be warned that log is not stationary. Soldiers must keep hands and fingers away from parts of log resting on cradle. Soldiers should not rock or roll log while others are negotiatiing it. Ground under obstacle should be covered with sand, sawdust or shredded rubber to lessen impact in event of fall.

6" x 6" or 6" dia log 8' 6" long, 2' 6" in ground.

12" dia log, 24' long

2" x 4" braces 2-2d nails each end.

2' 10"
4'
2'
6'
2' 6"
18'

Ground covered with sawdust, sand, or wood/rubber chips.

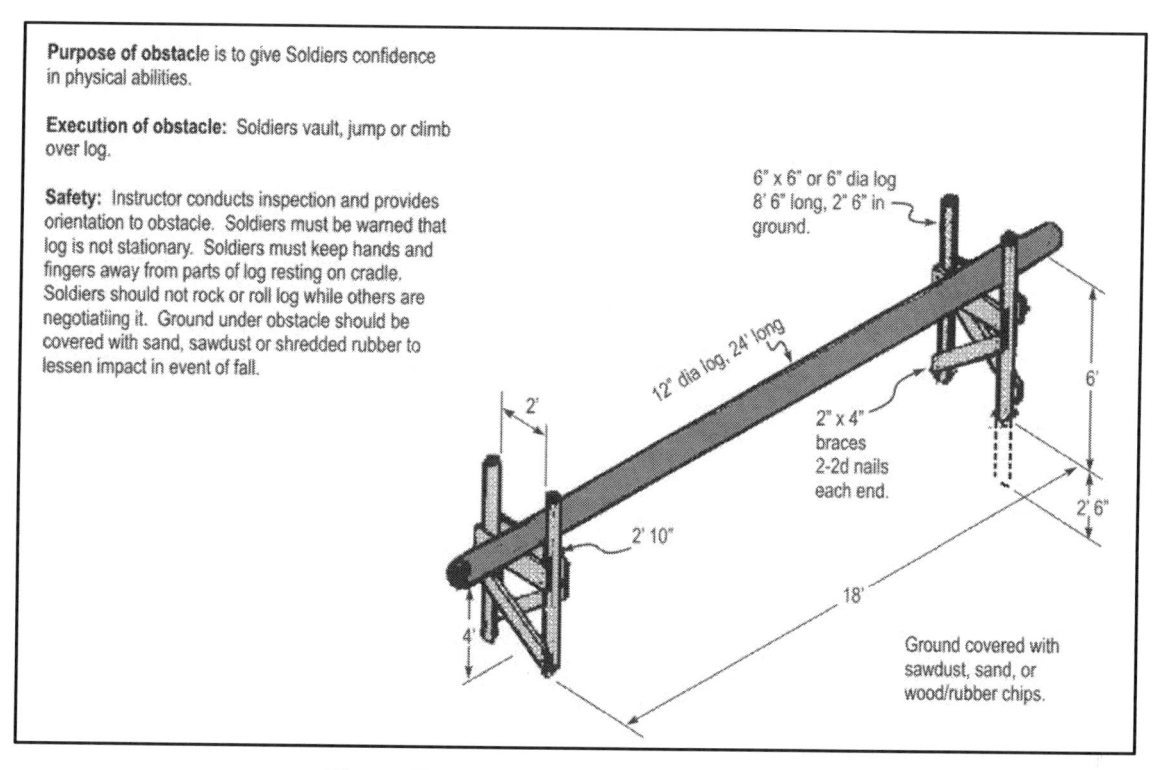

Figure E-28. Belly buster (course sketch)

Purpose of obstacle is to give Soldiers confidence in physical abilities.

Execution of obstacle: Soldiers move forward under wire, on their backs while raising wire with their hands to clear their bodies. Continuing to the end of the wire obstacle.

Safety: Instructor conducts inspection and provides orientation to obstacle. Wire should lay loosely on the ground. Crawling surface should be sand or sawdust, free of sharp objects. Direction of negotiating crawl may be reversed from time to time to maintain more level crawling surface.

Use tree or stump if handy.

Wrap barbed wire around post, leave lying loose on ground.

Barbed wire

4" dia log or 4" x 4" 3' long 12" in ground.

Ground covered with sawdust, sand.

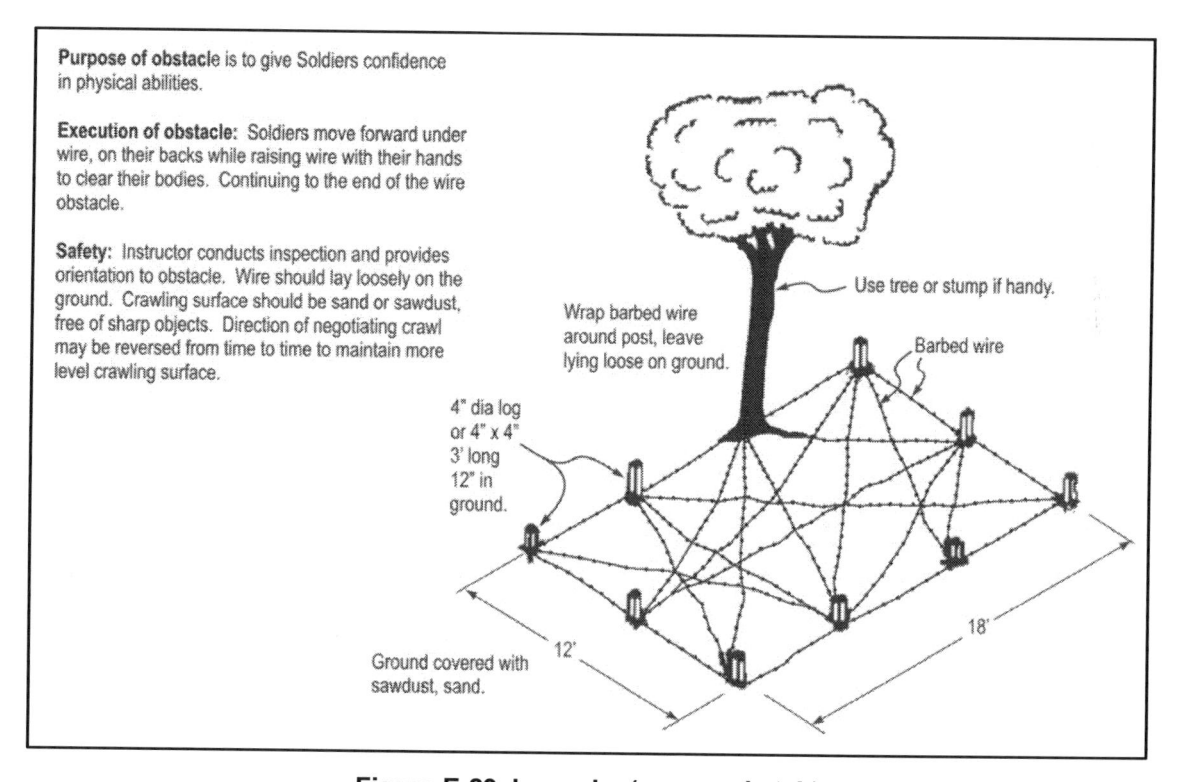

12'
18'

Figure E-29. Low wire (course sketch)

Obstacle Negotiations

Purpose of obstacle is to give Soldiers confidence in physical abilities.

Execution of obstacle: Soldiers step over bar; they either alternate legs or use the same leg each time while making an effort not to use their hands. (Shorter Soldiers may be required to use hands).

Safety: Instructor conducts inspection and provides orientation to obstacle. Ensure obstacle does not have sharp edges or splinters

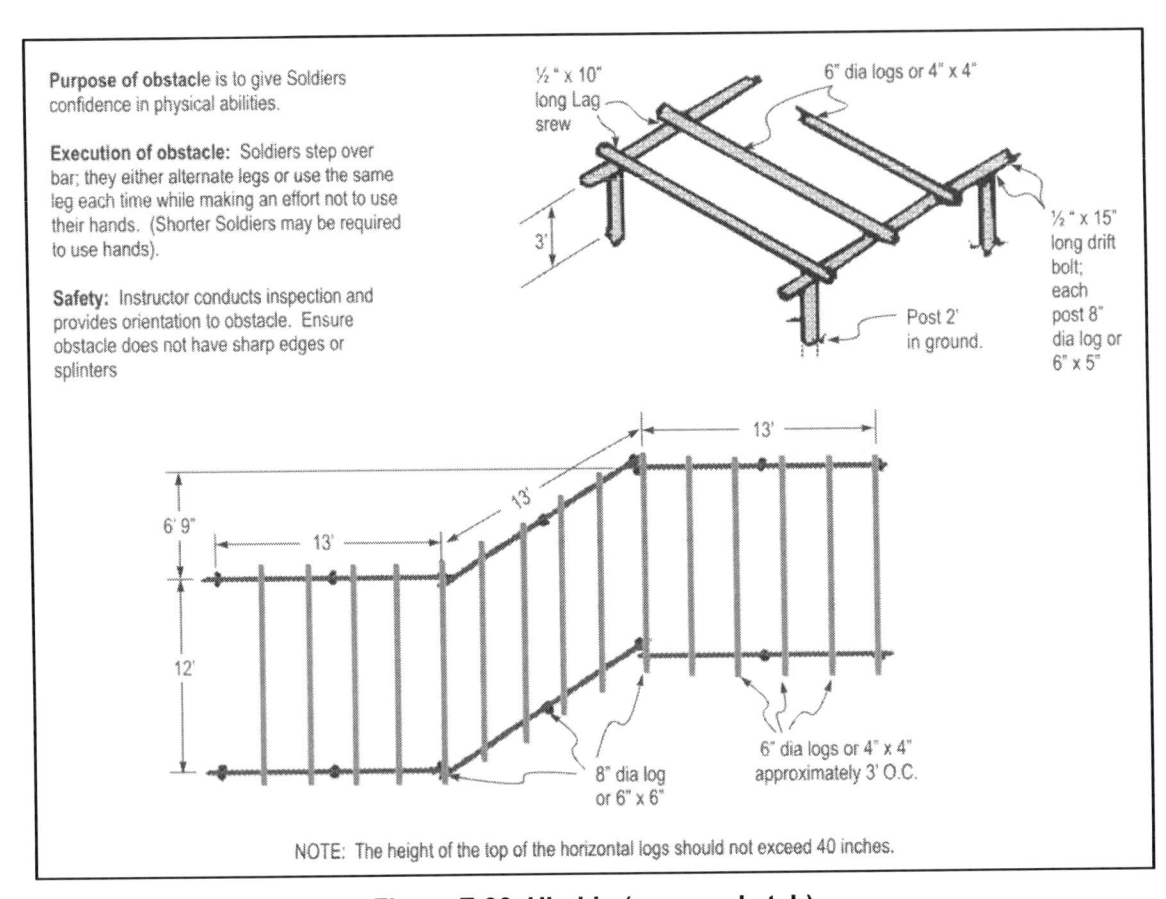

- ½" x 10" long Lag srew
- 6" dia logs or 4" x 4"
- 3'
- Post 2' in ground.
- ½" x 15" long drift bolt; each post 8" dia log or 6" x 5"

- 13'
- 13'
- 6' 9"
- 13'
- 12'
- 8" dia log or 6" x 6"
- 6" dia logs or 4" x 4" approximately 3' O.C.

NOTE: The height of the top of the horizontal logs should not exceed 40 inches.

Figure E-30. Hip-hip (course sketch)

Appendix E

Purpose of obstacle is to give Soldiers confidence in physical abilities.

Execution of obstacle: Soldiers approach the underside of climbing ladder and go down other side to the ground.

Safety: Instructor conducts inspection and provides orientation to obstacle. Ground under near side of obstacle must be covered with sawdust, sand, shredded tire or similar material to lessen impact of fall. Wood surface must be free of nails and splinters. Support braces will be padded. Spotters will be used between the support post.

4" dia logs or 4" x 4" spaced approximately 20" apart

10" dia logs or 6" x 6"

Place post 2' in ground.

Padding

45°

60°

13'

13' 6"

9'

Ground covered with sawdust, sand, or wood/rubber chips.

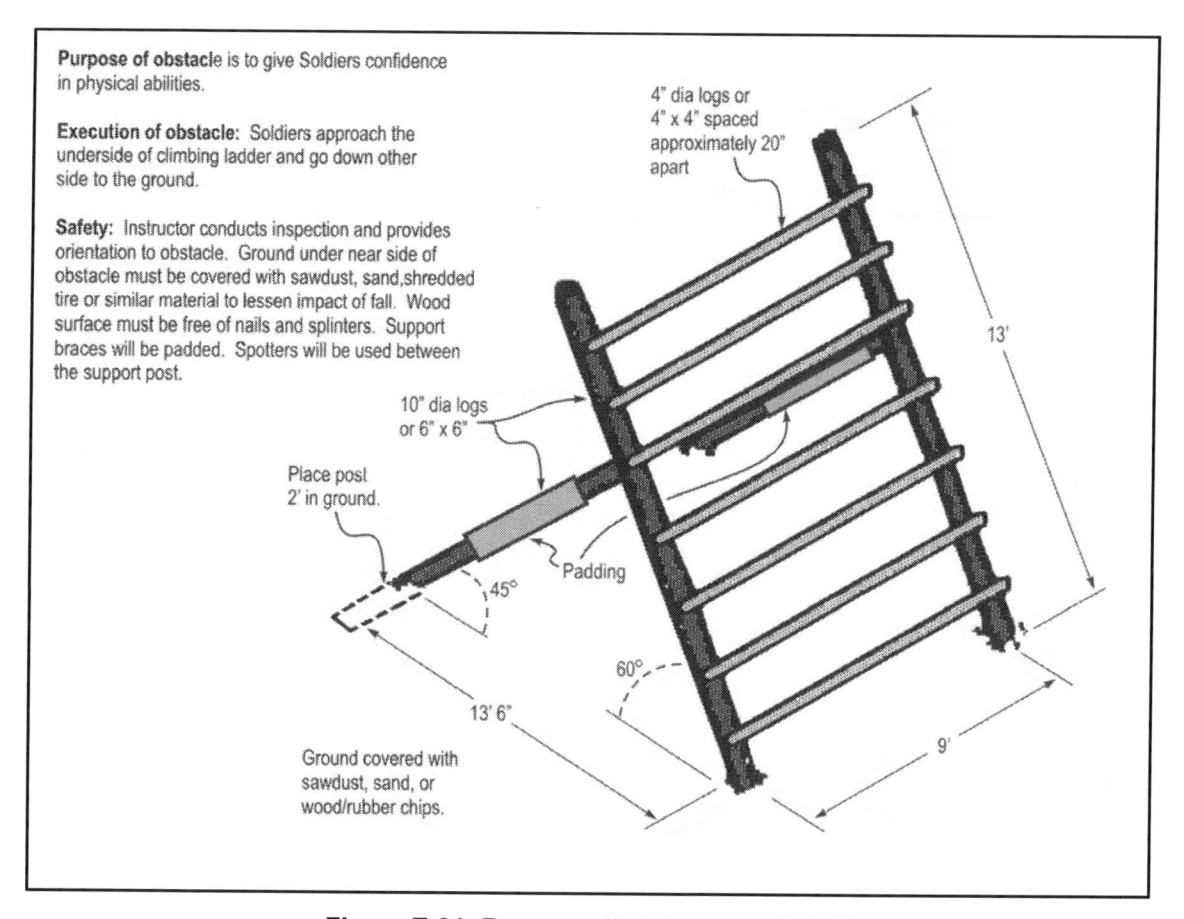

Figure E-31. Reverse climb (course sketch)

E-34   FM 7-22   26 October 2012

Obstacle Negotiations

Purpose of obstacle is to give Soldiers confidence in physical abilities.

Execution of obstacle: Soldiers move from one end of the obstacle to the other by weaving their bodies under one bar and over the next.

Safety: Instructor conducts inspection and provides orientation to obstacle. Ground under obstacle must be covered with sawdust, sand, shredded tire or similar material to lessen impact of fall. Wood surface must be free of nails and splinters. Spotters should be used in center. Safety pads will be used under the apex.

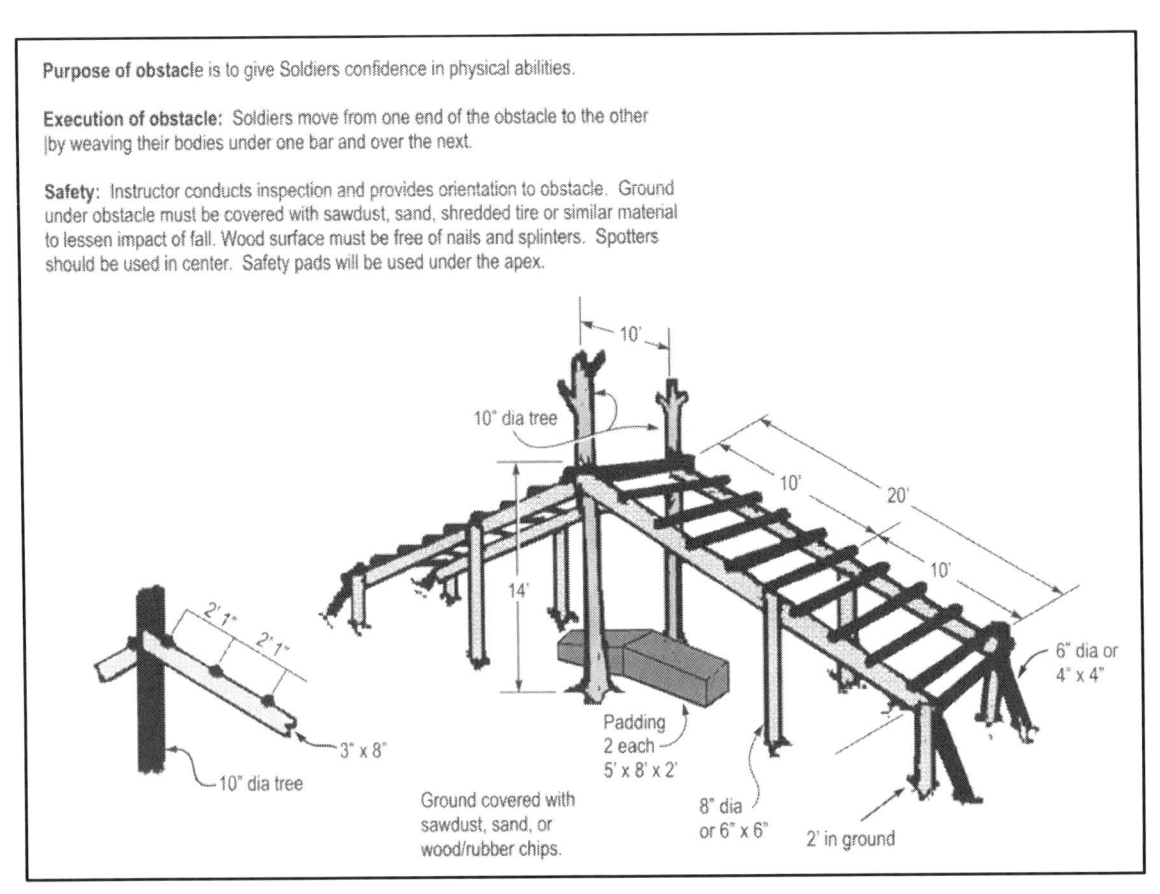

Figure E-32. Weaver (course sketch)

Purpose of obstacle is to give Soldiers confidence in physical abilities.

Execution of obstacle: Soldiers move from one end of the obstacle to the other by weaving their bodies under one bar and over the next.

Safety: Instructor conducts inspection and provides orientation to obstacle. Ground under obstacle must be covered with sawdust, sand, shredded tire or similar material to lessen impact of fall. Wood surface must be free of nails and splinters. Spotters should be used in center. Safety pads will be used under the apex.

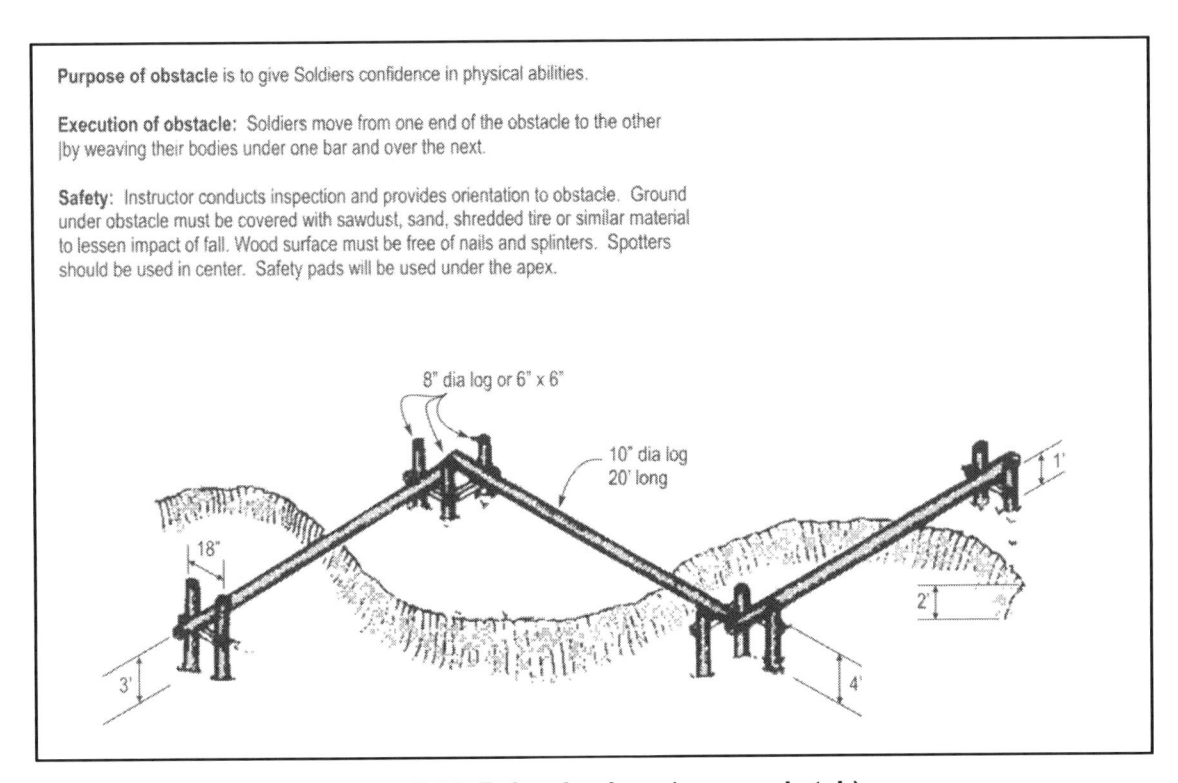

Figure E-33. Balancing logs (course sketch)

Appendix E

Purpose of obstacle is to give Soldiers confidence in physical abilities.

Execution of obstacle: Soldiers jump from one log to another until obstacle is negotiated from near to far side.

Safety: Instructor conducts inspection and provides orientation to obstacle. Wood surface must be free of sharp edges and should not be slippery) it may be necessary to rough up tops of logs/stumps to ensure traction or use 1-inch nails driven into the tops).

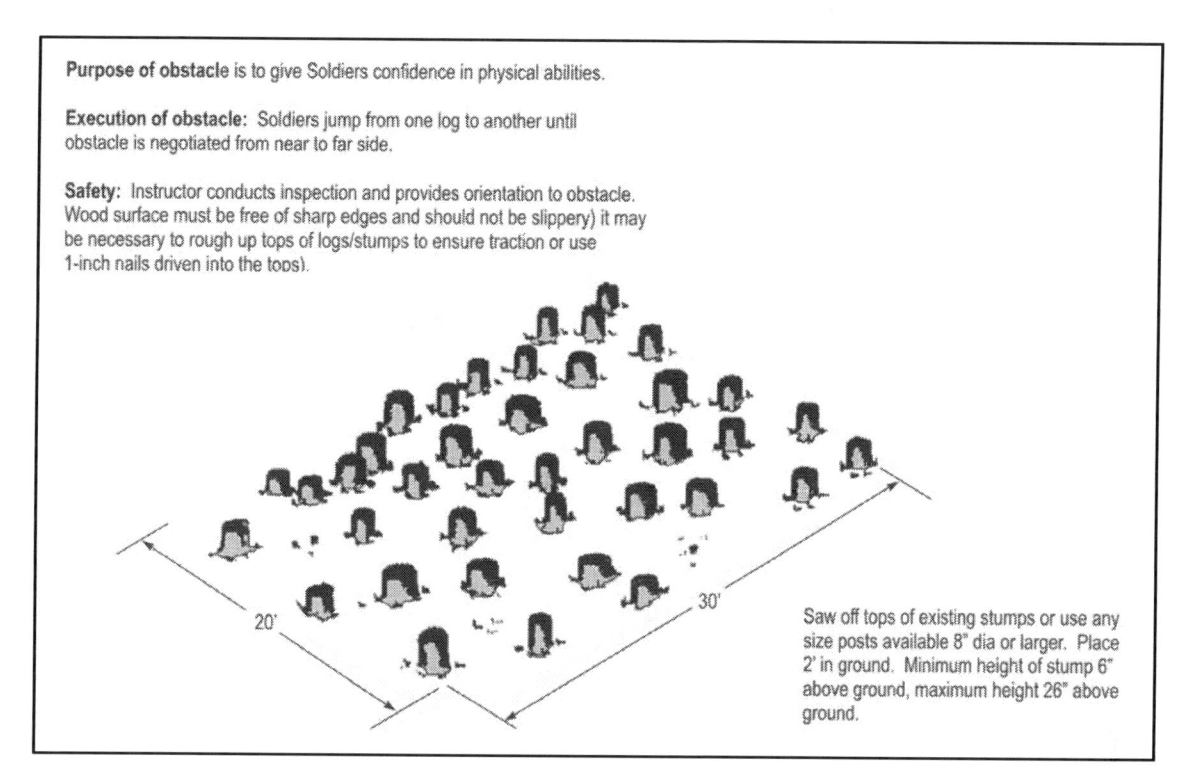

20' 30'

Saw off tops of existing stumps or use any size posts available 8" dia or larger. Place 2' in ground. Minimum height of stump 6" above ground, maximum height 26" above ground.

Figure E-34. Island hopper (course sketch)

Glossary

Acronym/Term	Definition
ACH	advanced combat helmet
ACU	Army combat uniform
AGR	ability group run
AI	assistant instructor
AIT	Advanced Individual Training
AKO	Army Knowledge Online
APFT	Army Physical Fitness Test
ARFORGEN	Army Force Generation
ARPL	Assistant Reconditioning Program Leader
AWCP	Army Weight Control Program
BCT	Basic Combat Training
BOLC A	Basic Officer Leader Course A
BOLC B	Basic Officer Leader Course B
B/G	black and gold
CB	combatives
CD	conditioning drill
CDOC	conditioning obstacle course
CFOC	confidence obstacle course
CL	climbing drill
C-METL	core mission essential task list
CRM	composite risk management
D-METL	directed mission essential task list
ETM	endurance training machines
4C	4 for the core
FM	field manual
FTX	field training exercise
GD	guerrilla drill
HR	hill repeats
HSD	hip stability drill
IDT	individual duty for training
IMT	initial military training
IPFU	individual physical fitness uniform
IOTV	improved outer tactical vest
METL	mission essential task list
MMD	military movement drill
MMRB	military medical review board
MOS	military occupational specialty
NCO	noncommissioned officer
NCOIC	noncommissioned officer in charge
OIC	officers in charge
OPTEMPO	operating tempo
OSUT	one station unit training
PD	preparation drill
POI	program of instruction
PPPT	pregnancy postpartum physical training
PRT	physical readiness training
PSD	push-up and sit-up drill
PTRP	physical training and rehabilitation program
RC	reserve component
RD	recovery drill

Glossary

RESET	returning from extended deployment
RPL	reconditioning program leaders
RR	release run
R/W/B	red, white, and blue
SR	shuttle run
SSD	shoulder stability drill
STC	strength training circuit
STM	strength training machine
USAPHC	United States Army Public Health Command
USAR	United States Army Reserve
WTBD	warrior tasks and battle drills

References

SOURCES USED

These are the sources quoted or paraphrased in this publication.

AR 40-501, *Standards of Medical Fitness*, 14 December 2007.
AR 350-1, *Army Training and Leader Development*, 18 December 2009.
AR 600-9, *The Army Weight Control Program*, 27 November 2006.
AR 670-1, *Wear and Appearance of Army Uniforms and Insignia*, 3 February 2005.
TC 3-21.5, *Drill and Ceremonies*, 20 January 2012.
TC 3-25.150, *Combatives*, 24 September 2012.
ADP 7-0, *Training Units and Developing Leaders*, 23 August 2012.
FM 21-18, *Foot Marches*, 1 June 1990.
Army Engineer Drawings 28-13-95, *Confidence Course Layout Plan*, and
Folio Number 1, "Training Facilities" Corps of Engineers Drawing Number 28-13-95.
   U.S. Army Corps of Engineers, 441 G Street NW, Washington, D.C. 20314-1000.
USAPHC Guide Series 255 A-E, *U.S. Army Pregnancy Postpartum Physical Training Program*.
   http://phc.amedd.army.mil/topics/healthyliving/pft/Pages/ArmyPregnancyPostpartumPhysicalTraining
      Program.aspx

Websites:
U.S. Army Research Institute of Environmental Medicine,
   http://www.usariem.army.mil/
   http://www.usariem.army.mil/Pages/download/altitudeacclimatizationguide.pdf
   http://www.usariem.army.mil/Pages/download/heatacclimatizationguide.pdf
United States Army Public Health Command
   http://phc.amedd.army.mil

DOCUMENTS NEEDED

These documents must be available to the intended users of this publication. *DA Forms are available on the APD website (www.apd.army.mil).*
DD Forms are available on the OSD website (www.dtic.mil/whs/directives/infomgt/forms/formsprogram.htm).

DA Form 705, *Army Physical Fitness Test Scorecard*.
DA Form 2028, *Recommended Changes to Publications and Blank Forms*.
DA Form 3349, *Physical Profile*.
DD Form 689, *Sick Slip, Individual*.

Index

4C, 5-5, 5-6, 5-14, 5-23, 5-24, 5-25, 5-26, 6-11, 6-16

AGR, 5-7, 5-8, 5-11, 5-14, 5-15, 5-21, 5-22, 5-23, 5-24, 5-25, 5-26, 5-27, 5-28, 5-29, 5-30, 10-3, 10-5, 10-18, 10-19

AR 350-1, 1-1, 1-2, 1-7, 2-2, 3-1, 4-5, 5-16, 5-18, 5-20, 5-35, 5-36, 6-1, 6-2, 6-8, A-1

ARFORGEN, 1-8, 2-1, 2-2, 2-7, 4-3, 4-4, 4-6, 4-7, 4-9, 5-18, 5-19, 5-35
    model, 1-8

Army Physical Fitness Training Program, 1-1, 1-2, 1-7, 1-9, 2-1, 5-1

BOLC A, 2-2, 4-1, 4-2, 5-7

BOLC B, 4-2, 5-14, 5-18

clothing, A-1, A-2

C-METL, 1-2, 1-3, 1-9, 2-1, 2-2, 2-3, 3-2, 4-3, 4-4, 4-8, 4-9, 5-19

commands, 3-3, 6-11, 7-1, 7-2, 7-3, 7-4, 7-5, 7-6, 7-11, 7-12, 7-13, 7-14, 8-1, 8-2, 8-15, 8-16, 9-3, 9-4, 9-11, 9-12, 9-37, 9-40, 9-69, 9-70, 10-4, 10-5, 10-15, C-1
    execution, 1-2, 1-4, 1-8, 2-2, 2-3, 3-3, 4-3, 5-1, 5-7, 5-36, 6-4, 6-8, 6-11, 6-71, 6-72, 6-76, 6-78, 6-80, 6-82, 6-84, 6-89, 6-91, 6-95, 6-97, 6-101, 6-103, 6-105, 6-110, 6-112, 6-114, 7-1, 7-15, 9-1, 9-3, 9-11, 9-40, 9-70, 9-74, C-7, E-2
    preparatory, 3-3, 7-1

company formation, 7-4

counts, 6-17, 6-29, 6-30, 6-70, 6-71, 6-72, 6-73, 6-75, 6-81, 6-86, 6-88, 6-99, 6-109, 7-11, 7-12, 7-13, 8-4, 8-5, 9-4, 9-11, 9-19, 9-37, A-11

decompression, C-6

D-METL, 1-3, 1-9, 2-1, 2-2, 2-3, 2-7, 3-2, 4-3, 4-4, 4-7, 4-8, 4-9, 5-19, 5-20, 5-35

drills, 1-2, 1-3, 1-4, 1-8, 1-9, 2-1, 2-2, 2-3, 2-4, 2-5, 3-1, 3-2, 3-3, 4-2, 4-4, 4-6, 4-7, 5-1, 5-2, 5-5, 5-6, 5-14, 5-15, 5-16, 5-17, 5-18, 5-19, 5-35, 5-36, 6-4, 6-6, 6-11, 6-31, 6-69, 6-1, 7-5, 7-13, 7-14, 9-1, 9-3, 9-4, 9-11, 9-37, 9-38, 9-39, 9-40, 9-41, 9-69, 9-74, 10-2, 10-5, 10-11, B-1, C-2, C-12
    climbing, 5-7, 5-8, 5-17, 5-21, 6-4, 6-70, 6-71, 6-72, 6-73, 6-74, 7-13, 9-2, 9-38, 9-40, 9-41, 9-48, 9-56, B-2, B-3, C-12, E-3, E-7, E-9
    conditioning, 2-3, 4-4, 4-6, 4-7, 5-6, 5-8, 5-14, 5-17, 5-21, 5-35, 6-70, 6-71, 6-72, 6-73, 6-74, 7-13, 9-2, 9-3, 9-4, 9-11, C-13, E-2, E-5, E-12
    endurance and mobility, 1-4, 1-7, 2-1, 2-3, 2-7, 3-1, 5-2, 5-4, 5-5, 5-6, 5-8, 5-15, 5-17, 5-19, 5-21, 6-2, 6-4, 6-7, 6-10, 6-16, 6-74, 6-83, 8-2, 9-2, 9-3, 9-4, 9-11, 9-17, 9-21, 9-37, 9-38, 9-58, 9-69, 10-2, 10-6, 10-11, 10-15, 10-16, 10-18, 10-19, 10-20, 10-22, D-1, E-12
    guerrilla, 2-3, 2-4, 2-5, 4-4, 5-15, 5-17, 5-19, 5-21, 7-5, 7-13, 8-1, 8-15, 9-2, 9-38, 9-39, 9-41, 9-70
    hip stability drill, 5-5, 6-11, 10-6, 10-11
    military movement, 5-8, 5-17, 5-21, 6-70, 6-71, 6-72, 6-73, 6-74, 7-13
    preparation, 1-2, 1-3, 4-3, 4-4, 5-6, 5-15, 5-36, 6-11, 6-16, 6-74, 7-11, 7-13, 8-1, 8-2, 8-3, 8-23, 9-54, 10-5, 10-6, 10-11, A-2, A-3, A-4, C-3
    push-up, 5-5, 5-7, 5-8, 5-17, 5-21, 6-70, 6-74, 6-94, 8-14, 8-15, 9-2
    reconditioning, 2-1, 2-3, 2-7, 4-4, 6-1, 6-3, 6-4, 6-5, 6-6, 6-7, 6-8, 6-9, 6-69, 6-70
    sit-up, 5-8, 5-17, 5-21, 6-70, 9-2, A-7
    strength and mobility, 4-6, 5-1, 5-4, 5-6, 5-8, 5-15, 5-17, 5-19, 5-21, 6-10, 6-11, 6-16, 6-25, 6-31, 6-102, 7-13, 8-16, 9-1, 9-2, 9-9, 9-39, 9-54, 9-56, 9-74, D-1

electronic devices, A-1

event scorer, A-1, A-2, A-11, A-13, A-14, A-15, A-17

event supervisors, A-2, A-7
    responsibility, A-3

exercise
    duration, 1-7, 2-3, 2-4, 4-5, 5-1, 5-2, 5-16, 5-19, 6-9, 6-10, 6-69, 9-1, 10-1, 10-2, 10-6, 10-11, 10-15, 10-16, 10-18, 10-19, 10-20, 10-21, A-12, D-1

extended scale scoring, A-1

falling without injury, E-5

foot marching, 5-15, 5-20, 5-36, C-14

formation, 3-3, 5-36, 6-23, 7-1, 7-2, 7-3, 7-4, 7-5, 7-6, 7-13, 7-14, 7-15, 8-1, 8-15, 9-3, 9-11, 9-37, 9-69, 10-2, 10-4, 10-5, 10-6, 10-16, 10-19

injury, 1-7, 2-1, 2-3, 2-5, 2-7, 4-1, 4-2, 4-4, 4-8, 5-1, 5-2, 5-4, 5-5, 5-15, 5-19, 6-1, 6-2, 6-3, 6-4, 6-5, 6-6, 6-7, 6-8, 6-69, 6-70, 6-116, 8-23, 9-39, 9-66, 9-69, 9-71, 10-4, 10-6, 10-20, C-1, C-5, C-7, C-12, D-2, E-1, E-12
    back, 6-73
    foot and ankle pain, 6-71
    knee pain, 6-70
    lower leg, 6-72
    reconditioning, 6-2
    risk, 5-3
    shoulder, 6-74

kettlebell, 2-2, 9-54, 9-56, 9-67, C-13

leaders, 1-2, 1-6, 1-7, 1-8, 1-9, 3-1, 3-2, 3-3, 3-4, 4-4, 4-6, 4-8, 5-5, 5-35, 6-1, 6-2, 10-18

leadership, 9-3, 9-11, 9-37, 9-40, 9-69
    enthusiasm, 3-2
    motivation, 3-1, 3-2
    personal appearance, 3-1
    physical qualifications, 3-1

Index

traits, 3-1

mission-essential task list (METL), 4-3, 4-6

OSUT, 2-2, 4-1, 4-2, 4-6, 5-4, 5-7, 5-8, 5-9, 5-14, 5-18, 6-5

overreaching, 5-2

overtraining, 5-2, 5-3

phase
  initial conditioning, 2-2
  sustaining, 1-7, 1-8, 2-1, 2-2, 2-3, 2-7, 3-3, 4-2, 4-4, 4-6, 5-1, 5-2, 5-6, 5-7, 5-14, 5-15, 5-16, 5-18, 5-19, 5-20, 5-35, 6-1, 6-3, 6-4, 6-5, 7-14, 8-2, 8-16, 9-1, 9-3, 9-4, 9-11, 9-20, 9-39, 9-40, 9-41, 9-48, 9-69, 10-2, 10-4, 10-5, 10-6, 10-11, 10-15, 10-16, 10-18, 10-19, 10-20, B-1
  toughening, 1-8, 2-2, 2-7, 3-3, 4-1, 4-2, 4-4, 5-1, 5-2, 5-6, 5-7, 5-14, 5-15, 5-18, 5-20, 6-3, 6-4, 6-11, 6-16, 6-25, 7-14, 8-2, 8-16, 9-1, 9-3, 9-4, 9-11, 9-20, 10-2, 10-4, 10-5, 10-6, 10-11, 10-15, 10-16, 10-18, 10-19, B-1

physical readiness training, 3-1

platoon, 3-1, 5-36, 6-6, 7-2, 7-3, 7-4, 7-5, 7-13, 7-14, 8-16, 9-3, 9-11, 9-37, 9-54, 9-56, 9-69, 10-4, 10-19
  in column, 7-3, 7-4, 10-4

profile, 4-4, 4-5, 4-8, 5-20, 5-35, 6-1, 6-2, 6-3, 6-4, 6-5, 6-7, 6-8, 6-9, 6-10, 6-25, 6-31, 6-66, 6-68, 6-69, 6-70, 6-71, 6-72, 6-73, 6-74, 6-76, 6-80, 6-82, 6-84, 6-87, 6-91, 6-93, 6-97, 6-99, 6-101, 6-103, 6-106, 6-112, 6-116, 10-2, 10-20, 10-21, 10-22, A-1, A-12

PRT System, 1-4, 1-7, 1-8, 2-1, 2-2, 2-3, 2-5

reconditioning, 2-1, 2-3, 2-7, 3-3, 4-2, 4-4, 4-7, 4-8, 5-35, 6-2, 6-3, 6-4, 6-5, 6-6, 6-7, 6-8, 6-9, 6-10, 6-11, 6-16, 6-25, 6-31, 6-116, 10-2, 10-20

recovery, 5-3, 5-6, 5-8, 5-14, 5-17, 5-18, 5-21, 5-22, 5-23, 5-24, 5-25, 5-26, 5-27, 5-28, 5-29, 5-30, 5-31, 5-36, 6-1, 6-2, 6-5, 6-7, 6-70, 6-71, 6-72, 6-73, 6-74, 6-106, 7-14, 8-1, 8-15, 8-16, 8-23, E-20

running, 2-1, 2-2, 2-3, 2-4, 2-5, 2-6, 3-3, 4-1, 4-4, 5-1, 5-2, 5-3, 5-6, 5-8, 5-14, 5-15, 5-16, 5-17, 5-19, 5-20, 5-21, 5-36, 6-4, 6-6, 6-8, 6-9, 6-69, 6-70, 6-71, 6-72, 6-73, 7-14, 8-15, 9-54, 9-74, 10-1, 10-2, 10-3, 10-4, 10-5, 10-6, 10-8, 10-11, 10-15, 10-16, 10-18, 10-19, 10-20, 10-22, A-3, A-4, A-18, C-5, C-14, D-1, E-1, E-2, E-4, E-5, E-9, E-10
  course, A-3
  form, 10-5
  speed, 1-5, 2-5, 5-15, 5-20, 5-36, 6-70, 6-71, 6-72, 6-73, 7-11, 10-4, 10-15
  sustained, 6-70, 6-71, 6-72, 6-73, 7-14, 10-4

scorer, A-11, A-12
  responsibility, A-3

spotter, 9-39

stopwatch, 9-54, 10-4, 10-15, A-12, A-14, A-17

support personnel, A-1, A-2, A-3, A-11, A-13, A-14, A-15, A-17

test personnel
  number of, A-2

timer, A-2, A-3, A-11, A-12, A-13, A-14, A-15, A-16, A-17, E-12

training
  Active and Reserve Component, 4-2
  advanced individual, 4-1
  basic combat, 4-1, 5-6, 6-5, 1
  Basic Officer Leader Course, 4-1, 5-6, 5-14
  concurrent, 1-8
  frequency, 1-7, 5-2, 5-15, 6-9, 6-10, 10-15, 10-18, 10-20, 10-21
  initial military, 2-2, 4-1, 5-3
  multi-echelon, 1-7
  multiple sessions, 5-4
  off-ground, 2-6
  one station unit, 4-1, 5-6, 5-14, 6-5
  on-ground, 2-6
  pregnancy postpartum physical, 4-4
  to standard, 1-6
  to sustain, 1-7

unit commander, 4-3, 4-7
  responsibility, 10-19

Weight Control Program, 4-5, 4-6, 5-35, 6-1